VOLUME EIGHTY

ADVANCES IN PARASITOLOGY

VOLUME EIGHTY

ADVANCES IN PARASITOLOGY

Edited by

S.I. HAY

Spatial Epidemiology and Ecology Group
Tinbergen Building, Department of Zoology,
University of Oxford, South Parks Road,
Oxford, UK

RIC PRICE

Centre of Tropical Medicine,
University of Oxford,
Oxford, UK

J. KEVIN BAIRD

Eijkman-Oxford Clinical Research Unit
Jalan Diponegoro No. 69
Jakarta, Indonesia

ELSEVIER

AMSTERDAM • BOSTON • HEIDELBERG • LONDON
NEW YORK • OXFORD • PARIS • SAN DIEGO
SAN FRANCISCO • SINGAPORE • SYDNEY • TOKYO

Academic Press is an Imprint of Elsevier

Academic Press is an imprint of Elsevier
The Boulevard, Langford Lane, Kidlington, Oxford, OX51GB, UK
32, Jamestown Road, London NW1 7BY, UK
Radarweg 29, PO Box 211, 1000 AE Amsterdam, The Netherlands
225 Wyman Street, Waltham, MA 02451, USA
525 B Street, Suite 1900, San Diego, CA 92101-4495, USA

First edition 2012

ISBN: 978-0-12-397900-1
ISSN: 0065-308X

Printed and bound in UK
12 13 14 15 10 9 8 7 6 5 4 3 2 1

CONTENTS

CONTRIBUTORS

Nicholas M. Anstey
Global Health Division, Menzies School of Health Research and Charles Darwin University, Darwin, Australia; Department of Infectious Diseases, Division of Medicine, Royal Darwin Hospital, Darwin, Australia

J. Kevin Baird
Eijkman-Oxford Clinical Research Unit, Jalan Diponegoro No. 69, Jakarta, Indonesia; Nuffield Department of Medicine, Centre for Tropical Medicine, University of Oxford, Oxford, UK

Katherine E. Battle
Department of Zoology, University of Oxford, South Parks Road, Oxford, UK

Nicholas M. Douglas
Global Health Division, Menzies School of Health Research and Charles Darwin University, Darwin, Australia; Department of Infectious Diseases, Royal Darwin Hospital, Darwin, Australia

Iqbal R.F. Elyazar
Eijkman-Oxford Clinical Research Unit, Jakarta, Indonesia

Peter W. Gething
Department of Zoology, University of Oxford, South Parks Road, Oxford, UK

Carlos A. Guerra
Fogarty International Center, National Institutes of Health, Bethesda, MD, USA

Simon I. Hay
Department of Zoology, University of Oxford, Oxford, South Parks Road UK; Fogarty International Center, National Institutes of Health, Bethesda, MD, USA

Rosalind E. Howes
Department of Zoology, University of Oxford, South Parks Road, Oxford, UK

Mallika Imwong
Department of Molecular Tropical Medicine and Genetics, Faculty of Tropical Medicine, Mahidol University, Bangkok, Thailand

George K. John
Centre for Tropical Medicine, Nuffield Department of Clinical Medicine, University of Oxford, Oxford, UK

Jason D. Maguire
Portsmouth Naval Hospital, Portsmouth, Virginia USA

Catherine L. Moyes
Department of Zoology, University of Oxford, South Parks Road, Oxford, UK

Jeanne R. Poespoprodjo
Papuan Community Health and Development Foundation, Timika, Papua, Indonesia; Department of Pediatrics, Timika General Hospital, Timika, Papua, Indonesia

Ric N. Price
Centre for Tropical Medicine, Nuffield Department of Medicine, University of Oxford, Oxford, UK; Global Health Division, Menzies School of Health Research, Charles Darwin University, Darwin, Australia

Lorenz von Seidlein
Global Health Division, Menzies School of Health Research, Charles Darwin University, Darwin, Northern Territory, Australia

G. Dennis Shanks
Army Malaria Institute, Enoggera, Queensland, Australia; Department of Zoology, University of Oxford, Oxford, UK

Marianne E. Sinka
Department of Zoology, University of Oxford, South Parks Road, Oxford, UK

Nicholas J. White
Mahidol Oxford Research Unit, Faculty of Tropical Medicine, Mahidol University, Bangkok, Thailand

PREFACE

We have long harboured an ambition to bring together a special issue of *Advances in Parasitology* on the epidemiology of *Plasmodium vivax*. The editors initially set out to present a contemporary synthesis of what is known about *P. vivax*, calling upon the authors of each chapter to (i) retrieve what has been 'lost' from 'history', (ii) summarize objectively the current state of knowledge including the reasons for the 'hiatus' in interest and; (iii) identify research gaps/directions/priorities to gently temper the prevailing 'hubris' with respect to control and elimination. The sub-title of the volumes is thus not hard to decipher. While we have largely succeeded in these objectives, we failed spectacularly in limiting these reviews to a single volume. This is a reflection of the sheer magnitude of information reviewed by our 35 contributing authors and has necessitated splitting the 12 chapters into two related volumes: Part A (volume 80, which you are reading) and Part B (volume 81, to be published in early 2013). These volumes are split thematically as outlined below.

Part A is composed of six chapters and deals principally with the most practical dimensions of vivax malaria, describing the epidemiology, clinical consequences, treatment, and control strategies shaped by the biology of the parasite. The first by Katherine Battle and colleagues sets the scene by discussing the global public health significance of *P. vivax*. The contemporary global spatial limits of *P. vivax* transmission are discussed alongside the most recent estimates of its endemicity and the distribution of its potential vector species. A very detailed collection of maps is made available for reference and a strong argument made that we have underestimated considerably the global burden of this disease. Nicholas White and Mallika Imwong provide the second foundation chapter discussing one of the most epidemiologically important aspects of *P. vivax* biology: the ability of the parasite to relapse. Much historical information is unearthed and combined with recent findings to argue that the biology of relapse remains poorly understood and propose that the activation of latent hypnozoites may well be triggered by other systemic illness; the clinical and epidemiological corollaries of this are then explored. Nicholas Anstey and colleagues provide the third chapter that reviews risk factors, pathogenesis and clinical outcomes of *P. vivax* infection. Strong conclusions from this chapter are the importance of co-morbidity in determining the outcome of infection – including significant mortality. In Chapter 4, Kevin Baird and colleagues move on to the issues of diagnosis and treatment. Emphasis is given to the historic neglect of the treatment of this infection, and the challenges faced in achieving a radical cure and overcoming the emergence and spread of drug-resistant parasites. Nicholas Douglas and colleagues extend this theme in Chapter 5 evaluating the transmission dynamics of *P. vivax*, and highlighting the importance of preventing the relapsing infections of the parasite when reducing transmission potential. In the final chapter of the first volume, Dennis Shanks extends this theme further and reviews aspects of the biology and epidemiology of *P. vivax* pertinent to its control and elimination. Hence this volume compiles a body of evidence pertinent to those dealing with the management of vivax malaria emphasising the very serious public health and clinical threat that it poses.

Part B brings together a further six chapters which will be detailed more extensively in its own preface. In brief, the chapters investigate more fully aspects of the parasite life cycle, innate and inherited aspects that confer host resistance to *P. vivax* infection, G6PD deficiency, the genome of *P. vivax* and finally the lessons that can still be learned from the interpretation of the neurosyphilis literature.

We believe that all authors have succeeded in providing authoritative, inclusive and balanced reviews and that the presentation of this body knowledge as a collection is much more valuable than the sum of its parts. We are grateful to a host of enthusiastic peer reviewers who provided fast and comprehensive feedback that has been incorporated into this final version. This collection is testament to the fact that *P. vivax* is less neglected than it once was. It also highlights the fact that the basic epidemiology of this parasite remains poorly understood, and that much still needs to be achieved in our understanding of *P. vivax* before we can move feasibly on to its control and, ultimately, its elimination.

Simon I. Hay, Ric N. Price and J. Kevin Baird

CHAPTER ONE

The Global Public Health Significance of *Plasmodium vivax*

Katherine E. Battle*, Peter W. Gething*, Iqbal R.F. Elyazar†, Catherine L. Moyes*, Marianne E. Sinka*, Rosalind E. Howes*, Carlos A. Guerra‡, Ric N. Price§,¶, J. Kevin Baird†,§, Simon I. Hay*,‡

*Department of Zoology, University of Oxford, South Parks Road, Oxford, UK
†Eijkman-Oxford Clinical Research Unit, Jalan Diponegoro No. 69, Jakarta, Indonesia
‡Fogarty International Center, National Institutes of Health, Bethesda, MD, USA
§Centre for Tropical Medicine, Nuffield Department of Medicine, University of Oxford, Oxford, UK
¶Global and Tropical Health Division, Menzies School of Health Research, Charles Darwin University, Darwin, NT, Australia

Contents

Advances in Parasitology, Volume 80
ISSN 0065-308X, http://dx.doi.org/10.1016/B978-0-12-397900-1.00001-3

Abstract

Plasmodium vivax occurs globally and thrives in both temperate and tropical climates. Here, we review the evidence of the biological limits of its contemporary distribution and the global population at risk (PAR) of the disease within endemic countries. We also review the most recent evidence for the endemic level of transmission within its range and discuss the implications for burden of disease assessments. Finally, the evidence-base for defining the contemporary distribution and PAR of *P. vivax* are discussed alongside a description of the vectors of human malaria within the limits of risk. This information along with recent data documenting the severe morbid and fatal consequences of *P. vivax* infection indicates that the public health significance of *P. vivax* is likely to have been seriously underestimated.

1. INTRODUCTION

Malaria is a highly significant global public health problem. Its greatest burden is imposed on the world's poorest countries (Sachs and Malaney, 2002). It is the third leading cause of death from infectious diseases for children under age of five worldwide (Black et al., 2010) and the fourth leading cause for all ages (WHO, 2008). After decades of neglect, malaria control research and financing has experienced a resurgence in recent years (RBMP, 2008) and targets have been raised by international initiatives aiming for the goal of elimination (Feachem et al., 2009; Chitnis et al., 2010a; Moonen et al., 2010; Tanner and Hommel, 2010; Tatem et al., 2010; Alonso et al., 2011). While elimination goals ostensibly address all human malarias, the allocation of funding and resources between the two parasites of greatest significance, *Plasmodium falciparum* and *Plasmodium vivax*, has been highly disparate. From 2007 to 2009, only 3.1% of all malaria research and development funding was targeted at *P. vivax* (PATH, 2011). *Plasmodium vivax* is epidemiologically and biologically different to *P. falciparum* and it is not, therefore, possible to assume that control methods developed for falciparum malaria are transferable to vivax malaria (Luxemburger et al., 1994; Bockarie and Dagoro, 2006; Baird, 2010; Bousema and Drakeley, 2011). Evidence from *P. vivax* infections in carefully monitored populations show that vivax malaria should no longer be thought of as a benign and rarely fatal disease but one that can lead to severe disease and death (Baird, 2007; Price et al., 2007b; Anstey et al., 2009). The reader is referred to an accompanying review in this thematic volume of *Advances in Parasitology* (Chapter 3, Volume 80), which addresses the clinical severity of *P. vivax*. The clinical importance of *P. vivax*, along with its wide geographic distribution extending well beyond the limits of falciparum malaria (Guerra et al.,

2006a, 2006b, 2010; Gething et al., 2011b) and the challenges it presents for control (Sattabongkot et al., 2004; Wells et al., 2010), has led to a call for greater attention to be paid to understanding the global distribution and burden of this neglected parasite (Baird, 2007; Price et al., 2007b; Mueller et al., 2009a).

A key information gap impeding *P. vivax* control and progress towards elimination has been the lack of geographical estimates of risk (Mendis et al., 2001; Price et al., 2007b; Malaria Eradication Research Agenda, 2011b). Assessment of geographic variations in levels of endemicity of the parasite is essential to estimate the burden of the disease and measure the impact of control and the feasibility of elimination (Pampana, 1969; Hay et al., 2008; Tatem et al., 2010). The Malaria Atlas Project (MAP) was founded to address this evidence gap (Hay and Snow, 2006; Hay et al., 2009; Gething et al., 2011a), with an initial focus on *P. falciparum*, and the resulting global malaria distributions have been used to assess the adequacy and equity in control funding (Pigott et al., 2012), to inform international policy and resource allocation (Anonymous, 2009; Feachem et al., 2009; McLaughlin et al., 2009; World Bank, 2009; Zanzibar Malaria Control Program, 2009; DFID, 2010; Global Partnership to Roll Back Malaria et al., 2010) and to estimate the global burden of disease (Patil et al., 2009; Gething et al., 2010a; Hay et al., 2010b). A suite of modelled spatial data on *P. vivax* transmission and endemicity (Guerra et al. 2010; Gething et al. 2012) and *P. vivax* vectors (Sinka et al., 2010a, 2010b, 2011, 2012) add to this evidence base for strategic disease-control planning, implementation and monitoring. Here, for the first time, we bring all these mapped data on *P. vivax* together in one place. We explain how these products were generated, their limitations and their value, and provide an overview for each malaria-endemic region of the world. We also consider those areas where our lack of geographical knowledge is particularly acute and would benefit most from concerted future research efforts. A full review of the methodologies used to generate each mapped product is also provided at the end of this chapter.

2. THE GLOBAL DISTRIBUTION OF *P. VIVAX* INFECTIONS

Plasmodium vivax has the widest geographical distribution of the human malarias with an estimated 2.49 billion individuals living at risk of infection in 2010 (Gething et al., 2012). Biological features of *P. vivax* that distinguish it from *P. falciparum* present unique challenges to the control of the parasite (Mendis et al., 2001; Sattabongkot et al., 2004; Wells et al., 2010); in elimination settings, *P. vivax* is often the 'last parasite standing' (Garnham, 1951; Yekutiel, 1960; Pampana, 1969; Wernsdorfer et al., 2009; Tatem et al., 2010). *Plasmodium vivax* gametocytes are present earlier in the progression of a primary or recrudescent infection than *P. falciparum* (Mendis et al., 2001; McKenzie et al., 2002), such that the majority of patients have sufficient gametocytaemia to allow transmission before the infection is diagnosed or treated (Ratcliff et al., 2007; Awab et al., 2010; Douglas et al., 2010). *Plasmodium vivax* gametocytes are transmitted more efficiently to *Anopheles* mosquito vectors (Boyd and Kitchen, 1937; Collins et al., 2002) than those of *P. falciparum* and are transmissible at lower parasite densities (Sattabongkot et al., 2004). Within the mosquito, vivax sporozoites develop faster than *P. falciparum* and with slightly wider viable temperature ranges, allowing for a greater geographical distribution (Gething et al., 2011b). In addition, due to vector bionomics in regions where *P. vivax* is most prevalent, methods of control that are broadly effective in reducing *P. falciparum* transmission, such as insecticide-treated bed nets (ITNs), show far less success in the control of *P. vivax* (Luxemburger et al., 1994; Bockarie and Dagoro, 2006). *Plasmodium vivax* malaria is typically carried with lower levels of parasitaemia, making it relatively difficult to diagnose (Mendis et al., 2001). However, there is evidence that, despite lower blood parasite loads, *P. vivax* immunity is acquired more rapidly than *P. falciparum* and may result in an earlier age-prevalence peak in areas of high transmission (Mueller et al., 2009a, 2009b). Detailed reviews regarding the biology (Chapter 2, Volume 80), control (Chapter 6, Volume 80), and acquired immunity (Chapter 3, Volume 81) of *P. vivax* are available elsewhere in this thematic issue of *Advances in Parasitology*.

Perhaps the most important feature of *P. vivax* biology is its ability to relapse in the weeks and months following a primary parasitaemia, via a dormant liver stage known as the hypnozoite (James, 1931; Coatney, 1976; Garnham, 1989; Prudencio et al., 2006; Chen et al., 2007; Imwong et al., 2007; White, 2011). It has long been known that there is significant geographical variation in the rate at which a 'strain' of *P. vivax* may relapse (Coatney and Cooper, 1948; Winckel, 1955; Garnham et al., 1975). The exact mechanism through which hypnozoite relapses are triggered is unknown (Cogswell, 1992; Prudencio et al., 2006; Baird, 2009; Mueller et al., 2009a). One theory is that the mechanism is an adaptive trait of the parasite to sequester or 'hibernate' during times when climatic conditions would be inhospitable to the *Anopheles* vector of the disease (Shute et al., 1976; Baird and Rieckmann, 2003; White, 2011). This theory is supported by observations that temperate strains of the parasite tend to exhibit longer relapse intervals than tropical strains (Garnham et al., 1975; Shute et al., 1976; Cogswell, 1992; Collins and Jeffery, 1996; Adak et al., 1998; Baird et al., 2007; Imwong et al., 2007). Primaquine is currently the only widely available drug with activity on the hypnozoite stage capable of preventing relapse (Baird and Hoffman, 2004; Galappaththy et al., 2007), but is associated with haemolysis in individuals with glucose-6-phosphate dehydrogenase (G6PD) deficiency (Beutler, 1994; Baird and Hoffman, 2004; Cappellini and Fiorelli, 2008). Primaquine is contraindicated in pregnant women because of the risk of haemolytic anaemia in the foetus of unknown G6PD status (WHO, 2010a). The reader is referred to reviews that address relapse (Chapter 2, Volume 80), treatment (Chapters 4 and 5, Volume 80) and G6PD deficiency (Chapter 4, Volume 81), which are provided elsewhere in this special issue *Advances in Parasitology*. Relapse also has implications for understanding the burden of *P. vivax* malaria based on prevalence rates derived from malariometric surveys and cartographic studies (Patil et al., 2009; Hay et al., 2010b).

To accurately illustrate *P. vivax* endemicity, it is necessary to incorporate the distribution of the Duffy-negative phenotype. The varying prevalence of Duffy negativity in populations throughout the world is a significant determinant of the distribution of *P. vivax* (Livingstone, 1984). Duffy-negative individuals are, for the most part, refractory to *P. vivax* infection and the phenotype is found at highest frequencies in Africa, whereas it is relatively rare elsewhere (Howes et al., 2011). A detailed review of the effect of Duffy negativity on the epidemiology of *P. vivax* (Chapter 2, Volume 81) is provided elsewhere in this series. The influence of Duffy negativity on *P. vivax* transmission reinforces the need to differentiate strategies employed to generate and interpret maps of *P. vivax* endemicity from those used for *P. falciparum* (Hay et al., 2009; Gething et al., 2011a).

Until recently, little work had been done to define the geographic limits and risk of *P. vivax* infection. The only map including vivax malaria endemicity was that of Lysenko from 1968 (Lysenko and Semashko, 1968). As shown in Fig. 1.1, Lysenko defined endemicity as the parasite rate (PR) in children aged between 2 and 10 years old (hypoendemic <10%, mesoendemic 11–50%, hyperendemic 51–75%), with the exception of the holoendemic class (>75%) where the PR was defined in the 1-year age cohort. The map was derived from a synthesis of historical records and malariometric indices of all four human malaria parasites: disease and vector presence and absence records; spleen, parasite, sporozoite and biting rates; sickle cell incidence; and others (Hay et al., 2004; Gething et al., 2010b). Lysenko interpolated the data globally to determine the distribution of

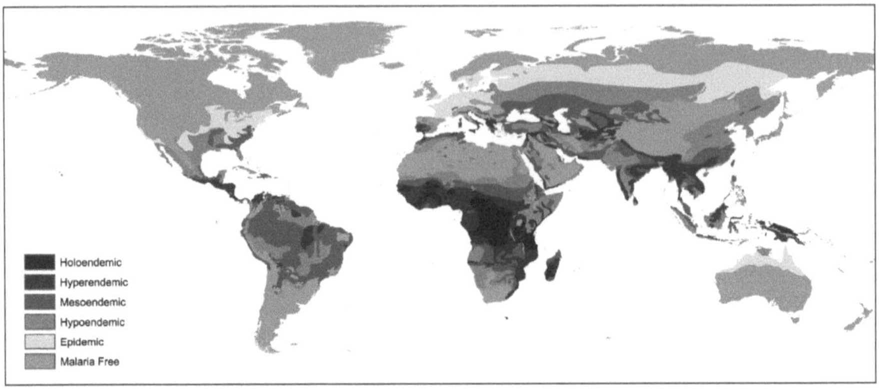

Figure 1.1 The Lysenko map of global malaria endemicity. The map was digitised from its original source (Lysenko and Semashko, 1968) where endemicity classes were defined by parasite rate (PR) in children between 2 and 10 years old (hypoendemic <10%, mesoendemic 11–50%, hyperendemic 51–75%), with the exception of the holoendemic class (>75%) where the parasite rate was defined in the 1-year age cohort. The 'epidemic' class is restricted to the temperate regions in these maps and it should be noted that this term is used differently today. For a colour version of this figure, the reader is referred to the online version of this book.

malaria at the peak of its historic distribution (*circa* 1900) using expert opinion, temperature ranges and rainfall isohyets (Lysenko and Semashko, 1968; Lysenko and Beljaev, 1969; Kaneko et al., 1998).

Lysenko's endemicity map was not specific to one human malaria parasite. The map was created before the advent of geographic information systems (GIS) and lacks geographic precision – as exemplified by a significant misplacement of the Nile River. Relative to the remote sensing data available today, the environmental data Lysenko and Semashko used to refine the endemicity boundaries were both limited and crude. The evidence base used was not described in detail and, since the map was not generated via a formal statistical method, there was no measure of uncertainty provided. It is therefore difficult to determine how robust the distribution estimates of Lysenko were.

3. SPATIAL DISTRIBUTION OF *P. VIVAX* MALARIA, POPULATIONS AT RISK AND ITS VECTORS

Here, we bring together maps at the regional level (Fig. 1.2) of the limits of transmission, endemicity and estimates of the populations at risk, along with the distribution of *P. vivax* vectors.

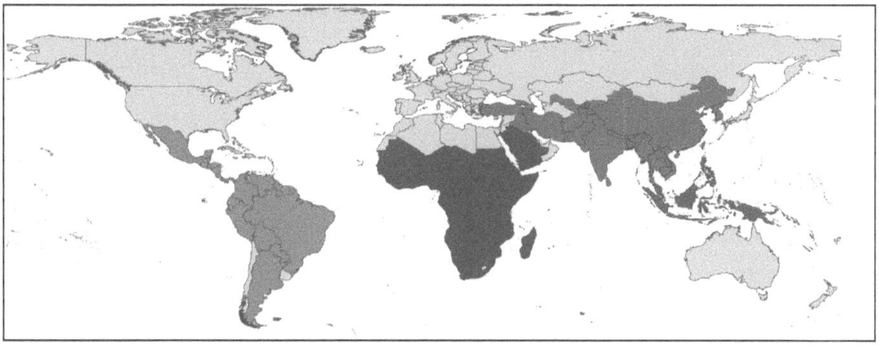

Figure 1.2 Regional tiles of *Plasmodium vivax* endemic countries. The regional tiles shown were used to stratify the modelling of *P. vivax* endemicity. Pink = the Americas; blue = Africa+; green = Asia; orange = Asia-Pacific; grey = non endemic for *P. vivax*. For interpretation of the references to colour in this figure legend, the reader is referred to the online version of this book.

3.1. *P. vivax* Malaria Limits and Endemicity

Figure 1.3 shows estimated limits of unstable (light grey) and stable (dark grey) transmission of *P. vivax* in Panel 1 (A1–D1) and, within Panel 2 (A2–D2), levels of endemicity (blue to yellow to red, red being >7% prevalence). The map is accompanied by two measures of uncertainty in our endemicity estimates: one absolute measure and the second weighted by population density in each location (Fig. 1.4).

A detailed review of the data and methodology used to create this map is given at the end of this chapter in Section 6. In brief, we have used 9970 spatiotemporally unique records of parasite prevalence (Fig. 1.3, Panel 1), environmental covariates and a Bayesian geostatistical model with seasonal and age-standardisation components to generate a smooth map of point estimates of parasite prevalence in all areas with stable transmission (Gething et al., 2012). Throughout the world, endemicity was predicted within a relatively narrow range, with the point estimate of the *P. vivax* PR age-standardised to the 1–99 year age range ($PvPR_{1-99}$) rarely exceeding 7%. Here, we highlight important aspects of the methodological approach taken that improve on preceding global estimates of *P. vivax* malaria risk, along with key assumptions and constraints:

Temporal influence. The maps presented (Figure 1.3, Panel 2) are of endemicity in 2010 but the data used to inform the model ranged from 1985 to 2010. The model was designed to down-weight older data, so that whilst they can help inform the contemporary estimates, particularly in areas lacking contemporary data, these had less influence on our estimates of prevalence. This was reflected by higher uncertainty those areas reliant on older data.

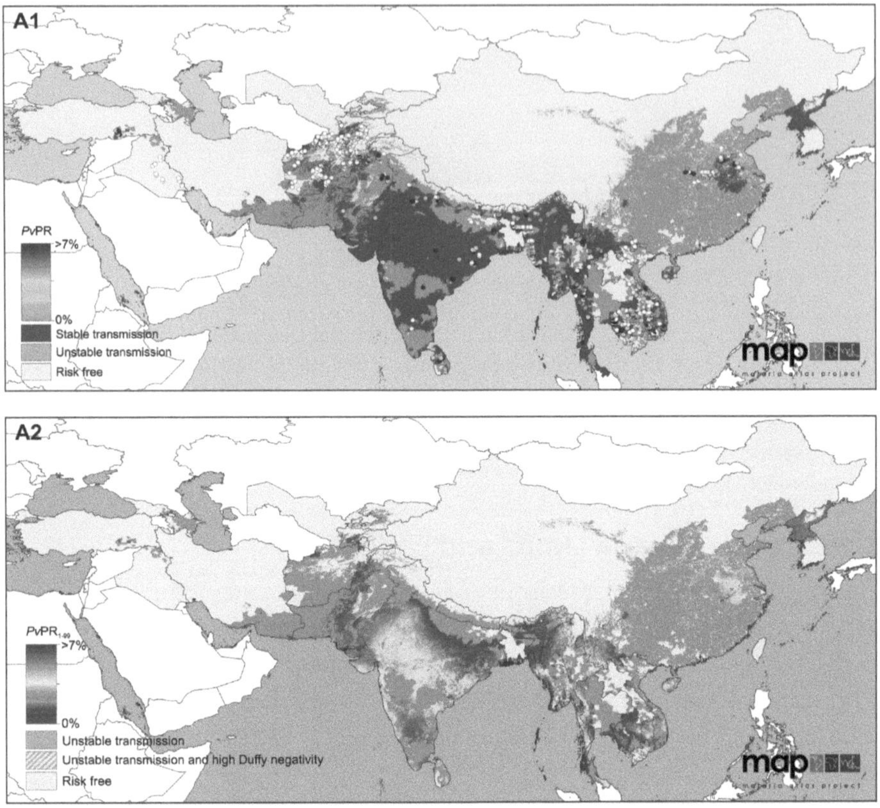

Figure 1.3 The spatial distribution of *Plasmodium vivax* malaria endemicity in 2010 in Asia, Asia-Pacific, the Americas and Africa+. The spatial distribution of *P. vivax* malaria endemicity is shown at the regional levels: Asia (A), Asia-Pacific (B), the Americas (C) and Africa+ (D). Panel 1 in A–D shows the 2010 spatial limits of *P. vivax* malaria risk defined by *Plasmodium vivax* annual parasite incidence (*Pv*API) with further medical intelligence, temperature and aridity masks. Areas were defined as stable (dark grey areas, where *Pv*API ≥ 0.1 per 1000 per annum), unstable (medium grey areas, where *Pv*API < 0.1 per 1000 p.a.) or no risk (light grey, where *Pv*API = 0 per 1000 p.a.). Only the *P. vivax* malaria endemic countries (*Pv*MECs) in each region are shaded in. The community surveys of *P. vivax* prevalence conducted between January 1985 and June 2010 are plotted. The survey data are presented as a continuum of light blue to red (see map legend), with zero-valued surveys shown in white. Panel 2 in each region shows the model-based geostatistics (MBG) point estimates of the annual mean $PvPR_{1-99}$ for 2010 within the spatial limits of stable *P. vivax* malaria transmission, displayed on the same colour scale. Areas within the stable limits (Panel 1) that were predicted with high certainty (>0.9) to have a $PvPR_{1-99}$ less than 1% were classed as unstable. Areas in which Duffy negativity gene frequency is predicted to exceed 90% (Howes et al., 2011) are shown in hatching for additional context. For interpretation of the references to colour in this figure legend, the reader is referred to the online version of this book.

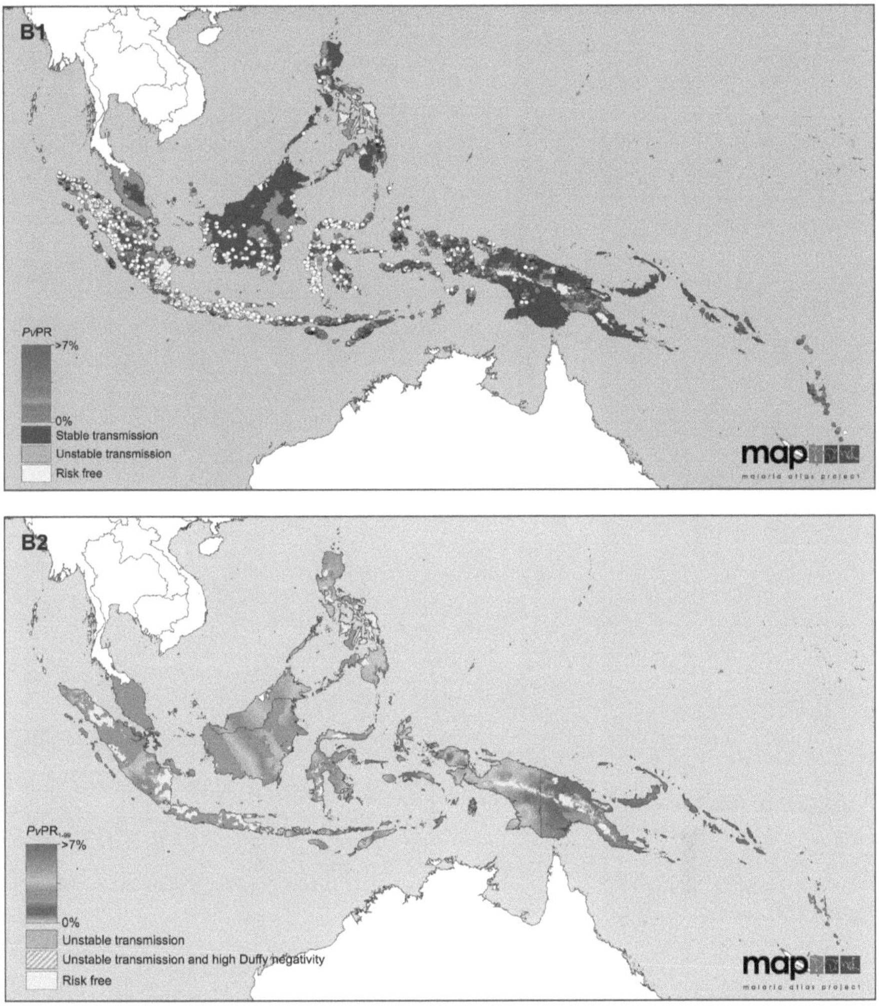

Figure 1.3, cont'd

Seasonal fluctuations. The estimates provide the average *P. vivax* endemicity over a year. Seasonal fluctuations of parasite prevalence were included in the model structure but aggregated in the outputs. Seasonal estimates are not currently available and would be confounded by relapses that can occur outside the transmission season (see the section 'Relapse of *Plasmodium vivax* blood infections' below).

Smoothness of the global map. Malaria endemicity can be highly heterogeneous over small distances and, therefore, our model produced candidate maps that are also highly heterogeneous. An artefact of presenting one map that summarises the full model output using either a mean or a median

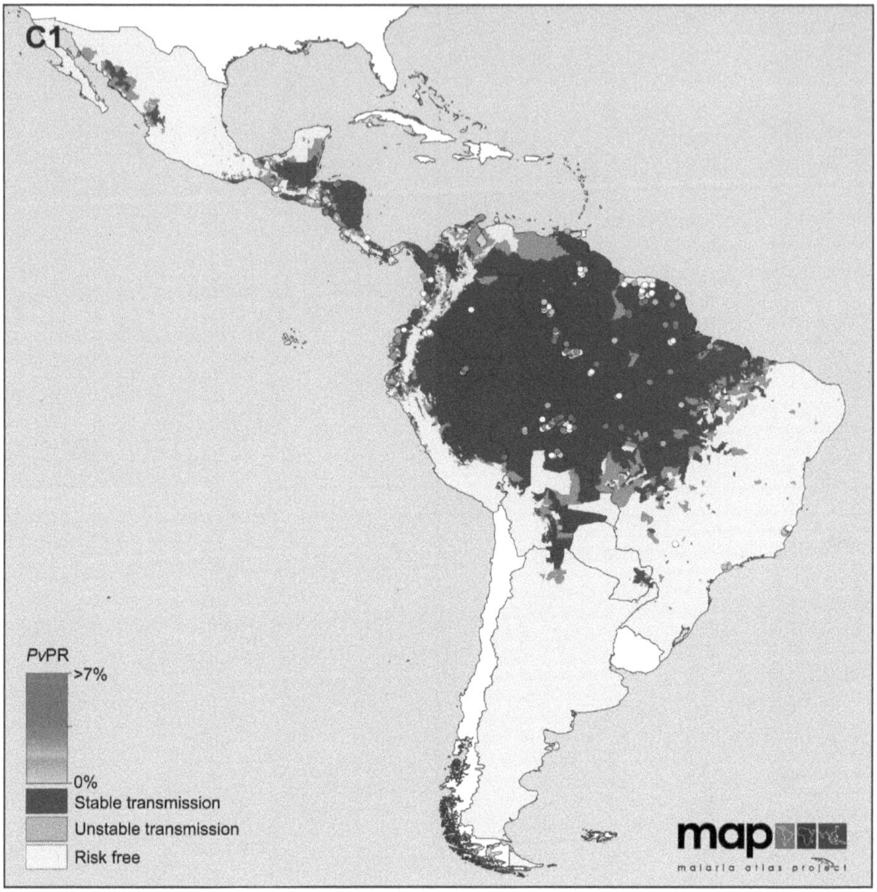

Figure 1.3, cont'd

of the estimates for each location was that this process smoothed out the patchy nature of malaria endemicity.

Use of parasite prevalence data. It is important that our estimates were based on real data, with a mathematical model that compensated for the fact that the real data were neither comprehensive nor universally contemporary. The most commonly available metric from almost all countries is prevalence (often referred to as parasite rate (PR)), estimated from surveys that are conducted using a common methodology (Hay et al., 2008). Prevalence data are relatively robust but have the disadvantage that required sample sizes become prohibitively large when prevalence rates are low.

Use of clinical incidence data. Annual clinical incidence per 1000 people (commonly referred to as annual parasite incidence (API) data) from public

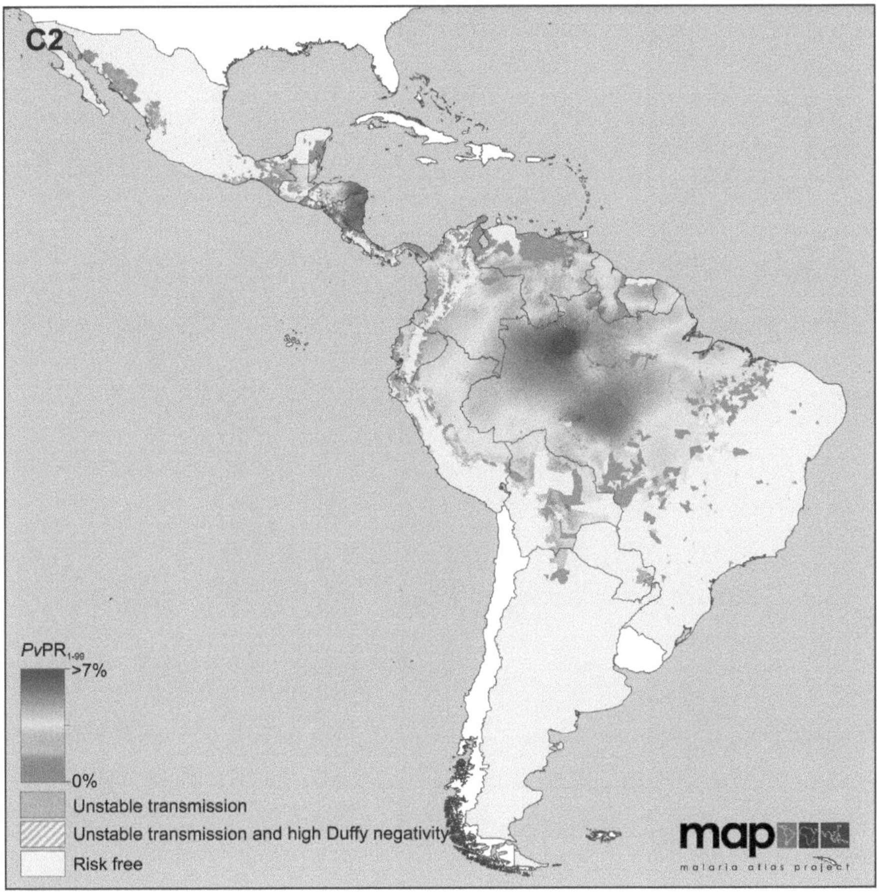

Figure 1.3, cont'd

health records is more sensitive at low parasite prevalence values, but is biased. For example, we do not know the proportion of people within every district who do not seek care for malaria at a public health facility and are therefore missed by the statistics. We used clinical incidence data to define the areas where we know very little, other than that malaria exists but is so low as to be essentially unquantifiable using available data. We referred to this level of risk as unstable malaria transmission and defined it as less than one case per 10,000 people per year (or <0.1 API). At very low levels of incidence, a small change is proportionally large so API is measured across an administrative (ADMIN) division and we averaged this figure over 4 years (providing data are available).

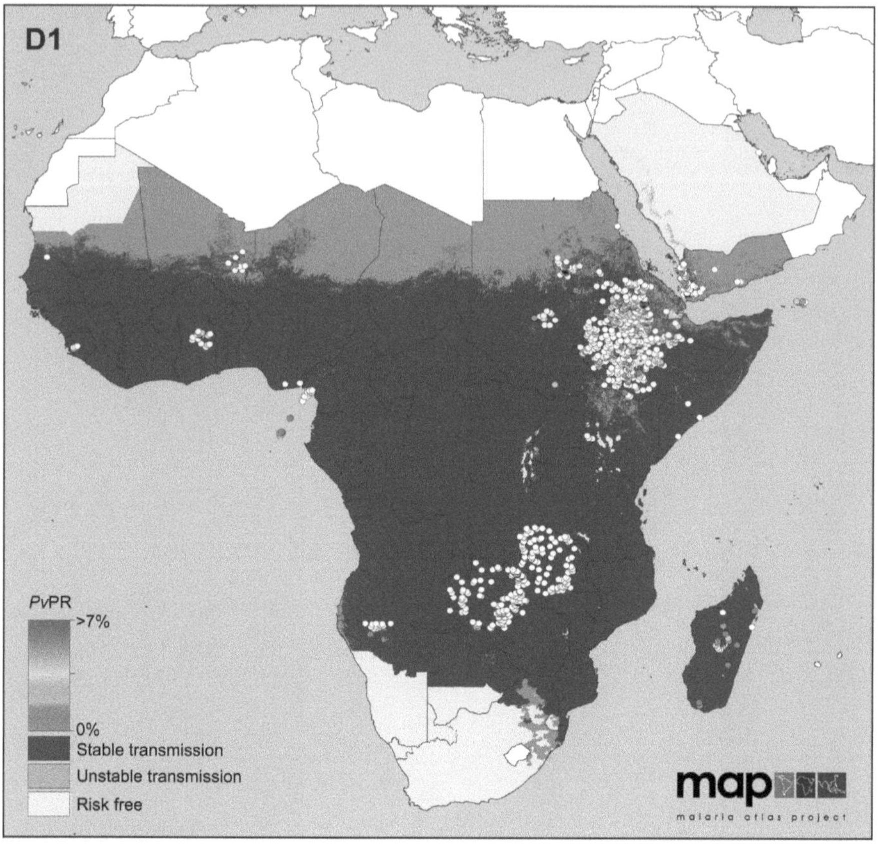

Figure 1.3, cont'd

Use of biological data to refine the limits defined by clinical incidence data. Remote sensing surfaces, showing temperature (Fig. 1.5) and aridity (Fig. 1.6), and biological models were used to identify areas where extreme temperature or aridity regimes would reduce or preclude *P. vivax* transmission. The limits of transmission are adjusted accordingly.

The effect of interventions. The PR data used came from many areas where interventions (ITNs, indoor residual spraying (IRS), etc.) have been implemented so that the effect of these interventions at these locations was included within the estimates produced by the model. This effect was not modelled separately and cannot be extracted from our results.

The impact of Duffy negativity. Since Duffy negative individuals are largely refractory to *P. vivax* infection (Miller et al., 1976), high population frequencies of this phenotype suppress endemicity, even where conditions are

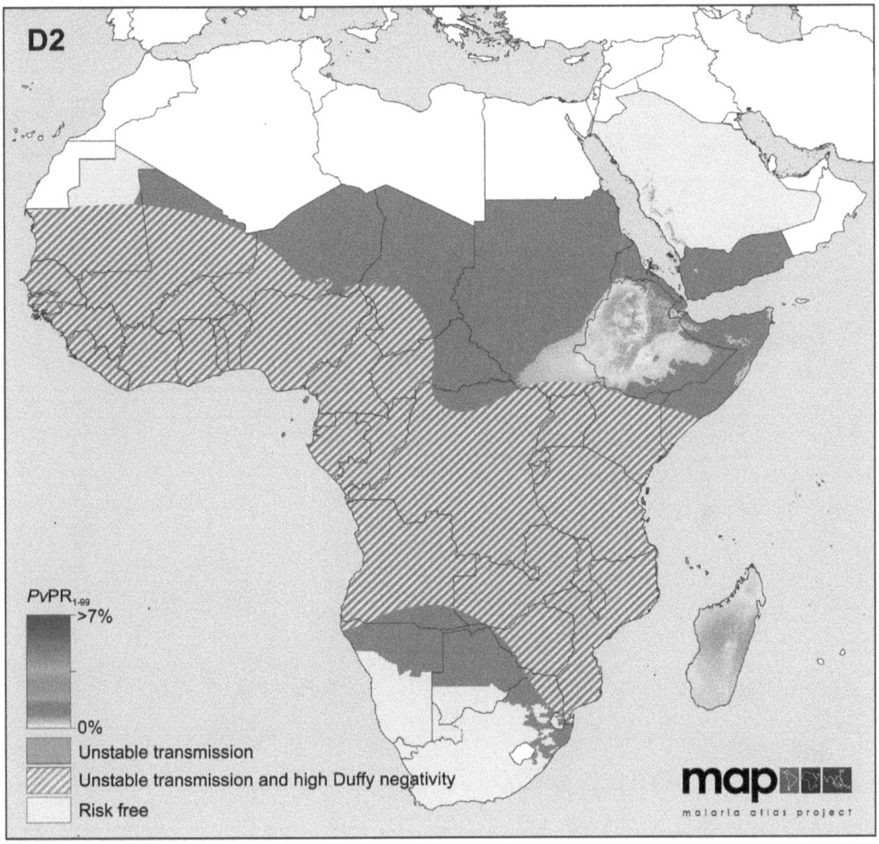

Figure 1.3, cont'd

otherwise well suited for transmission (Guerra et al., 2010). We incorporated recent global estimates of the frequencies of Duffy negativity (Howes et al., 2011) in our model. Regional maps of the predicted prevalence of Duffy negativity are shown in Fig. 1.7. The fraction of the population at each population that was Duffy-negative was excluded from the denominator in the prevalence data, such that any *P. vivax*-positive individuals are assumed to have arisen from the Duffy-positive population subset. Thus, in a location with 90% Duffy negativity, five positive individuals in a survey of 100 would give an assumed prevalence of 50% amongst Duffy positives. Correspondingly, prediction of prevalence was then restricted to the Duffy-positive proportion, with the final prevalence estimate re-converted to refer to the total population. This approach meant prevalence could never exceed the Duffy-positive proportion of a population and, where *P. vivax*

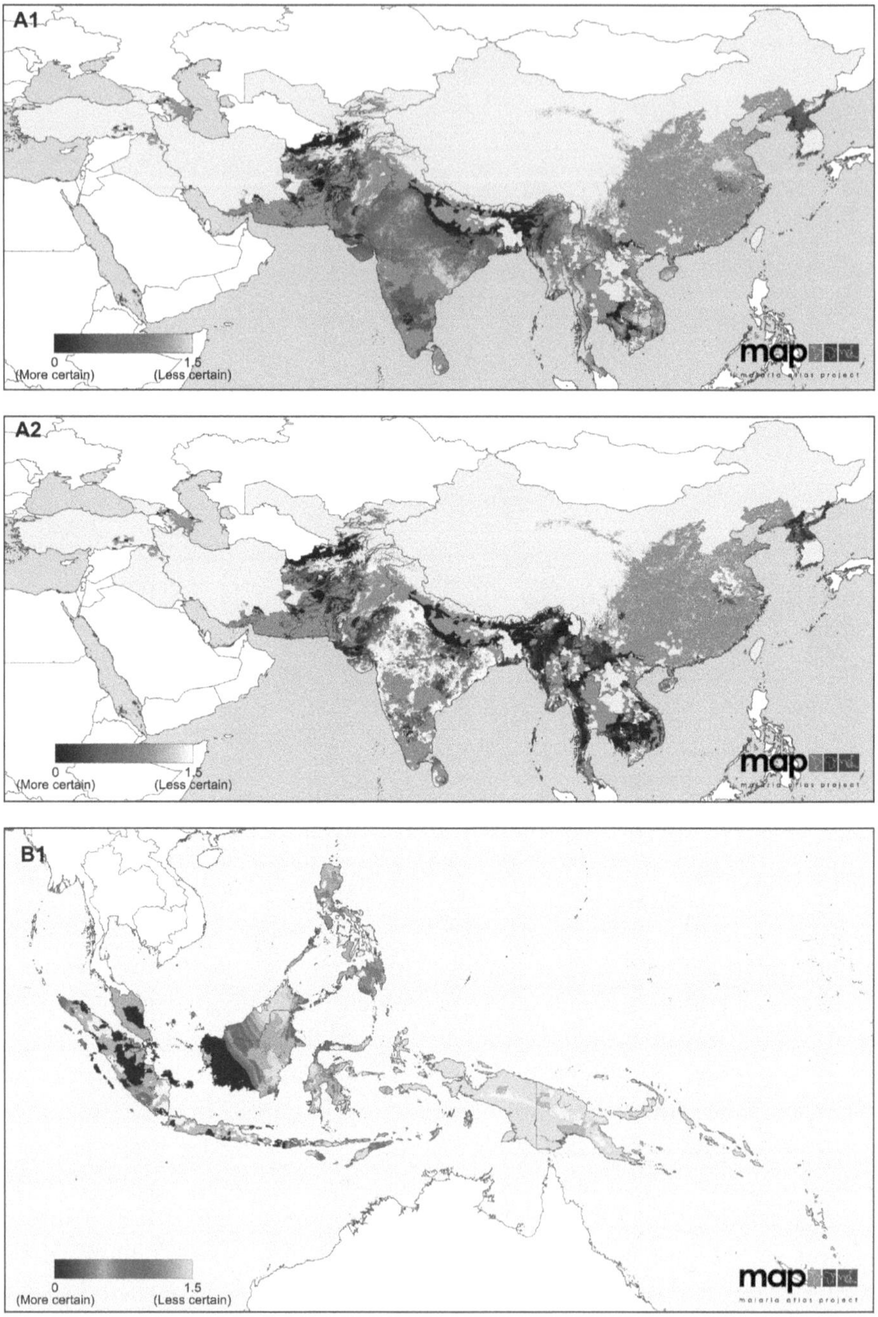

Figure 1.4 Uncertainty associated with predictions of *Plasmodium vivax* endemicity in Asia, Asia-Pacific, the Americas and Africa+.

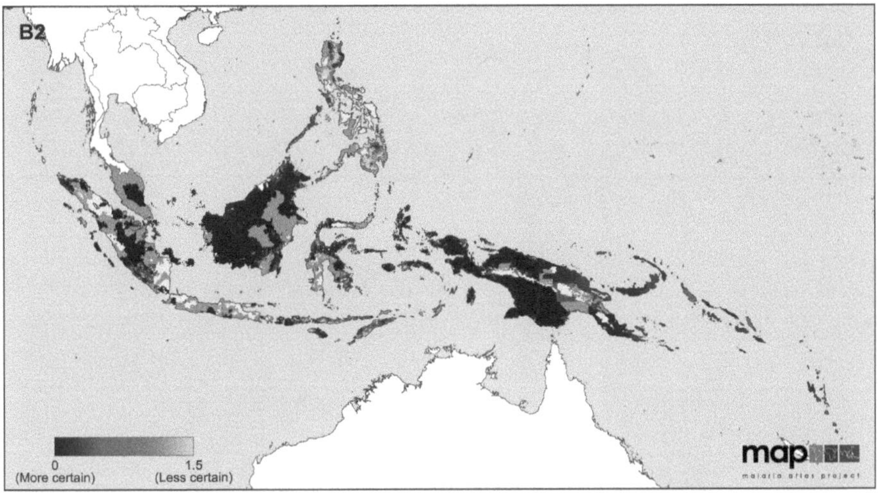

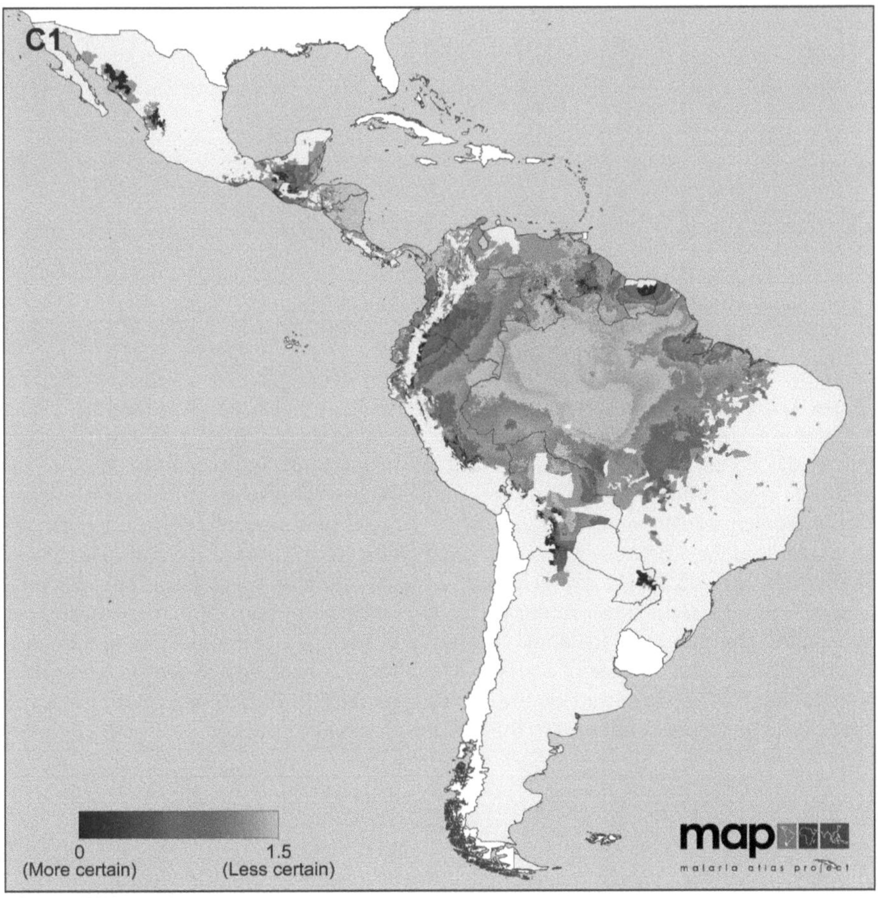

Figure 1.4, cont'd

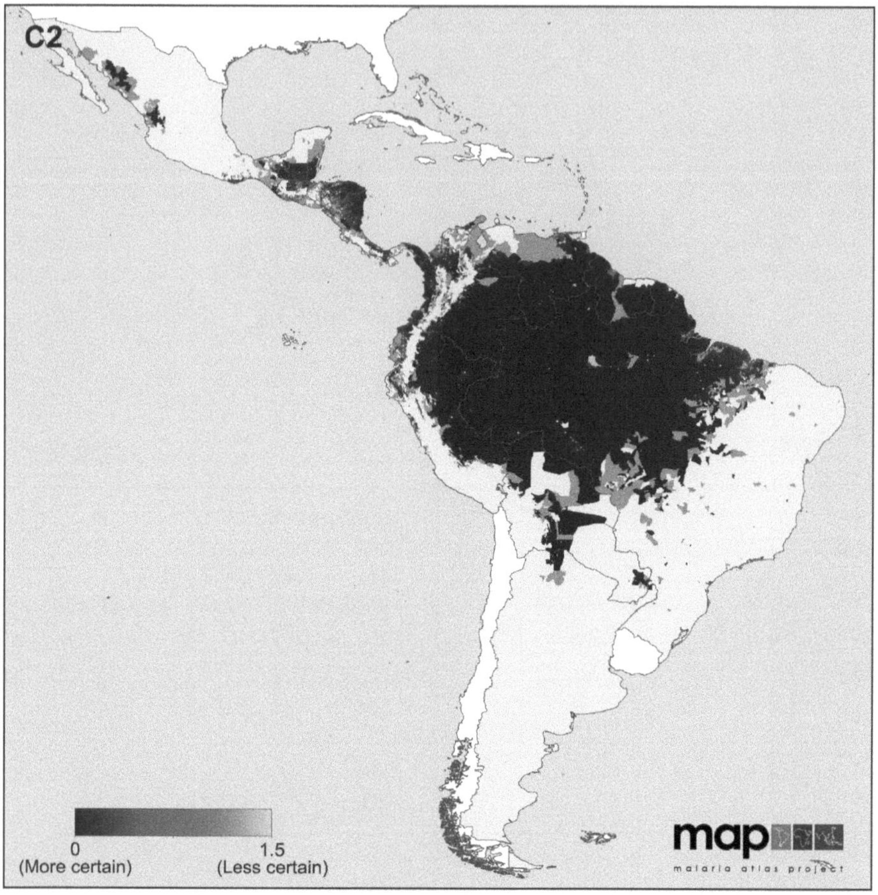

Figure 1.4, cont'd The uncertainty associated with *P. vivax* malaria endemicity predictions is shown at the regional levels: Asia (A), Asia-Pacific (B), the Americas (C) and Africa+ (D). Panel 1 in A–D shows the ratio of the posterior inter-quartile range to the posterior mean prediction at each pixel. Large values indicate greater uncertainty: the model predicts a relatively wide range of $PvPR_{1-99}$ as being equally plausible given the surrounding data. Conversely, smaller values indicate a tighter range of values have been predicted and, thus, a higher degree of certainty in the prediction. Panel 2 in each region shows the same index multiplied by the underlying population density and rescaled to 0–1 to correspond to Panel 1. Higher values indicate areas with high uncertainty and large populations. Areas of no risk within *Pv*MECs are shown in grey and countries not endemic for *P. vivax* or outside the named region are in white. For a colour version of this figure, the reader is referred to the online version of this book.

survey data were sparse across much of Africa, the prevalence predictions could borrow strength from the Duffy negativity map because predictions of prevalence were restricted to a narrower range of possible values.

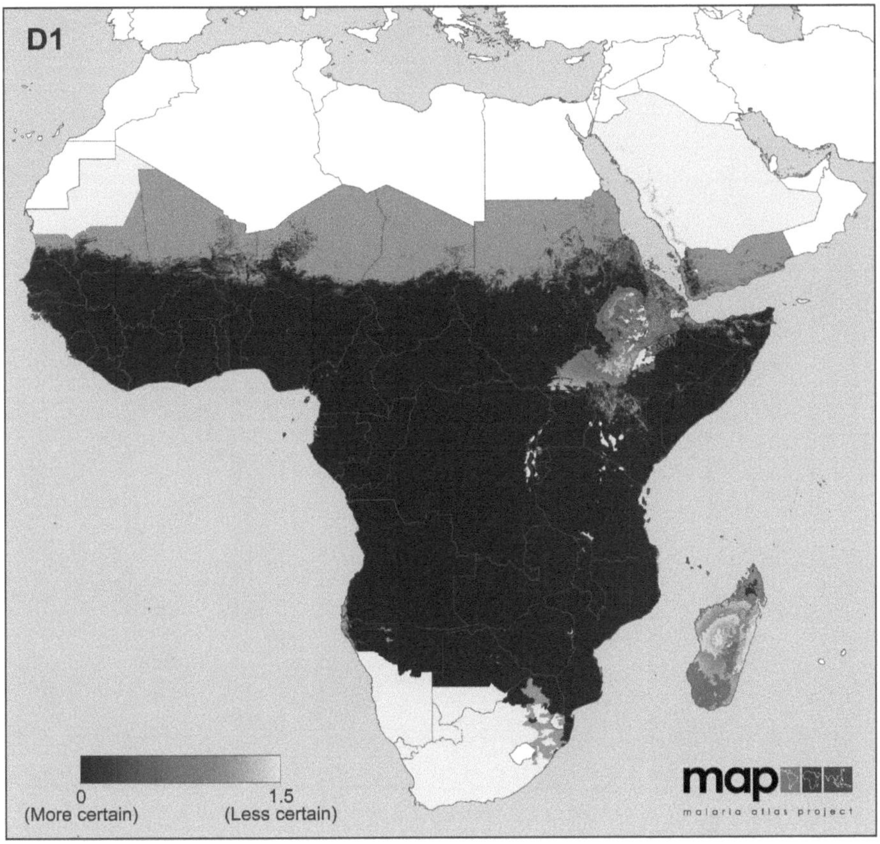

Figure 1.4, cont'd

Transmission in Africa. The predominance of Duffy negativity in Africa has led to a historical perception that *P. vivax* is absent from much of the continent (Rosenberg, 2007). Evidence exists, however, of autochthonous *P. vivax* transmission in nearly every African country (Guerra et al., 2010) and that *P. vivax* is capable of causing severe disease in the continent (Mahgoub et al., 2012). Therefore, we did not preclude any areas at risk before modelling endemicity. We initially assumed stable *P. vivax* transmission, unless the biological mask layers or clinical incidence (API) data confirmed otherwise. In some regions provisionally labelled as stable in this way, the endemicity model subsequently predicted extremely low prevalence, either because of survey data reporting zero infections or because of very high Duffy-negativity prevalence. Locations predicted with high certainty (probability >0.9) of being less than 1% prevalence were therefore re-assigned

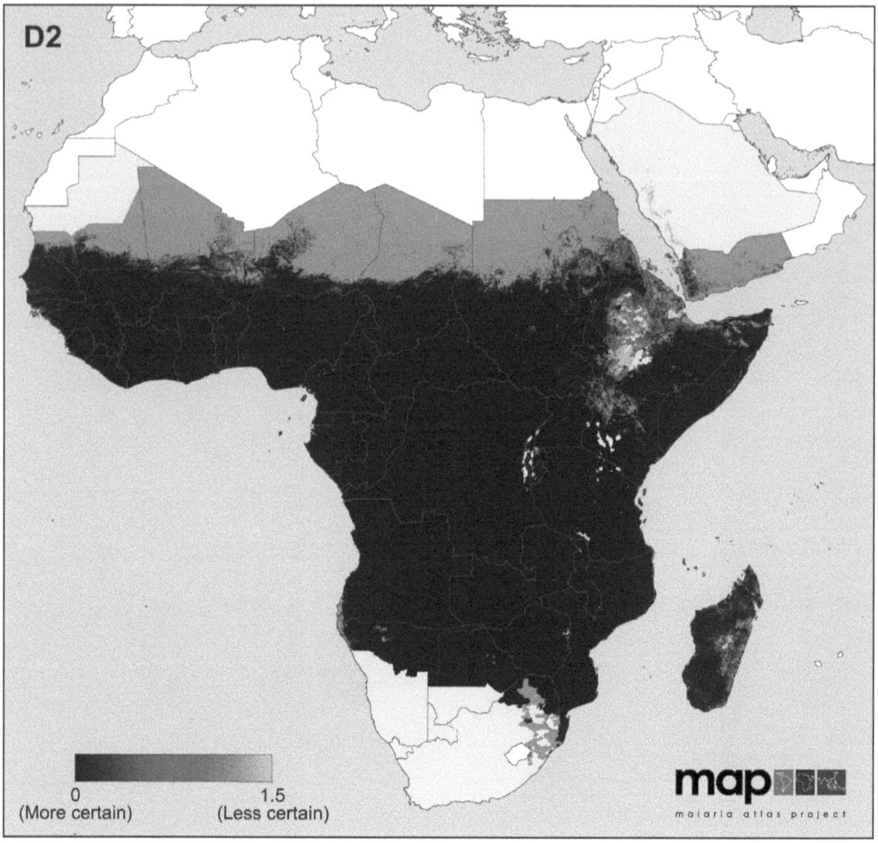

Figure 1.4, cont'd

to the unstable transmission class. This re-assignment is shown in the final limits presented in Fig. 1.3A1–D2.

Relapse of Plasmodium vivax *blood infections.* The prevalence surveys used detect parasites in the blood and not latent infections in the liver. Primary blood infections detected by these surveys are indistinguishable from relapses caused by previously dormant parasites from the liver. The estimates produced by the model were, therefore, the combined prevalence of primary infections and relapses.

The impact of low parasitaemia levels. Plasmodium vivax parasites are typically found in the blood at much lower densities than are found for *P. falciparum* and so are less likely to be detected by some diagnostic methods. We made no direct adjustments for these 'sub-patent' infections. It is important to note that *P. vivax* can cause fever and anaemia at lower

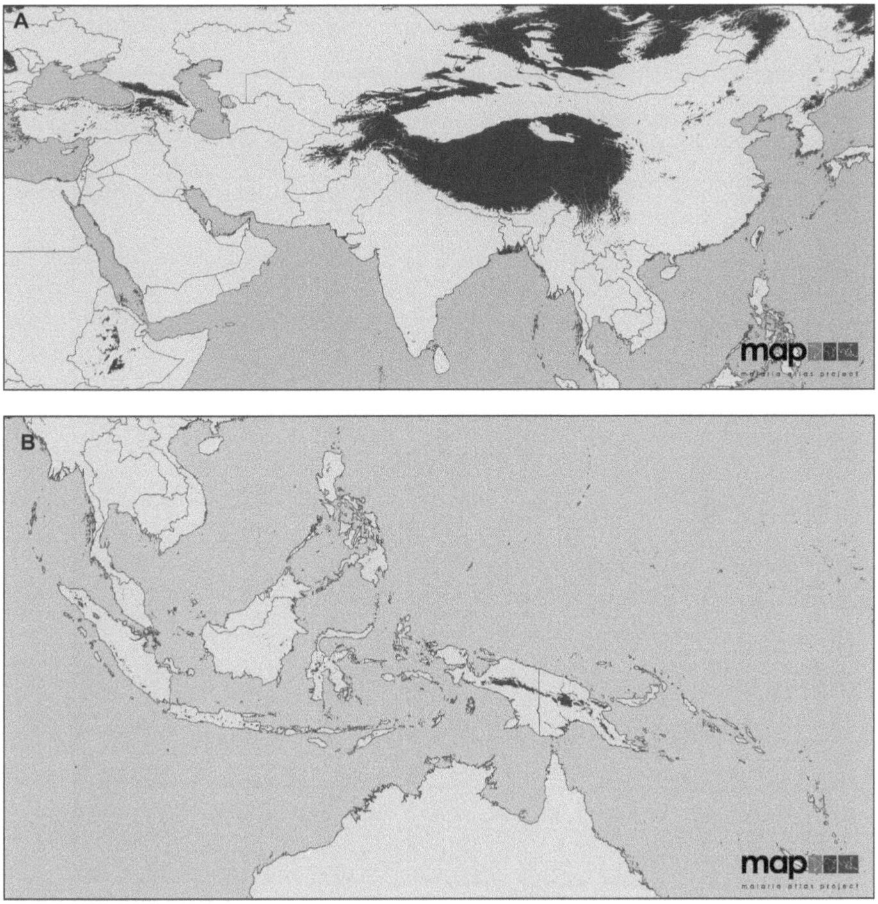

Figure 1.5 Environmental suitability for transmission of *P. vivax* as defined by temperature in Asia, Asia-Pacific, the Americas and Africa+. Areas shaded green in Asia (A), Asia-Pacific (B), the Americas (C) and Africa+ (D) are those in which no windows exist across an average year in which the annual temperature regime is likely to support the presence of infectious vectors. The temperature suitability model is described in full elsewhere (Gething et al., 2011b). For interpretation of the references to colour in this figure legend, the reader is referred to the online version of this book.

parasitaemia levels than *P. falciparum* (Mendis et al., 2001; Mueller et al., 2009a).

Uncertainty in the estimates of endemicity. The modelling framework allowed the uncertainty in predicted endemicity values to vary between locations, depending on the observed variation, density and sample size of surveys and the predictive utility of the suite of environmental covariates (described in detail in Section 1.6). The uncertainty associated with

Figure 1.5, cont'd

predictions is summarised by maps showing the ratio of the inter-quartile range (IQR) to its mean (Fig. 1.4A1–D1). The IQR is a simple measure of the precision with which each value is predicted, and standardisation by the mean produces an uncertainty index less affected by underlying prevalence levels and more illustrative of relative model performance. This index was then also weighted by the underlying population density to produce a second map indicative of those areas, where uncertainty is likely to be most operationally important (Fig. 1.4A2–D2).

3.2. Population at Risk of *P. vivax* Malaria

Table 1.1 shows our estimates of the population at risk (PAR) of infection and a detailed description of the methodology used to generate these

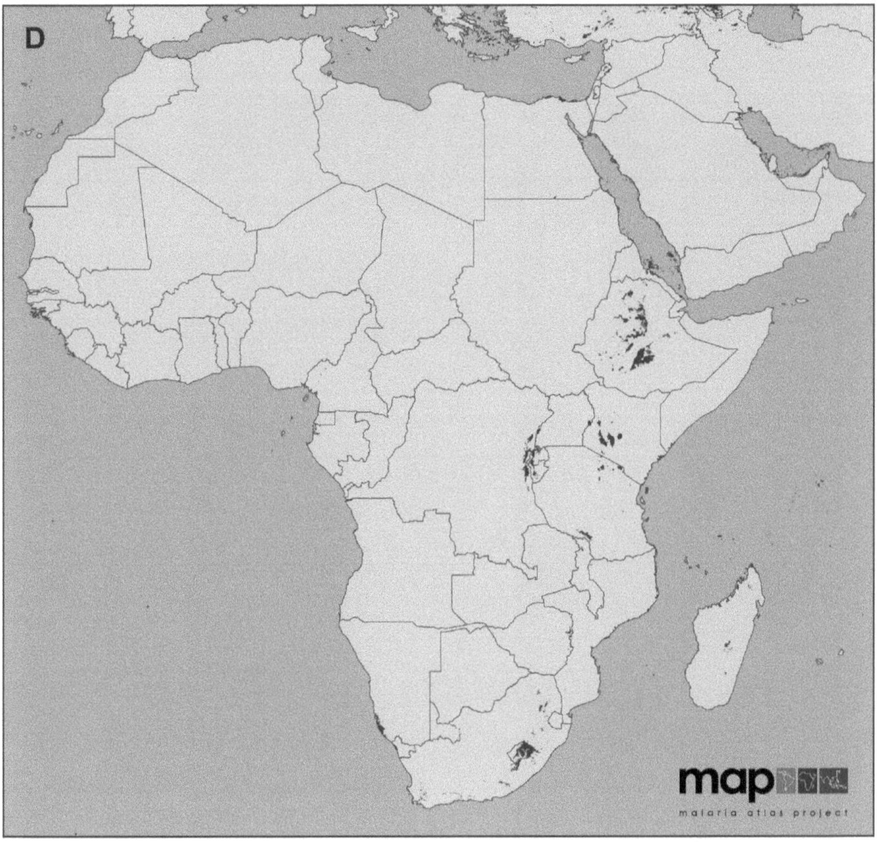

Figure 1.5, cont'd

estimates is given at the end of this chapter in Section 6. In brief, these estimates were generated by combining maps of the limits of transmission described above with 1 × 1 km resolution gridded population surfaces for 2010 projected from the Global Rural-Urban Mapping Project (GRUMP) year 2000 *beta* version population counts (Balk et al., 2006; CIESIN/IFPRI/WB/CIAT, 2007). Regional maps of 1 × 1 km gridded population estimates for 2010 are shown in Fig. 1.8. Fine-scale population data were used to ensure the detailed variations in risk level described in our limits maps are appropriately assigned to the underlying population.

3.3. Vectors of the *P. vivax* Parasite

Figure 1.9 shows the distribution of *Anopheles* species considered to be the dominant vector species/species complexes (DVS) of human malaria

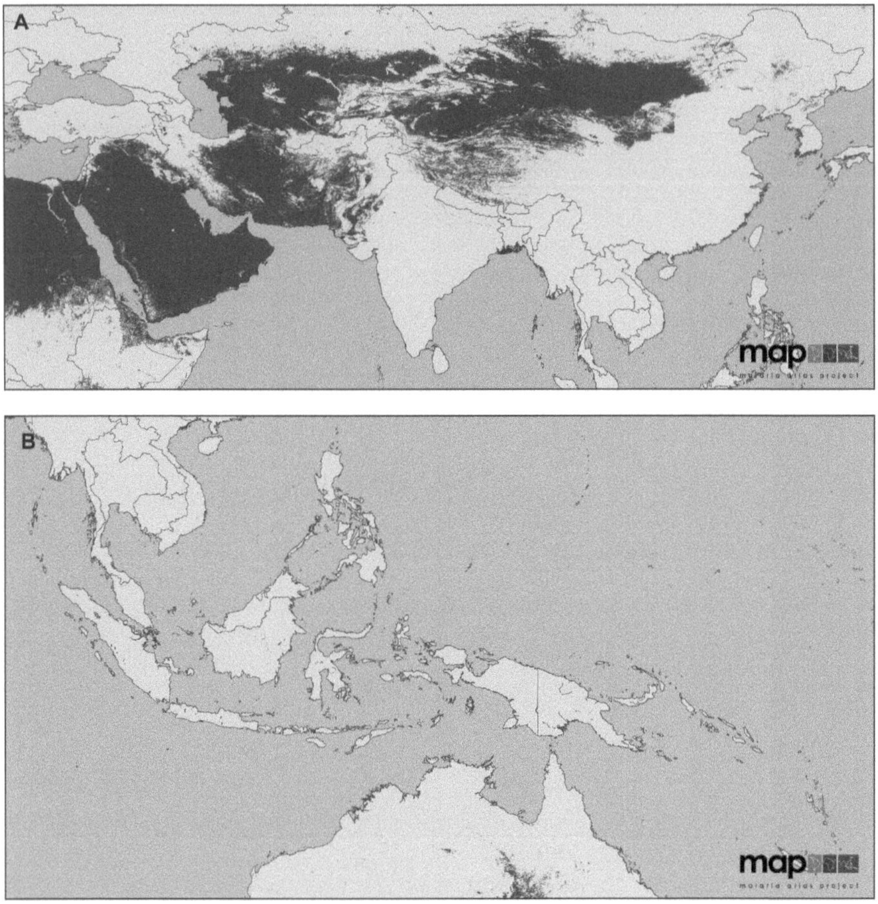

Figure 1.6 Environmental suitability for transmission of *P. vivax* as defined by extreme aridity in Asia, Asia-Pacific, the Americas and Africa+. Areas shaded in brown are those classified as bare areas by the GlobCover land cover product in Asia (A), Asia-Pacific (B), the Americas (C) and Africa+ (D), interpreted as lacking sufficient moisture to support populations of *Anopheles* necessary for transmission. For interpretation of the references to colour in this figure legend, the reader is referred to the online version of this book.

(*P. falciparum* and *P. vivax)* within each region of the world (Sinka et al., 2012), and a detailed description of the methods to generate these maps is given in Section 6.4. In brief, the distribution of each of the dominant species or species complex was individually predicted using occurrence data (mainly abstracted from published surveys), the Boosted Regression Trees (BRT) ecological niche modelling technique (Elith et al., 2008), environmental covariates, and species range expert opinion maps derived from

Figure 1.6, cont'd

consultation with vector experts (Hay et al., 2010c; Sinka et al., 2010a, 2010b, 2011). The resulting maps illustrate the predicted distribution of the DVS using a probability of occurrence metric. While the ecological covariates of the model are based on annualised means of temperature and precipitation, the maps cannot represent the season fluctuations that may occur in the distribution of the DVS, which may be significant in those that extend into temperate climes. Expert opinion was also used to classify the most important species/combinations of species per region (i.e. those with the highest impact) and their individual distributions were overlaid, with the most competent vectors species uppermost, to generate the maps shown. The combined vector map aims to aid vector control planning by showing which species need to be controlled in each area.

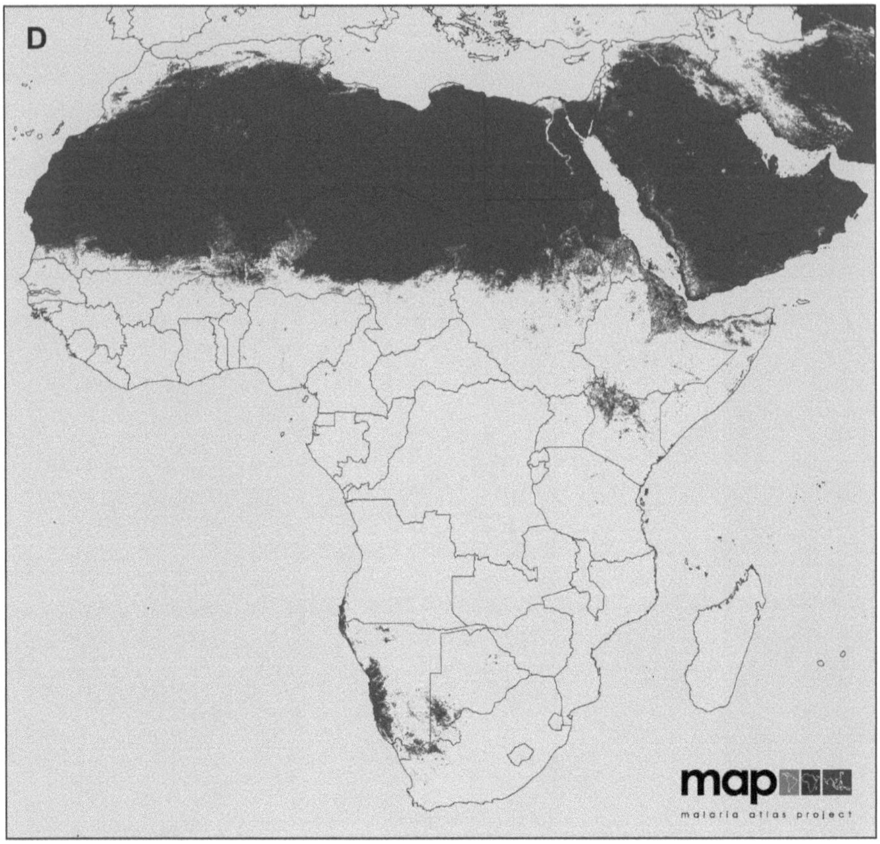

Figure 1.6, cont'd

A species complex is a group of closely related species that are often indistinguishable based on morphology alone. Reference to a complex may be indicated by referring to the species as '*sensu lato*' or '*s.l.*' (e.g. *Anopheles gambiae s.l.*) meaning 'in the broad sense', whereas '*sensu stricto*' or '*s.s.*' ('in the strict sense') indicates the species alone (often a sibling within a species complex has the same name as the complex). The presence of species complexes adds a level of complexity to vector control efforts. Sibling species that are morphologically indistinguishable and often sympatric within an area can have such varied bionomics that one sibling is rendered a dominant vector and the other a non-vector (Meek, 1995; Manguin et al., 2008). The proper identification of species and knowledge of their ranges, often rapidly altered by expanding agriculture and land use changes (Amerasinghe et al., 1991a; Amerasinghe and Indrajith, 1994; Lee, 1998; Singh and Mishra, 2000;

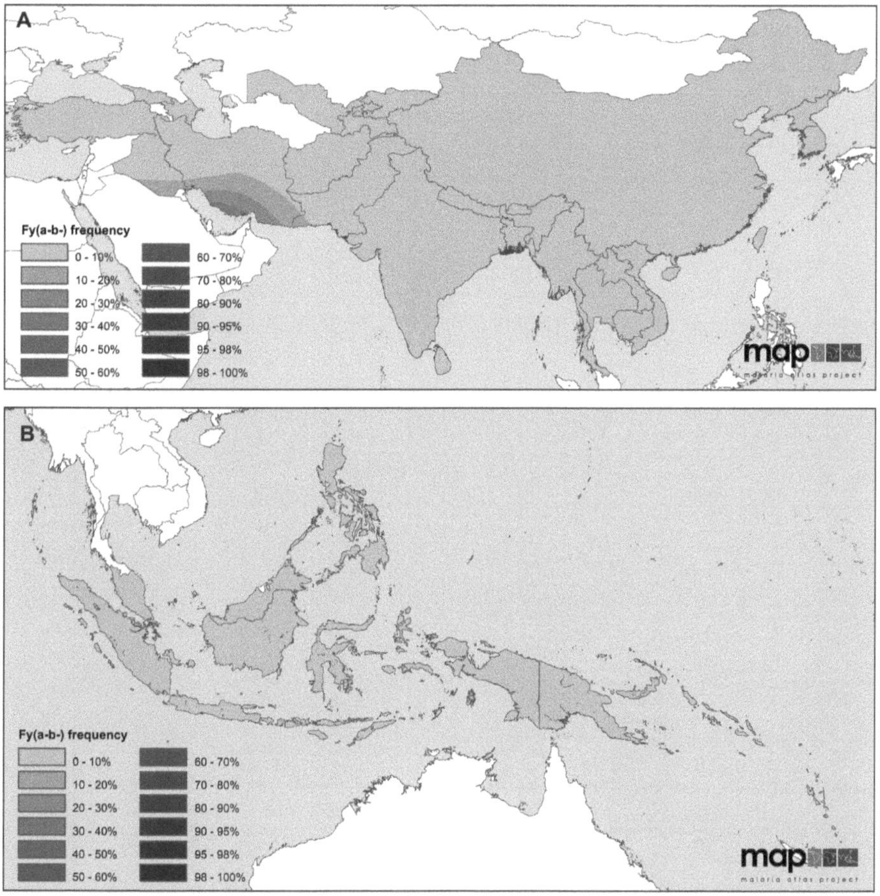

Figure 1.7 Distribution of the Duffy negative phenotype in Asia, Asia-Pacific, the Americas and Africa+. The prevalence of the Duffy-negative phenotype, Fy(a–b–), is shown in Asia (A), Asia-Pacific (B), the Americas (C) and Africa+ (D). The Duffy prevalence model is full described elsewhere (Howes et al., 2011). Only the *Pv*MECs in each region are shown in colour. For a colour version of this figure, the reader is referred to the online version of this book.

Vythilingam et al., 2005), is vital for appropriate allocation of vector control resources.

A search of the current literature has identified 71 species/species complexes with the potential ability to transmit *P. vivax* (Table 1.2), which includes all 41 species classified by Sinka et al. (2010a, 2010b, 2011) as DVS, 34 of which are shown in Fig. 1.9 (a number of species were not included in the global multi-species maps as they were of lesser importance when classified by region and were completely overlaid by higher impacting vectors).

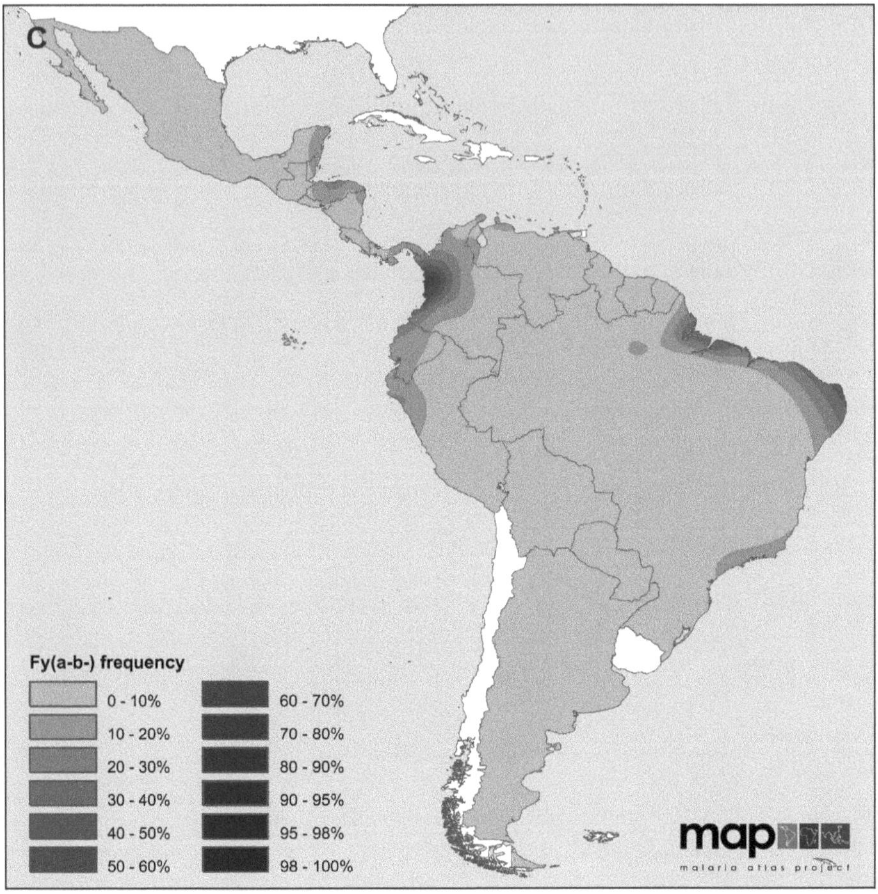

Figure 1.7, cont'd

This suggests, and the literature appears to corroborate, that all vectors capable of transmitting *P. falciparum* are also capable of transmitting *P. vivax*. Interestingly, the reverse (all *P. vivax* vectors being capable of transmitting *P. falciparum*) appears not to hold true.

4. REGIONAL SUMMARIES OF THE PUBLIC HEALTH SIGNIFICANCE OF *P. VIVAX* MALARIA

4.1. Asia

The Asia region (defined for these purposes as mainland Asia but excluding the Malaysian Peninsula; Fig. 1.2) has, amongst the highest endemicity estimates of *P. vivax* malaria globally (Fig. 1.3A2), a diverse range of

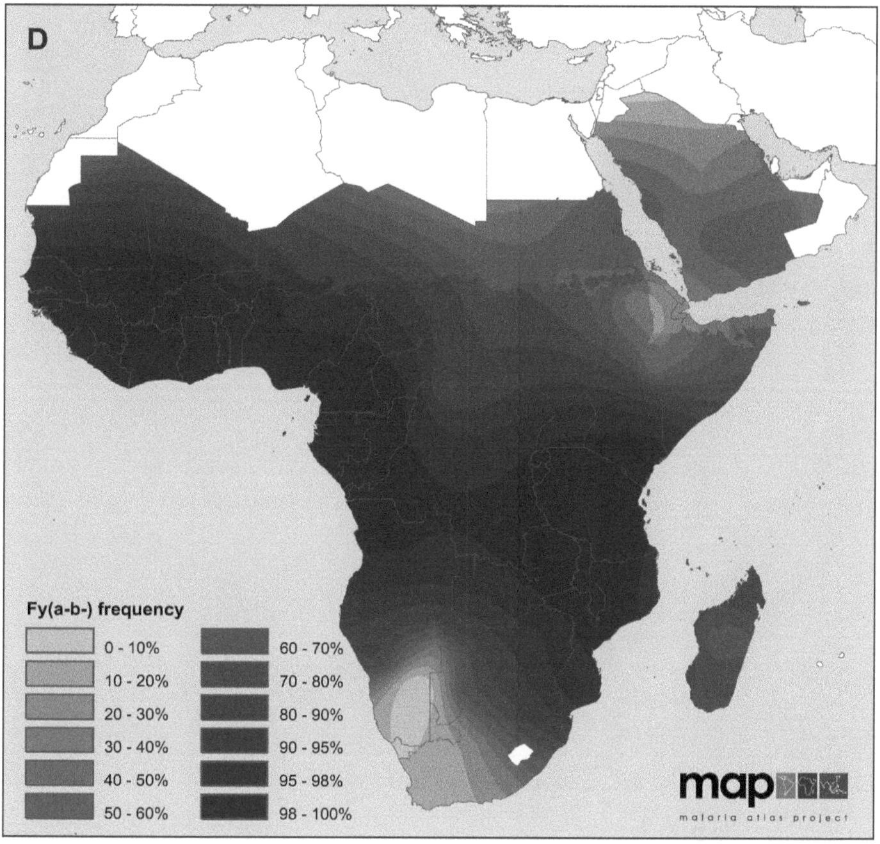

Figure 1.7, cont'd

Table 1.1 Area and Populations at Risk of *Plasmodium vivax* Malaria in 2010

	Area (million km^2)			Population (million)		
Region	**Unstable**	**Stable**	**Any risk**	**Unstable**	**Stable**	**Any risk**
America	1.38	8.08	9.46	87.66	49.79	137.45
Africa+	20.60	1.86	22.46	48.72	37.66	86.38
Asia	5.60	3.63	9.24	1236.92	812.55	2049.47
Asia-Pacific	0.96	1.78	2.74	150.17	64.90	215.07
World	28.55	15.34	43.90	1523.47	964.90	2488.37

Risk is stratified into unstable risk (*Pv*API < 0.1 per 1000 people p.a.) and stable risk (*Pv*API ≥ 0.1 per 1000 people p.a.). America = Central and South America; Africa+ = Africa, Yemen and Saudi Arabia; Asia = mainland Asia excluding the Malaysian Peninsula; Asia-Pacific = southern islands of Asia-Pacific and the Malaysian Peninsula.

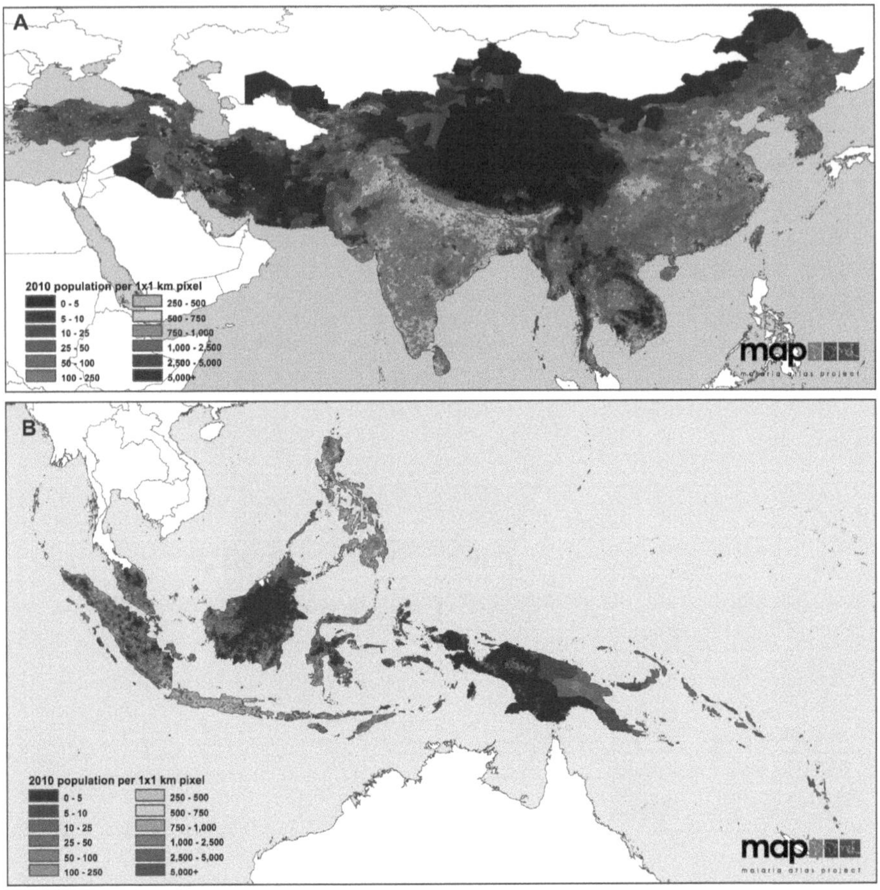

Figure 1.8 Population surfaces for 2010 in Asia, Asia-Pacific, the Americas and Africa+. Population estimates in Asia (A), Asia-Pacific (B), the Americas (C) and Africa+ (D) were estimated from the GRUMP beta 2000 data (Balk et al., 2006; CIESIN/IFPRI/WB/CIAT, 2007) and are shown here at 0.0083 × 0.0083 decimal degree resolution (0.0083 decimal degrees is approximately equal to 1 km at the equator). Only the *Pv*MECs in each region are shown in colour. For a colour version of this figure, the reader is referred to the online version of this book.

epidemiologically important vectors, and by far the largest populations at risk. A low prevalence of Duffy negativity and high population densities in many of the endemic countries in Asia make it an important global *P. vivax* transmission setting to understand and the vast majority (84%) of the global PAR of stable *P. vivax* transmission live in this region.

Defining the limits of transmission. API data were available for all countries except Uzbekistan and Democratic People's Republic of Korea (Korea

Figure 1.8, cont'd

DPR). Five countries reported data up to the year 2010 (Bhutan, Georgia, Nepal, Sri Lanka and Thailand) with the majority (14/23) providing data up to 2008. India and China's last year of reporting was 2007.

Transmission was estimated to span 9 million square kilometres of land in Asia (around 45% of the total land area in Asia). The majority of this area (61%; 5.60 million km^2) was at unstable risk. The areas at risk in India and China covered 2.97 and 2.82 million km^2 each and, therefore, comprised 63% of the total area at risk in Asia (Panel A1 of Fig. 1.3).

Estimating endemicity. There were 2665 records of prevalence data collected from the region. The three most data-rich countries were Viet Nam (n = 657), Afghanistan (n = 493) and Bangladesh (n = 365), as illustrated in

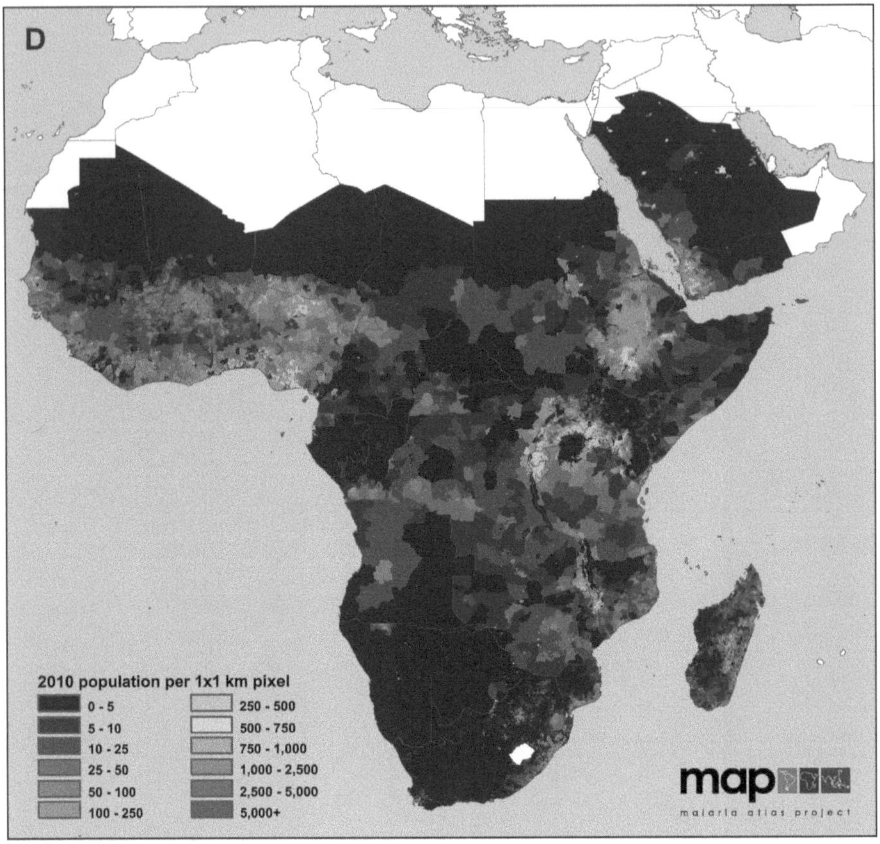

Figure 1.8, cont'd

Panel A1 of Fig. 1.3. The predicted prevalence estimates in stable transmission areas in Asia were highly heterogeneous. Areas with point estimates above 7% were found in small parts of India, Myanmar and Thailand. Regions with $Pv\mathrm{PR}_{1-99}$ estimates between 3.5 and 7% were found in large areas of India and pockets of China, Myanmar, Thailand, Cambodia and Lao People's Democratic Republic (Lao PDR). Uncertainty maps reveal that areas of high uncertainty correspond with high transmission areas. These areas also had sparse prevalence survey data. Relative to its size, India had very little prevalence data available. Myanmar and Thailand also had large regions with stable transmission but with few surveys to support the prevalence estimates. While predictions on the Korean peninsula were made with relatively high certainty (Fig. 1.4A1), there were no survey data available from Korea DPR. Known prevalence values from this region as well as from India, Myanmar,

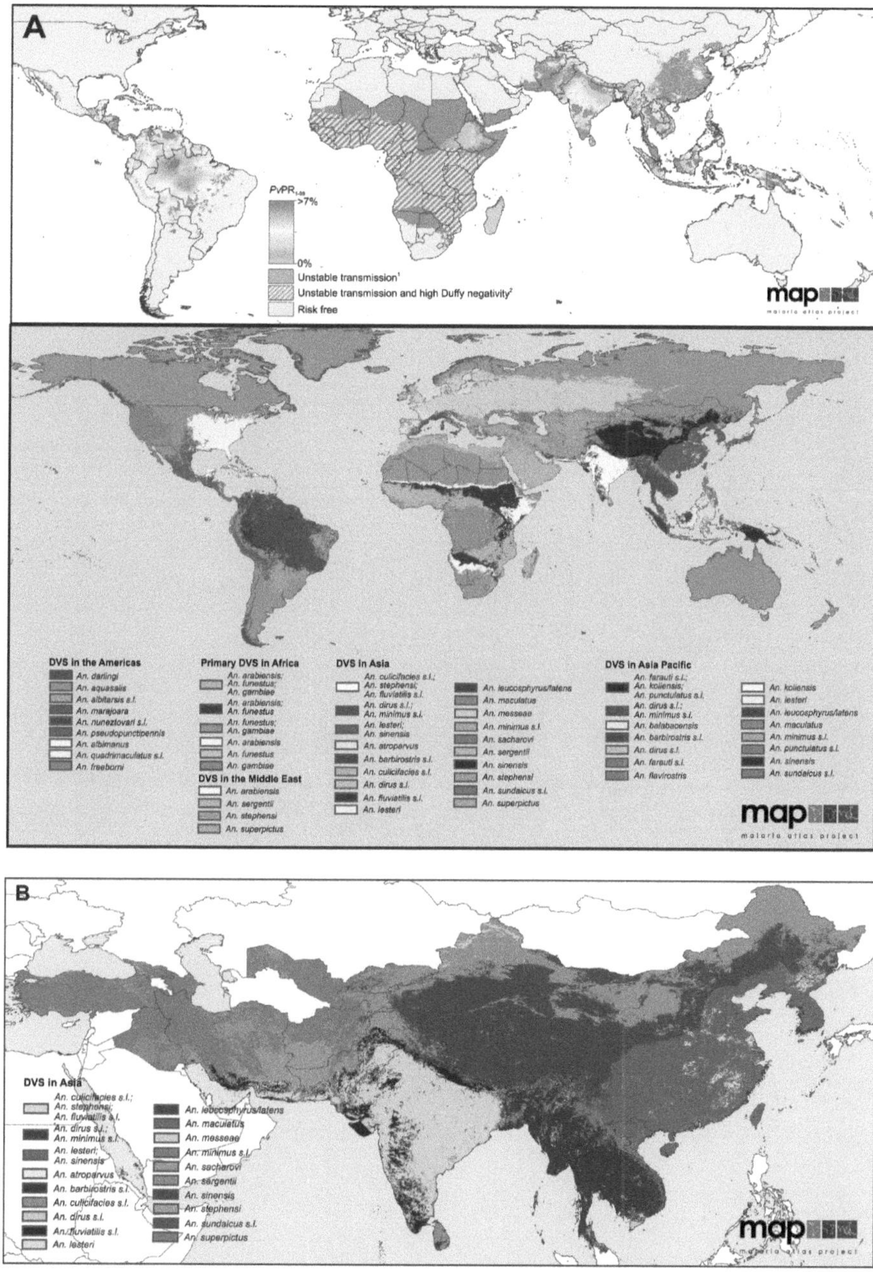

Figure 1.9 Distribution of dominant vector species globally and in Asia, Asia-Pacific, the Americas and Africa+.

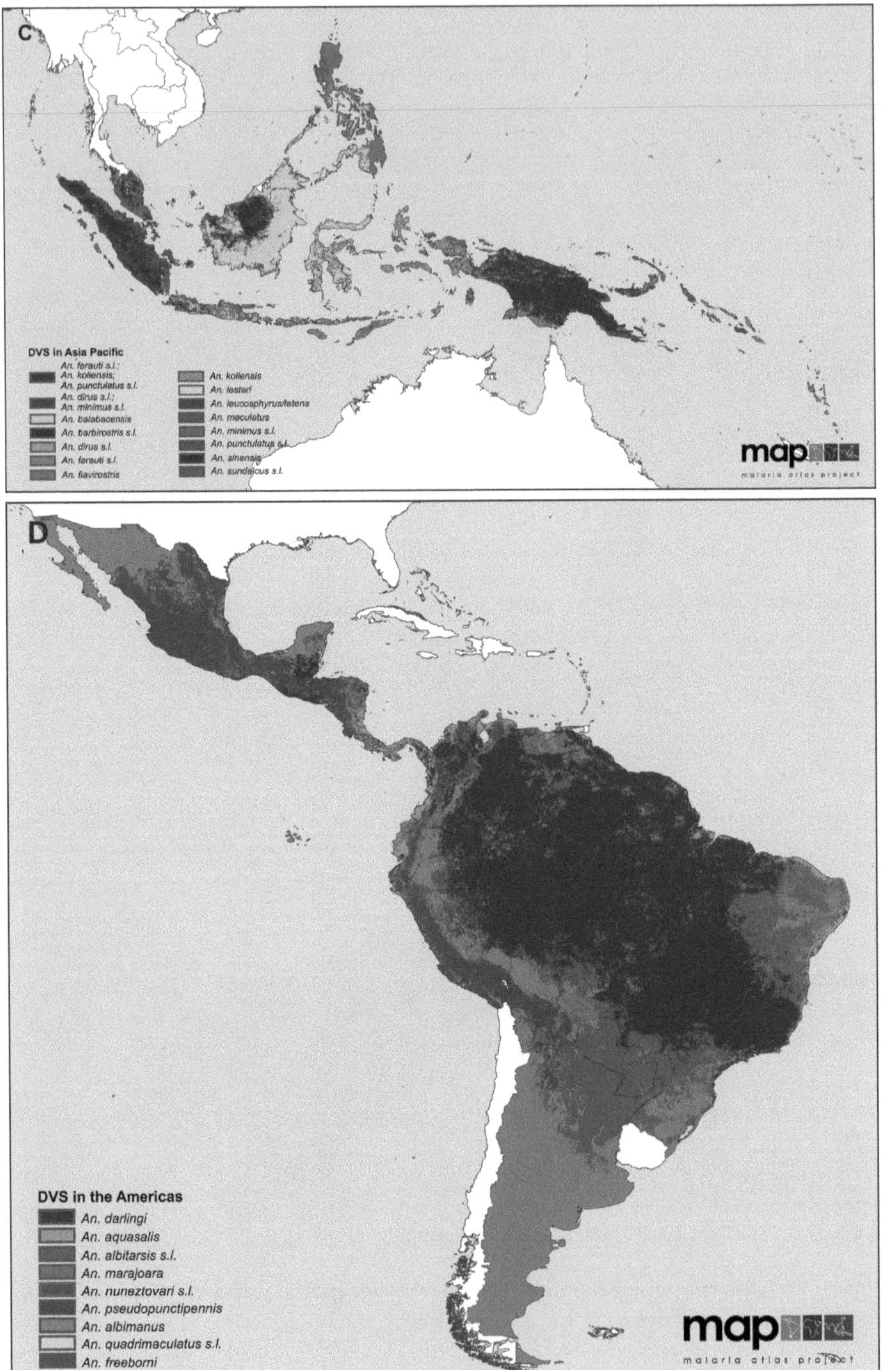

Figure 1.9, cont'd

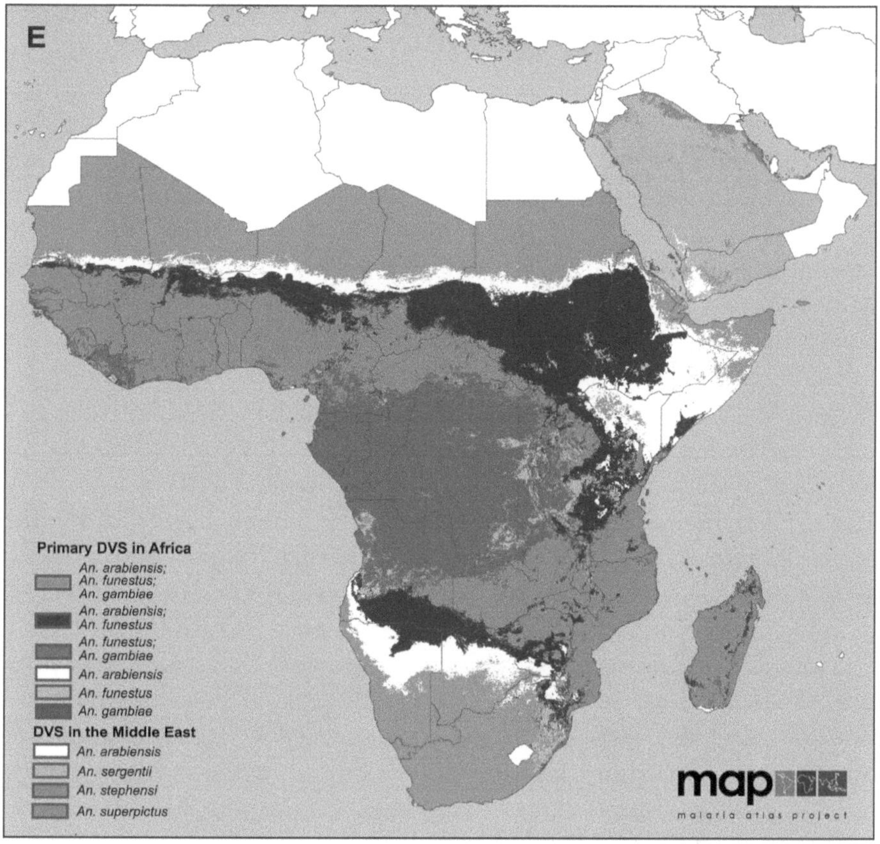

Figure 1.9, cont'd The global distribution of the main dominant vector species is shown in Panel A alongside the predicted endemicity of *P. vivax* in 2010. The distribution of the primary vector species are also shown at the regional level: Asia (B), Asia-Pacific (C), the Americas (D) and Africa+ (non-endemic countries are shaded white). The legend for Africa+ is disaggregated in to Africa and the Middle east because only the primary vector species in Africa are shown in the map. For colour version of this figure, the reader is referred to the online version of this book.

Thailand and parts of China with stable vivax transmission would improve the certainty of $Pv\text{PR}_{1-99}$ predictions for the area as a whole. Regions with low endemicity and high density of surveys, such as Afghanistan, Cambodia and the small region with stable transmission in Turkey, were predicted with high certainty. The population-weighted uncertainty map differs substantially for parts of this region (Fig. 1.4A2). Incorporating population estimates allows weighting to highlight areas where high uncertainty and large populations coincide. Thus, sparsely populated areas of Myanmar, Thailand,

Cambodia, Lao PDR, Viet Nam and Korea DPR displayed small values of this index despite uncertain predictions of prevalence. Conversely, values for highly populated areas across India, parts of Pakistan and the stable transmission region in China, were inflated substantially. Predictions in this region would be best improved, therefore, with more detailed prevalence data from these population centres.

Population at risk. The PAR in Asia was, by far, the largest of all of the regions due to large areas of risk and high population densities. There were over two billion people living at risk in this region in 2010, which constituted 82% of the total global PAR. The greatest PAR was found in India, China and Pakistan with 1.13 billion, 462 and 169 million at risk, respectively. India's PAR comprised more than half (55%) of the Asia-region PAR. Together, China and Pakistan made up another 30% (23 and 8%, respectively), indicating that 85% of the region's PAR was attributable to three nations. These nations also had the largest populations in this region. Compared to India and Pakistan, a smaller proportion of China's overall population was living at risk of *P. vivax*. Of China's estimated 1.4 billion people, 33% were at risk of *P. vivax*, whereas 95% of India's 1.2 billion and 90% of Pakistan's 188 million people were living at some level of risk. When we consider both levels of risk (unstable and stable transmission), we find that 60%, or 1.2 billion, of the PAR in Asia experienced unstable risk, indicating that over 800 million individuals in this region lived in stable transmission areas. Countries devoid of stable transmission areas were Iraq, Kyrgyzstan and Uzbekistan. Countries with the greatest population at stable risk in this region were again India (642.07 million), Pakistan (43.82 million) and China (31.92 million). Other countries that had large (>10 million) PAR living in stable transmission areas were Korea DPR (21.07 million), Myanmar (20.48 million), Thailand (17.11 million) and Bangladesh (16.04 million). Ninety eight percent of the population living in stable transmission areas was, therefore, found in 30% (7 out of 24) of the countries in Asia.

Vectors. Of the 35 potential *P. vivax* vector species found within Asia as a whole (Table 1.2), there are 19 DVS (Sinka et al., 2011) of which, at least seven are considered to be species complexes (Harbach, 2004). Nine DVS are found solely in the Asia region, whilst the remaining 10 have distributions extending into the Asia-Pacific region.

A total of 10,667 unique occurrence points were georeferenced for the 19 DVS across Asia. Records of species occurrence obtained from observations recorded to point (≤ 10 km^2) and wide area (10–25 km^2) locations (sites) were used for analyses. In some cases, more than one occurrence point was obtained for a single site. The greatest number of points were found for the *Anopheles culicifacies* complex (n = 1568) followed by the *Anopheles subpictus* (n = 1143) and *Anopheles barbirostris* (n = 1064) complexes. From the 15 Asian *P. vivax* malaria endemic countries (*Pv*MECs), where anopheline occurrence data were found, data were obtained from 5388 sites. The greatest number of sites were in Myanmar (n = 1830), India (n = 1529) and China (n = 355). There was only one reported occurrence located in Turkey.

The predicted distributions of the DVS in Asia resulted in a complex multi-species map, where the main vector species overlap over large areas of land and exist independently in small areas (Fig. 1.9B). It is beyond the scope of this review to describe the individual distributions and bionomics of the 19 potential DVS in this region, however, a complete discussion of the ranges and bionomics of the global DVS may be found elsewhere (Sinka et al., 2010a, 2010b, 2011, 2012). The DVS described in the following paragraphs are those that were deemed to be primary malaria vectors by the project's technical advisory group (TAG), which are therefore shown prominently in Fig. 1.9A, and showed conclusive evidence for the ability to transmit *P. vivax* (Table 1.2). The potential primary DVS that met these criteria were the *An. culicifacies* complex (Culicifacies Complex; *An. culicifacies s.l.*), the *Anopheles dirus* complex (Dirus Complex; *An. dirus s.l.*), the *Anopheles minimus* complex (Minimus Complex; *An. minimus s.l.*), *Anopheles sinensis*, *Anopheles stephensi* and *Anopheles superpictus*.

Anopheles culicifacies s.l. is a prominent vector across the Indian subcontinent, found in sympatry with *An. stephensi* and *Anopheles fluviatilis s.l.* Species B of the Culicifacies Complex is considered a non-vector of *P. falciparum*, currently attributed to highly zoophilic behaviour. However, there is evidence shown in Table 1.2 that the species is also partially refractory to *P. vivax* and, hence, perhaps to malaria parasites overall (Adak et al., 2005; Vijay et al., 2011). Evidence of wild-infected populations and experimental infections suggest that at least species A and C of the complex are primary vectors of malaria on the Indian subcontinent. Bangladesh, Myanmar, Thailand, Cambodia, Lao PDR and Viet Nam are dominated by the Dirus and Minimus Complexes (Fig. 1.9B). The range of the *An. sinensis* complex covers the majority of China, where it is shown to co-exist with *Anopheles lesteri* along the eastern areas of the country and the Korean peninsula. However, the overlapping distributions of these two vector species may be an artefact of mis-identification in some areas rather than true sympatry

Table 1.2 Known and potential vector species of *Plasmodium vivax*

Species, species complex* or group	Distribution and bionomics reviewed by MAP†	*P. vivax* vector (Yes/No)‡	Notes and reference(s)
Asia			
An. atroparvus van Thiel	Yes	Yes(?)	*An. atroparvus* has been shown to be experimentally infected with *P. vivax* by humans (Daskova and Rasnicyn, 1982) and (Collins et al., 1980; Collins et al., 2009). This species has been shown to be refractory to *P. falciparum* infection (de Zulueta et al., 1975; Daskova and Rasnicyn, 1982; Romi et al., 2001).
An. baimaii Sallum & Peyton	Yes (part of the *An. dirus* complex)	Yes(?)	*P. vivax* and *P. falciparum* are commonly found in *An. baimaii* (Sallum et al., 2005; Obsomer et al., 2007), but original sources do not differentiate the members of the Dirus Complex (Prakash et al., 2001).
An. culicifacies complex	Yes	Yes	Naturally *P. vivax*-infected *An. culicifacies* were found in Madhya Pradesh (Species C and D) (Subbarao et al., 1992) and Uttar Pradesh (Species A) (Subbarao et al., 1988) in India, and in Sri Lanka (Amerasinghe et al., 1991b). *An. culicifacies* have been experimentally infected with *P. vivax* in from infected monkeys (Collins et al., 2009) and laboratory colonies established from wild caught *An. culicifacies* (Species A, B and C) were infected with vivax from blood drawn from infected humans (Adak et al., 1999). Species A and C showed relatively high oocyst infections but Species B did not. Sporozoite infections were relatively high in Species A compared to Species C and negligible in Species B. There is evidence that Species B may be refractory to *P. vivax* infection (Adak et al., 2006; Vijay et al., 2011).
An. fluviatilis complex	Yes	Yes(?)	Sporozoites were detected in specimens collected in Iran and the species is said to transmit both *P. vivax* and *P. falciparum* (Eshghi et al., 1976). Both oocytes and sporozoites were found in an *An. fluviatilis* (Species T) laboratory colony infected by human volunteers under controlled conditions (Adak et al., 2005).

Hyracanus Group	No	Yes	Wild *P. vivax*-infected Hyracanus Group mosquitoes were found in Assam, India (Rattanarithikul et al., 1996; Prakash et al., 2004).
An. messeae Falleroni	Yes	Yes(?)	*An. messeae* was implicated as a vector of *P. vivax* in eighteenth and nineteenth century Finland (Hulden et al., 2008; Hulden, 2009) and the authors refer to evidence of *An. messeae* as a vector in Russia (Sokolova and Snow, 2002), but only indirect evidence linking malaria incidence with vector density was provided.
An. nimpe Nguyen, Tran & Harbach	No (part of the Hyracanus Group)	Yes(?)	*An. nimpe* was implicated as a vector (Nguyen et al., 2000) of *P. vivax* and *P. falciparum* after being sampled in coastal Viet Nam, but direct evidence of infection was not available.
An. philippinensis-nivipes complex	No	Yes	Wild-infected *An. philippinensis-nivipes s.l.* specimens were found to be positive for *P. vivax* in Assam, India (Prakash et al., 2004).
An. pulcherrimus Theobald	No	Yes	*An. pulcherrimus* is thought to be the main vector in Afghanistan (Brooker et al., 2006). *Plasmodium vivax* (VK210 and VK247 subtypes) was detected in specimens collected in Afghanistan (Rowland et al., 2002) and *P. vivax* circumsporozoite proteins were detected though enzyme-linked immunosorbent assay (ELISA) in wild-infected wild-infected mosquitoes (the vast majority of which were *An. pulcherrimus*), but the authors do not explicitly state that *P. vivax* was found in *An. pulcherrimus* (Faulde et al., 2007).
An. sacharovi Favre	Yes (part of the Maculipennis Subgroup)	Yes	*P. vivax* circumsporozoite antigens were detected in wild populations sampled in Turkey (Simsek et al., 2010). *P. vivax* oocysts and sporozoites were detected in the salivary glands of a laboratory colony of *An. sacharovi* fed on vivax infected humans (Kasap, 1990).
An. sawadwongporni Rattanarithikul & Green	No (part of the Maculatus Group)	Yes	*P. vivax* (VK247) was detected in specimens collected in Thailand (Coleman *et al.*, 2002).
An. sergentii (Theobald)	Yes	Yes	*P. vivax* circumsporozoite proteins were detected in *An. sergentii* specimens sampled from desert oases in Egypt (Kenawy et al., 1990).

Continued

Table 1.2 Known and potential vector species of *Plasmodium vivax*—*cont'd*

Species, species complex* or group	Distribution and bionomics reviewed by MAP†	*P. vivax* vector (Yes/No)‡	Notes and reference(s)
An. stephensi Liston	Yes	Yes	*P. vivax* (VK210 and VK247) was detected in specimens collected in Afghanistan (Rowland et al., 2002). *An. stephensi* has been experimentally infected with *P. vivax* from infected monkeys (Basseri et al., 2008; Collins et al., 2009) and humans (Adak et al., 2005); both oocysts and sporozoites found were found in the mosquitoes.
An. superpictus Grassi	Yes	Yes	*P. vivax* (VK247) was detected in specimens collected in Afghanistan (Rowland et al., 2002). Laboratory colonies *An. superpictus* have been infected with *P. vivax*-infected humans; oocysts and sporozoites were in salivary glands (Kasap, 1990). This species seems to be generally implicated as a vivax vector in other articles.
An. varuna Iyengar	No (part of the Aconitus Subgroup)	Yes	Wild-infected mosquitoes were found to be positive for *P. vivax* in Assam, India (Prakash et al., 2004).
Asia and Asia-Pacific			
An. aconitus Dönitz	Yes	Yes	*P. vivax* sporozoites have been detected (by ELISA) in wild captured *An. aconitus* in Sri Lanka (Amerasinghe et al., 1991b). Laboratory studies revealed that *An. aconitus* forms B and C were susceptible to *P. vivax* and *P. falciparum*, and that *An. aconitus* form C was susceptible only to *P. vivax* (Junkum et al., 2005).
An. annularis van der Wulp	Yes	Yes	*P. vivax* sporozoites have been detected (by ELISA) in wild captured *An. annularis* in Sri Lanka (Amerasinghe et al., 1991b) and India (Prakash et al., 2004).
An. barbirostris complex	Yes	Yes	*P. vivax* circumsporozoite proteins were found in specimens collected in Thailand with infectivity rates of 0.24% (Rattanarithikul et al., 1996) and 4.8% (Frances et al., 1996).

An. campestris Reid	No	Yes	*P. vivax* (VK210) was detected in specimens collected in Thailand (Coleman et al., 2002).
An. dirus complex	Yes	Yes	*P. vivax* circumsporozoite proteins have been detected in an *An. dirus* specimen collected in Thailand (Baker et al., 1987). The vectorial capacity of *An. dirus* for *P. vivax* has been estimated (Prakash et al., 2001). Experimental infection from humans (Wirtz et al., 1985), monkeys (Collins et al., 2009) and membrane feeding has been shown (Coleman et al., 2004; Junkum et al., 2005).
An. donaldi Reid	No	Yes	*P. vivax* circumsporozoite proteins have been detected in *An. donaldi* specimens sampled in Malaysia (Seng et al., 1999). *An. donaldi* was successfully infected with *P. vivax* and *P. falciparum* sporozoites in the lab and was capable of transmitting both species to man (Hardin et al., 1973; Harrison and Scanlon, 1975).
An. hodgkini Reid	No	Yes	*P. vivax* was detected in specimens collected in Thailand (VK247) (Coleman et al., 2002).
An. karwari (James)	No	Yes	*P. vivax* circumsporozoite proteins in *An. karwari* specimens collected in Thailand (Frances et al., 1996).
An. kochi Dönitz	No	Yes	Wild *P. vivax*-infected *An. kochi* mosquitoes were found in Assam, India (Prakash et al., 2004).
An. lesteri Baisas & Hu (synonymous with *An. anthropophagus* Xu & Feng)	Yes	Yes(?)	*An. lesteri* (*An. anthropophagus*) were experimentally infected with *P. vivax* from humans in Republic of Korea (Shin et al., 2002). Human laboratory feedings showed *An. lesteri* to be a highly competent vector (Joshi et al., 2009).
An. letifer Sandosham	No	Yes(?)	Specimens collected in Sarawak, Malaysia were found to be sporozoite positive and although the species of the parasite was not identified, >90% of the malaria infections in the region at the time were due to *P. vivax* (Chang et al., 1997). Implied transmission, but with no direct evidence, is described in other references (Rahman et al., 1997; Fryauff et al., 1998).

Continued

Table 1.2 Known and potential vector species of *Plasmodium vivax*—*cont'd*

Species, species complex* or group	Distribution and bionomics reviewed by MAP†	*P. vivax* vector (Yes/No)‡	Notes and reference(s)
An. leucosphyrus Dönitz & *An. latens* Sallum & Peyton	Yes (sister species in the Leucosphyrus Complex)	Yes	*P. vivax* circumsporozoite proteins have been detected in *An. leucosphyrus* specimens sampled in Malaysia (Seng et al., 1999). However, the species sampled was most likely *An. latens*, which was commonly misidentified as its sister species *An. leucosphyrus* (Sallum et al., 2005).
An. ludlowae (Theobald)	No	?	No references were found to confirm that *An. ludlowae* transmits *P. vivax*, but it is a potential vector in areas where vivax malaria is found (e.g. the Philippines and Indonesia) (Lien et al., 1977; Wooster and Rivera, 1985).
An. minimus complex	Yes	Yes	*P. vivax* was detected in specimens collected in Thailand (VK210 and VK247) (Coleman et al., 2002) and India (VK247) (Prakash et al., 2004). *An. minimus* has also been experimentally infected with *P. vivax* from monkeys (Rattanarithikul et al., 1996; Collins et al., 2009).
An. nigerrimus Giles	No	Yes	*P. vivax* (VK210) circumsporozoite proteins have been detected in *An. nigerrimus* sampled in China (Alam et al., 2010) and *An. nigerrimus* has been incriminated as a vector there (WHO, 1986).
An. sinensis Wiedemann	Yes (part of the Hyracanus Group)	Yes	*P. vivax* (VK210 and VK247) was detected in specimens collected in the Republic of Korea (Lee et al., 2002). Deliberate infection was also observed in laboratory conditions (Shin et al., 2002).
An. subpictus complex	Yes	Yes	*P. vivax* circumsporozoite proteins have been detected in *An. subpictus* sampled in Sri Lanka (Amerasinghe et al., 1991a).
An. sundaicus complex	Yes	Yes(?)	*An. sundaicus* is described as a vector of primarily *P. falciparum* malaria, implying that it will also transmit *P. vivax* (Am et al., 1993) and *P. vivax* oocysts and sporozoites have been found in infected laboratory colonies (Adak et al., 2005).

An. tessellatus Theobald	No	Yes	*P. vivax* circumsporozoite proteins were detected in *An. tessellatus* species sampled in Sri Lanka (Amerasinghe et al., 1991a).
An. vagus Dönitz	No	Yes	*P. vivax* sporozoites have been detected (by ELISA) in wild captured *An. aconitus* in Sri Lanka (Amerasinghe et al., 1991a) and India (Prakash et al., 2004).
Asia-Pacific			
An. balabacensis Baisas	Yes (part of the *An. leucosphyrus* complex)	Yes(?)	Sporozoites were found in specimens collected in Central Java, Indonesia (Barcus et al., 2002) and Palawan, Philippines (Schultz, 1992). Laboratory reports indicate transmission of *P. vivax* to non-human primates (Collins et al., 1980).
An. farauti complex	Yes	Yes	*P. vivax* circumsporozoite proteins were found in specimens collected in Papua New Guinea (Burkot et al., 1988).
An. flavirostris (Ludlow)	Yes	?	*An. flavirostris* has been described as the 'principal vector' in the Philippines, where *P. vivax* and *P. falciparum* malaria occur (Foley et al., 2003) and has been found with sporozoites in Malaysia (Hii et al., 1985) and the Philippines (Oberst et al., 1988), but the parasite species was not given.
An. koliensis Owen	Yes (part of the Punctulatus Group)	Yes	*P. vivax* circumsporozoite positivity of 0.76% was reported for wild-infected *An. koliensis* in Papua New Guinea (Attenborough et al., 1997).
An. maculatus Theobald/Maculatus Group	Yes	Yes	*P. vivax* (VK210 and VK247) was detected in *An. maculatus* specimens collected in Thailand (Coleman et al., 2002). *An. maculatus* has been experimentally infected with *P. vivax* from monkeys (Wirtz et al., 1985; Collins et al., 2009) with a reported infectivity of 4.8% (Collins et al., 1980).
An. punctulatus complex	Yes	Yes	*P. vivax* circumsporozoite antigens were detected in wild populations of the *An. punctulatus* complex sampled in Papua New Guinea (Burkot et al., 1988).

Continued

Table 1.2 Known and potential vector species of *Plasmodium vivax*—*cont'd*

Species, species complex* or group	Distribution and bionomics reviewed by MAP†	*P. vivax* vector (Yes/No)‡	Notes and reference(s)
The Americas			
An. albimanus Wiedemann	Yes	Yes	*P. vivax* sporozoites have been detected (by ELISA) in wild captured *An. albimanus* in Mexico (Ramsey et al., 1994) and in specimens experimentally infected from humans (Wirtz et al., 1985). Experimental feeding on monkeys revealed that *An. albimanus* was much more susceptible to 'New World' *P. vivax* (*P. vivax* specimens from Central and South America; 21.2% infection rate), than 'Old World' *P. vivax* (0.4% infection rate) (Li et al., 2001). Other laboratory studies using monkeys showed infectivity rates of only 0.6–0.7% (Collins et al., 1980).
An. albitarsis complex	Yes	Yes	*P. vivax* infection was detected (by dissection to find presence and ELISA to determine *Plasmodium* species) in wild captured *An. albitarsis* in Northern Brazil (de Arruda et al., 1986). Mosquito oocyst infection was observed in laboratory conditions after feeding on infected human volunteers; sporozoites were found in the salivary glands approximately two weeks after feeding (Klein et al., 1991). *P. vivax* infection was also seen in mosquitoes experimentally fed on infected monkeys (Collins et al., 1985).
An. aquasalis Curry	Yes	Yes	*P. vivax* sporozoites have been detected by ELISA in wild captured *An. aquasalis* in Brazil (Povoa et al., 2003). *An. aquasalis* has been experimentally infected with *P. vivax* from infected humans (da Silva et al., 2006).
An. argyritarsis Robineau-Desvoidy	No	Yes(?)	*An. argyritarsis* naturally infected with *P. vivax* have been found in Argentina and the species is identified as a principal vector in parts of Brazil. However, there is some debate whether the species sampled were *An. argyritarsis* or misidentified *An. darlingi* specimens (Linthicum, 1988).

An. bellator Dyar & Knab	No	Yes(?)	*An. bellator* has been found with *P. vivax* oocysts or both oocysts and sporozoites in Brazil (Deane, 1986) and have been experimentally infected from an infected human (Rozeboom and Laird, 1942).
An. benarrochi Gabaldón	No	Yes(?)	Naturally *P. vivax* infected *An. benarrochi* were observed in Peru (Flores-Mendoza et al., 2004), however when mosquito oocyst infection was observed in laboratory conditions after feeding on infected human volunteers, no sporozoites were found in the salivary glands approximately two weeks after feeding (Klein et al., 1991).
An. braziliensis (Chagas)	No	Yes(?)	Naturally *P. vivax* infected *An. braziliensis* were observed in Brazil (da Silva-Vasconcelos et al., 2002), however when mosquito oocyst infection was observed in laboratory conditions after feeding on infected human volunteers; no sporozoites were found in the salivary glands approximately two weeks after feeding (Klein et al., 1991).
An. cruzii Dyar & Knab	No	Yes	*P. vivax* (VK247) infectivity rates of 0.086–0.179% in wild-infected mosquitoes were reported in Brazil (Branquinho et al., 1997).
An. darlingi Root	Yes	Yes	*An. darlingi* specimens were found to be naturally infected with *P. vivax* in Amapa (Conn et al., 2002) and Para States (by dissection to find presence and ELISA to determine *Plasmodium* species) (de Arruda et al., 1986) in Brazil and in French Guiana (Girod et al., 2008). Mosquito oocyst infection was observed in laboratory conditions after feeding on infected human volunteers; sporozoites were found in the salivary glands approximately two weeks after feeding (Klein et al., 1991).
An. deaneorum Rosa-Freitas	No (part of *An. albitaris* complex)	Yes	*An. deaneorum* is a sibling of *An. albitarsis* complex and may be an important vector in Amazonian Brazil (Conn et al., 2002). This species can be experimentally infected by both *P. vivax* and *P. falciparum* (Klein et al., 1991; Senise et al., 2006).
An. freeborni Aitken	Yes	Yes(?)	*An. freeborni* has been experimentally infected with *P. vivax* from humans (Burgess and Young, 1950) and monkeys (Collins *et al.*, 2009).

Continued

Table 1.2 Known and potential vector species of *Plasmodium vivax*—*cont'd*

Species, species complex* or group	Distribution and bionomics reviewed by MAP†	*P. vivax* vector (Yes/No)‡	Notes and reference(s)
An. hermsi Barr & Guptavanj	No	Yes(?)	*An. hermsi* has been implicated in outbreaks of *P. vivax* in California (Maldonado et al., 1990; Ginsberg, 1991) and *An. hermsi* has been experimentally infected with vivax from monkeys (Collins et al., 2009).
An. marajoara Galvão & Damasceno	Yes (part of the *An. albitarsis* complex)	Yes	Specimens of *An. marajoara* (a sibling of *An. albitarsis s.l.*) collected in Amapa State, Brazil were found to be naturally infected with vivax and may be a superior vector for this parasite over *An. darlingi* in this location (Conn et al., 2002).
An. mediopunctatus Lutz	No	Yes(?)	Mosquito oocyst infection was observed in laboratory conditions after feeding on infected human volunteers and *P. vivax* sporozoites were found in the salivary glands approximately two weeks after feeding (Klein et al., 1991), however other literature states that this species was present in low numbers (de Arruda et al., 1986) or was a non-vector (Vittor et al., 2006).
An. nuneztovari complex	Yes	Yes	*P. vivax* infection was detected in wild captured *An. nuneztovari* by dissection and ELISA was used to determine *Plasmodium* species in Para State, Northern Brazil (de Arruda et al., 1986).
An. oswaldoi Peryassú	No	Yes	*An. oswaldoi* was incriminated as a vector of *P. vivax* in Southern Colombia (Quinones et al., 2006) and Northern Brazil (de Arruda et al., 1986). Mosquito oocyst infection was observed in laboratory conditions after feeding on infected human volunteers and *P. vivax* sporozoites were found in the salivary glands approximately two weeks after feeding (Klein et al., 1991).

An. pseudopunctipennis complex	Yes	Yes(?)	*An. pseudopunctipennis* is implied as the vector of *P. vivax* in Bolivia (Lardeux et al., 2007) and has been experimentally infected with *P. vivax* with variable infectivity rates from different vivax strains (Rodriguez et al., 2000). While natural infections with *P. vivax* have been reported from Argentina and Mexico (Warren et al., 1980), there is also evidence of *Plasmodium* sporozoites being absent from *An. pseudopunctipennis* populations in Belize (Achee et al., 2000) and El Salvador (Warren et al., 1980) and experiments have shown *An. pseudopunctipennis* to be refractory to *P. vivax* (VK210) (Gonzalez-Ceron et al., 2007).
An. punctimacula Dyar & Knab	No	Yes	Experimental man-to-mosquito transmission has been confirmed, and naturally infected *An. punctimacula* have been found in Panama (Simmons, 1937; Quinones et al., 2006; Ulloa et al., 2006; Loaiza et al., 2008).
An. quadrimaculatus complex	Yes	Yes(?)	Species of the *An. quadrimaculatus* are implied to be vectors of *P. vivax* malaria (Jensen et al., 1996) and has been infected in the laboratory from *P. vivax*-infected monkeys (Collins et al., 2009).
An. triannulatus Neiva & Pinto	No	Yes	*P. vivax* was detected in wild captured *An. triannulatus* in Para State, Northern Brazil (de Arruda et al., 1986). Mosquito oocyst infection was observed in laboratory conditions after feeding on infected human volunteers and *P. vivax* sporozoites were found in the salivary glands approximately two weeks after feeding (Klein et al., 1991).
Africa+			
An. arabiensis Patton	Yes (part of the *An. gambiae* complex)	Yes	*P. vivax* sporozoites have been detected (by ELISA) in wild captured *An. arabiensis* in Southern Ethiopia (Taye et al., 2006) and Madagascar (Fontenille et al., 1990). *An. arabiensis* have been experimentally infected with *P. vivax* from infected monkeys (Collins et al., 2009).
An. funestus Giles	Yes	Yes	*P. vivax* (VK247) circumsporozoite proteins were found in *An. funestus s.s.* specimens collected in Kenya (Ryan et al., 2006).

Continued

Table 1.2 Known and potential vector species of *Plasmodium vivax*—*cont'd*

Species, species complex* or group	Distribution and bionomics reviewed by MAP†	*P. vivax* vector (Yes/No)‡	Notes and reference(s)
An. gambiae Giles/ *An. gambiae* complex	Yes	Yes	*P. vivax* (VK247) circumsporozoite proteins were found in *An. gambiae s.s.* specimens collected in Kenya (Ryan et al., 2006) and *An. gambiae* has been infected with *P. vivax* in laboratory settings from infected monkeys (Wirtz et al., 1985; Collins et al., 2009).
An. labranchiae Falleroni	Yes	Yes	*An. labranchiae* has been stated to be a vector of *P. vivax* (Lindsay and Thomas, 2001; Romi et al., 2001), and a vectorial capacity for *P. vivax* in *An. labranchiae* has been estimated (Romi et al., 1997; Romi et al., 2001).
An. melas Theobald	Yes (part of the *An. gambiae* complex)	?	No natural *P. vivax* infection has been found (Moreno et al., 2004; Bigoga et al., 2007), but that is perhaps because the indigenous populations, rather than the mosquito, are refractory to *P. vivax* (Moreno et al., 2004).
An. merus Dönitz	Yes (part of the *An. gambiae* complex)	?	No natural *P. vivax* infection has been found; *An. melas* sampled in Tanzania were found not to be infected with *P. vivax* (Temu et al., 1998).
An. moucheti Evans	Yes	?	There is no indication of *P. vivax* infection, but this may be due to lack of the parasite within its distribution (Antonio-Nkondjio et al., 2005).
An. multicolor Cambouliu	No	Yes(?)	No *P. vivax* parasites were found in *An. multicolor* collected in desert oases in Egypt (El Said et al., 1983; Kenawy et al., 1990; Morsy et al., 1995). However, the authors infer the potential for transmission and refer to experimental transmission of *P. vivax* by *An. multicolour* under laboratory conditions.

An. nili complex	Yes	?	There is no indication of *P. vivax* infection by ELISA in specimens sampled in Ethiopia (Nigatu et al., 1994), but this may be due to lack of the parasite within its distribution (Antonio-Nkondjio et al., 2005).
An. pharoensis Theobald	No	Yes	*P. vivax* circumsporozoite proteins were detected by ELISA in *An. pharoensis* sampled in southwest Ethiopia (Nigatu et al., 1994).

Vector species were identified during the preparation for a series of articles (Hay et al., 2010c; Sinka et al., 2010a, 2010b, 2011, 2012) regarding the dominant vector species of human malaria. The potential for 71 anopheline species and species complexes to transmit *P. vivax* are shown alphabetically by region.
*Harbach (2004).
†Sinka et al. (2010a, 2010b, 2011, 2012).
‡Yes = evidence of wild/naturally infected mosquitoes; Yes(?) = evidence of laboratory infected mosquitoes or secondary information implicating species as a vector; ? = no information found to confirm species can carry *P. vivax*, but nothing to state it cannot, or unconfirmed or unclear references to species *P. vivax* vector status.

(Sinka et al., 2012). In addition to its sympatric range with *An. culicifacies s.l.* and *An. fluviatilis s.l.* over India, *An. stephensi* was predicted to occur in Afghanistan and Pakistan, where it was sympatric with *An. superpictus*.

The sources for the following summary can be found in Sinka et al. (2011) and Sinka et al. (2010a).

The DVS cover a large range of varying ecological niches and the presence of a large number of species complexes in the region means that variation in behaviours are often seen. Indeed, behavioural variability within a species is also common, for example, *Anopheles annularis* has a range extending across India, down through South-East Asia, across many of the Indonesian islands and Timor Island. However, *An. annularis* only has a focal role in malaria transmission in selected areas of India. Elsewhere it is considered of little importance.

The Indian DVS have overlapping, sympatric ranges and include a number of species complexes. For example, there are five sibling species in the *An. culicifacies* complex (species A, B, C, D and E) and although species A, C, D and E are all reported to be vectors of malaria in India (Sinka et al., 2011), species E is the most important because of its highly anthropophilic and endophilic behaviour. Members of the complex are found across the plains and up into the highlands (Iyengar, 1954; Barik et al., 2009). Larval habitats include a variety of man-made and naturally occurring water bodies and tolerance of brackish water has been observed in species E (Roberts, 1996). Species C shows plasticity in its behaviour, being found in both forested and deforested areas. Varied bionomics are therefore observed both within and among members of this species complex.

The other primary Indian DVS are *An. fluviatilis s.l.* and *An. stephensi*. Variable behaviour is observed amongst the three members of the Fluviatilis Complex (species S, T and U). Within the complex, there are reports of anthropophilic, zoophilic and exo- and endophagic populations, biting at dusk and during the night. *Anopheles stephensi* is primarily zoophilic and endophagic, and both *An. fluviatilis s.l.* and *An. stephensi* rest primarily indoors. *Anopheles stephensi* is unusual amongst *Anopheles* species in that it appears to be able to use virtually any water condition/habitat as a larval site, hence, its success in urban areas. Less is known regarding the larval habitats of *An. fluviatilis s.l.*, but the complex is associated with slow-moving streams or river margins.

Amongst the northern Asia DVS, *An. sinensis s.l.* and *An. superpictus* are considered potential primary vectors of *P. vivax*. In Thailand, members of the *An. sinensis* complex are considered zoophilic and exophagic, and therefore, of little vectorial importance. In the Korean Peninsula, on the other hand, *An. sinensis s.l.* shows a propensity for biting humans and has been identified as a primary vector (Lee et al., 2001). The larvae of *An. sinensis* complex are associated with lowland, shallow, freshwater habitats. Adult females may hibernate to overwinter from October to April when the temperatures drop below 19 °C (Chow, 1970). Little is known regarding the biting habits of *An. superpictus*, particularly regarding host preference, but they are generally considered to be exophagic. The larvae are commonly found in shallow, flowing, fresh, sun-lit water. However, the species can adapt to human influences by using man-made pits and holes.

The primary DVS in the south-eastern part of Asia are the *An. dirus* and *An. minimus* complexes, both of which contain highly competent vector species. Within the Dirus Complex, which contains seven sibling species, again there is some variability in behaviours (depending on location and sibling species) but the main vectors of the complex (*An. dirus s.s.* and *Anopheles baimaii*) are both highly anthropophilic and will feed both indoors and out. Biting activity appears highly sibling/species dependent with *Anopheles scanloni* (a focal vector within the complex) biting at dusk, *An. dirus* biting in the evening (between 2000 h and 2300 h) and *An. baimaii* biting as late as 0200 h. In general, outdoor resting both before and after feeding has been reported for this species complex. The larvae of this complex inhabit small, temporary, freshwater bodies and are therefore most abundant during the rainy (monsoon) season. *Anopheles minimus s.l.* has three siblings, only two of which (*An. minimus s.s.* and *Anopheles harrisoni*) are considered viable contemporary vectors. Variability in reported behaviour may be a consequence of misidentification or lack of differentiation between these two species, although both are also considered opportunistic and plastic in their behaviour. *Anopheles minimus s.s.* is potentially more anthropophilic, but also adaptable depending on host availability. Overall, *An. harrisoni* is considered more exophagic, exophilic and zoophilic than its sibling. Minimus Complex larvae are typically associated with small- or moderate-sized streams with slow, clear, cool and partially shaded water. *Anopheles minimus s.s.* has been observed in a wider range of habitats from canopied forests to open fields. *Anopheles harrisoni*, on the other hand, is more commonly associated with agricultural areas. Further studies that utilise molecular assays to differentiate sibling species are needed to understand the varied behaviour and habitat preference of *An. minimus s.s.* and *An. harrisoni*.

It is clear that the behaviour of the Asian vectors is complicated, variable and, as yet, not well understood. However, many of the species do exhibit characters that are amenable to control, for example, those that bite indoors at night are conducive to control using ITNs and indoor resting facilitates the effectiveness of IRS.

Asia summary. Plasmodium vivax is an important public health problem across large parts of Asia. Six of the 10 largest PAR estimates of the 95 *Pv*MECs were in Asia and these countries accounted for 87% of the global PAR. Control of *P. vivax* in *Pv*MECs in Asia would, therefore, dramatically reduce the impact of this widespread disease. In Asia, in particular, densely populated areas are endemic with stable *P. vivax* transmission. This is due to environmental suitability as well as a vector species (*An. stephensi*) that is well suited to urban transmission. The limits and endemicity maps produced show that *P. vivax* has a wide geographic range in this region, across which the level of transmission is generally low, with small pockets of intense stable transmission (Fig. 1.3A1 and A2). In high-prevalence areas, such as those found in parts of India (Orissa State) and Myanmar, the clinical character of *P. vivax* has been found to behave more like *P. falciparum*, resulting in cases of severe disease and death (Mendis et al., 2001; Baird, 2007; Price et al., 2007a; Kochar et al., 2009; Mahgoub et al., 2012). Uncertainty estimates were high throughout India and much of Myanmar and population-weighted estimates emphasise the need to improve the estimates in India with more survey data (Fig. 1.4A1 and A2). Improved *P. vivax* surveillance in these regions would greatly improve prevalence predictions. By 2012, several countries in this region (Azerbaijan, Bhutan, China, Georgia, Korea DPR, Republic of Korea, Kyrgyzstan, Sri Lanka, Tajikistan, Thailand, Turkey, Uzbekistan and Viet Nam) had entered the malaria elimination phase (The Global Health Group and the Malaria Atlas Project, 2011). Improved control in neighbouring countries with intense, stable transmission, such as Myanmar, will improve the success of those nations already working towards elimination by reducing the potential for imported malaria from human migration (Tatem and Smith, 2010). As elimination efforts continue, spatially detailed mapping will be needed to capture focal areas of transmission that remain. Further and improved vector surveillance using molecular techniques to disaggregate species complexes is needed to illuminate which vectors are truly the primary vector species of *P. vivax* to allow for appropriate control measures to be deployed in this highly variable region.

4.2. Asia-Pacific

The Asia-Pacific region (defined for these purposes as the southern islands of Asia-Pacific and the Malaysian Peninsula; Fig. 1.2) has amongst the highest endemicity estimates of *P. vivax* malaria (Fig. 1.3B) and the most complex spectrum of vectors. This region has a relatively small land mass, which means it has a much lower estimated PAR but the range in epidemiological and entomological factors here mean it presents a unique and challenging control setting.

Defining the limits of transmission. Annual parasite index data were provided for all the *P. vivax*-endemic countries in Asia-Pacific. The *Pv*MECs in this region are Indonesia, Malaysia, the Philippines, Papua New Guinea, the Solomon Islands, Timor-Leste and Vanuatu.

Transmission was estimated to span 2.74 million km^2 of land in this region, which is primarily made up of islands. The area at risk in Asia-Pacific constituted 6% of the global area at risk. Indonesia, the largest country in the region, had the greatest area at risk, consisting of 1.71 million km^2 of land, which was 90% of the country's total area (1.90 million km^2) and 60% of the total area at risk for the region.

Estimating endemicity. There were 5277 records of prevalence data collected from Asia-Pacific, which made up more than half (53%) of the global *P. vivax* PR survey database. This was primarily due to a large data contribution from Indonesia, which provided 4457 (84%) data points in the Asia-Pacific region and 45% of the global dataset (Fig. 1.3B1).

As with mainland Asia, *P. vivax* endemicity in regions with stable transmission was predicted to vary greatly across the Asia-Pacific region as shown in Panel B2 of Fig. 1.3. $PvPR_{1-99}$ point estimates were predicted to exceed 7% in small parts of Indonesia and the Solomon Islands, and much of Papua New Guinea. Estimates towards the east of the region (Sumatra and Kalimantan) were generally low, while endemicity values of around 5% were predicted to the north (Malaysian Borneo and the Philippines). The vast number of surveys from Indonesia meant that the certainty of the predictions in that portion of the region was relatively high (Fig. 1.4B1). Uncertainty was greatest on New Guinea and parts of Borneo, where prevalence data were sparse. The Philippines, which also had little survey data available, was predicted with relatively low certainty. Again, the population-weighted uncertainty map was substantively different (Fig. 1.4B2). While the highly populated areas of Asia were predicted with less certainty, the high-density areas of Asia-Pacific, which also had a high number of prevalence surveys (Indonesia), showed an increase in certainty in the population-weighted estimates. Uncertainty was also greatly reduced following population-weighting in focal areas in Papua, Indonesia and Papua New Guinea, where population density is low. The Philippines had relatively low uncertainty even when weighted by population estimates, indicating that this region would benefit from increased prevalence surveillance data.

Population at risk. Relative to the geographic size of the region, Asia-Pacific had a large PAR due to pockets of intense transmission in highly populated endemic areas. It was estimated that there were 215 million people at risk in this region. This was only 11% of the PAR in mainland Asia. However, the proportion of the population that was at any risk of *P. vivax* relative to the total population of the *Pv*MECs in Asia-Pacific was comparable to the Asia region: the PAR was 59% of the total population in Asia-Pacific and 58% in Asia. The largest PAR in Asia-Pacific was in Indonesia, with an estimated population of 129.28 million individuals at risk of *P. vivax* out of a total population of 231.60 million (56%) in 2010. The PAR of stable and unstable transmission in Indonesia was 60% of the PAR of Asia-Pacific as a whole. The next largest PAR estimates were from the Philippines and Malaysia. The Philippines had 50.34 million people at risk or 54% of the total population of the Philippines and 23% of the regional PAR. Malaysia had 27.88 million (99.8% of the Malaysian population; 13% of the regional PAR) people at risk. Papua New Guinea had a PAR of 5.64 million (81% of the national population; 3% of the regional PAR) and Timor-Leste had 1.15 million (98% of the national population; 0.5% of the regional PAR). The Solomon Islands and Vanuatu had a PAR less than one million: 0.53 and 0.24 million (both nearly 100% of the total national populations and <0.3% of the regional PAR). Seventy percent of the total PAR in the region was exposed to unstable transmission. Unstable transmission predominated in Indonesia and Malaysia, where 79% (26.78 million) and 84% (4.45 million) of the PAR, respectively, were exposed to that level of transmission. In the Philippines, 53% (26.59 million) of the PAR lived in areas of stable transmission. Nearly all of the transmission in Vanuatu was stable (99.57%). The entire PAR in both Timor-Leste and the Solomon Islands experienced stable transmission. More than half of the population (59%) in Asia-Pacific is at risk of some level of vivax malaria transmission and the 215 million people at risk was 8.7% of the global PAR.

Vectors. Of the 26 potential vector species of *P. vivax* (Table 1.2), we predict that 16 are DVS that occur in Asia-Pacific (*Anopheles aconitus*, *An. annularis*, *An. balabacensis*, *An. barbirostris* complex, *An. dirus* complex, *An. farauti* complex, *An. flavirostris*, *An. koliensis*, *An. lesteri*, *An. leucosphyrus* & *An. latens* (sister species grouped together here due to known mis-identification problems within the current literature), Maculatus Group, *An. minimus* complex, *An. punctulatus* complex, *An. sinensis* complex, *An. subpictus* complex and *An. sundaicus* complex), six of which are found only in Asia-Pacific (*An. balabacensis*, *An. farauti* complex, *An. flavirostris*, *An. koliensis*, *An. leucosphyrus* & *An. latens*, and the Punctulatus Group) (Sinka et al., 2011). The predicted distributions of the DVS in this region are shown in Fig. 1.9B.

For the 16 Asia-Pacific DVS, 9052 unique georeferenced occurrence records were assembled. The most data-rich species/species complexes were the *An. farauti* (*n* = 1737), *An. subpictus* (*n* = 1143) and *An. barbirostris* (*n* = 1064) complexes; and the most data-poor were *An. leucosphyrus* & *An. latens* (*n* = 12). Occurrence data were reported from all seven *Pv*MECs in this region. Most of the data originated from Papua New Guinea (*n* = 1503 sites), followed by Indonesia (*n* = 890), Solomon Islands (*n* = 160), Malaysia (*n* = 145), the Philippines (*n* = 124), Vanuatu (*n* = 36) and Timor-Leste (*n* = 1). Given the heterogeneity in vector distribution in this region, improved data collection from many of these countries would improve the fidelity of the predicted distributions of this complicated region.

The distribution and behaviour of the DVS discussed below are those that met the criteria for having broad distributions in the region (Fig. 1.9C) and have been conclusively incriminated as vectors of *P. vivax* (Table 1.2). A complete discussion of their ranges and bionomics may be found elsewhere (Sinka et al., 2011, 2012). The species that were identified as specific vectors of *P. vivax* in Asia-Pacific are *An. barbirostris s.l.*, *An. farauti s.l.*, *An. koliensis*, *An. leucosphyrus* & *An. latens* and *An. punctulatus s. l.*

Anopheles barbirostris s.l. is predicted to occur across large areas of inland Sumatra and Borneo. It is typically a highland species, but is also found in coastal regions (e.g. western Timor). Adult females of this species complex tend to feed outdoors on animals during dusk and night time, bringing its significance as a vector of malaria into question. Further molecular analysis and identification of the sibling species is needed to determine if variations observed in biting behaviours are due to plasticity within a species or differences amongst different species in the complex. A variety of larval habitats are also observed for *An. barbirostris s.l.* Though they are typically considered swamp breeds, a variety of water depth, size, light intensity and movement has been observed. This variation may again be due to the differences among species within the complex rather than the plasticity of a single species.

Anopheles farauti s.l., *An. koliensis*, and *An. punctulatus s.l.* are in sympatry across most of the islands of New Guinea, New Britain and into the Solomon Islands. All three are found within the Punctulatus Group. There are two sibling species in the Punctulatus Complex: *An. punctulatus s.s.* and *Anopheles* sp. near *punctulatus*.

The latter of the two species is relatively uncommon, whereas *An. punctulatus s.s.* is reportedly widespread and an important vector on the island of New Guinea. The distribution of *An. punctulatus s.s.* spans eastern Indonesia, Papua New Guinea and the Solomon Islands. Adult females feed readily on humans outdoors or occasionally indoors. Those that feed indoors may rest there, but are more likely to exit to find an outdoor resting spot. Peak biting times vary based on the geographic location. The larvae of *An. punctulatus s.s.* thrive in temporary pools that result from disturbed ecology. The habitats tend to be small, shallow, sunlit pools of water that may be turbid, but never brackish.

The distribution of the eight cryptic species of the Farauti Complex is largely dependent on the species' tolerance of salinity such that some members are coastal species (e.g. *An. farauti s.s.*) while others are restricted to inland areas with freshwater larval habitats (e.g. *Anopheles hinesorum*). The feeding habits of *An. farauti s.l.* also vary; host preference changes based on the availability of hosts and resting may be dependent on where feeding occurred. Peak biting times fluctuate with location, but it should be noted that daytime biting may occur. The larval habitats of members of the complex are typically in natural rain-fed water bodies, but artificial containers such as water drums or coconut shells may also be used. The habitats range in salinity and light intensity due to the wide geographic distribution of the species complex.

The predicted distribution of *An. koliensis* is similar to that of the Punctulatus Complex. The adult females tend to be anthropophilic, but will feed on other animal hosts. Although females may go indoors in search of a host, resting typically occurs outdoors. Feeding can occur throughout the night, with variable peak times based on location. The larval habitats of *An. koliensis* are intermediate between those of *An. farauti s.l.* and *An. punctulatus s.l.* and are generally associated with permanent freshwater bodies in open grassland such as irrigation ditches.

Anopheles leucosphyrus and *An. latens* are sister species of the Leucosphyrus Complex. Both species are found in forested areas with the former on the island of Sumatra and the latter on the southern portions of the Thai and Malaysian Peninsula and parts of Borneo. Both species are considered to be important vectors of malaria and were long thought of as the same species until molecular and cross-mating studies distinguished between them (Baimai *et al.*, 1988). Little is known of the bionomics of *An. leucosphyrus s.s.* because much of the literature on '*An. leucosphyrus*' is now known to refer to *An. latens*. The species has been shown to bite humans both inside and outside, however. *Anopheles latens* bites throughout the night with peak times that vary based on the seasonal climate and location. Its larval habitats include shaded temporary pools found in the forest such as swamps, stump ground holes or wheel tracks.

The bionomics of the DVS in Asia-Pacific are as variable as in mainland Asia. This is due to variation observed both within species complexes and among separate species. Such variations increase the challenge of developing a single universal vector control strategy across the region, however, a combination of methods could prove effective (Lindsay et al., 2004).

Asia-Pacific summary. The vivax malaria problem in Asia-Pacific is complex and presents a variety of challenges to control. Indonesia and the Philippines are densely populated over large land masses (Fig. 1.8B) and *P. vivax* is predicted to be endemic across most of their national territories. They have the fourth and fifth largest PAR estimates globally with 129 million at risk in Indonesia and 50 million in the Philippines. Some of the highest predicted $Pv\text{PR}_{1-99}$ values were also observed in this region (Fig. 1.3B1). As in the Asia regions, it is in these areas, where intense *P. vivax* transmission occurs (e.g. Indonesia and Papua New Guinea), where drug resistance is emerging and malaria cases of equivalent clinical severity to *P. falciparum* have been observed (Baird, 2004; Pukrittayakamee et al., 2004; Tjitra et al., 2008; Price et al., 2009). This and the large areas of unstable transmission make this region a worthy and vital candidate for research, control and eventually elimination efforts; four of the seven *Pv*MECs have already declared their status of working towards elimination: Malaysia, the Philippines, the Solomon Islands and Vanuatu (The Global Health Group and the Malaria Atlas Project, 2011). Uncertainty estimates, shown in Fig. 1.4B1 and B2, were high in areas that also displayed high $Pv\text{PR}_{1-99}$ estimates (Fig. 1.3B2), such as Papua New Guinea. While the population-weighted correction shows that parts of these regions are, for the most part, sparsely populated (Fig. 1.8B), improved surveillance to facilitate high-resolution mapping will be essential to tackle the final transmission hotspots of this region as elimination efforts progress. The vector situation in this region is complex and further research is needed to differentiate members of species complexes and incriminate those vectors that are vectors of *P. vivax* so that appropriate control measures may be taken.

4.3. Americas

The Americas have amongst the highest *P. vivax* endemicity values in the world but these typically occur in areas of very low population density, so the populations at risk are much lower than those found in Asia (Fig. 1.3C1 and Fig. 1.8C). The diversity of vector species is also less in this region (Fig. 1.9C).

Defining the limits of transmission. Plasmodium vivax annual clinical incidence data were available from all 19 countries that are considered to be endemic in the Americas. The most recent year of reporting was 2008 for the majority of the countries. Guyana, Nicaragua and Panama reported data up to 2007; Belize, El Salvador, French Guiana and Guatemala provided data up to 2006 and Colombia's last year of reporting was 2005.

Endemic areas in this region were estimated to span 9.46 million km^2, most of which were found in the Amazon basin. The majority of the areas at risk in the Americas (85%; 8.08 million km^2) were at stable transmission (Fig. 1.3C1). This indicated that, for the most part, the disease was either endemic or wholly absent in this region with limited areas at risk of unstable transmission. The transmission level was generally found to be inversely proportional to the population density: areas experiencing stable transmission were those of lower population density. For example, the largest area at risk of stable transmission in a single country was 4.40 million km^2 in Brazil, which was 54% of the stable transmission area for the Americas region. However, the population at stable risk in Brazil was 26% (13.02 million) of the region's total PAR of stable transmission. The Americas comprised 53% of the global land area at stable transmission but only 5% of the population at that level of risk.

Estimating endemicity. There were relatively few PR records and surveys for the Americas, with 388 unique records available for modelling. The three most data-rich countries in this region were Brazil (n = 175; 45%), Venezuela (n = 65; 17%) and Peru (n = 51; 13%), as shown in Fig. 1.3C1.

The predicted prevalence, or $Pv\text{PR}_{1-99}$, was highly variable throughout the Americas. Much of the areas of stable transmission had PRs between 3 and 5%, with isolated areas nearing 0% and others exceeding 7%. Regions of high prevalence (>7%) were found in Amazonia (Northeast Brazil) and Central America (Nicaragua and Honduras). The uncertainty in the $Pv\text{PR}_{1-99}$ predictions was relatively high overall due to the relatively sparse PR survey data in this region (Fig. 1.4C1). However, when population density was incorporated into the weighted uncertainty calculations, the resulting uncertainty index was greatly lowered because the areas with stable risk and high uncertainty predictions had low population densities (Fig. 1.4C2). This implies that the observed uncertainty has relatively little operational importance at the global scale, although highlights the current lack of precision for defining high-risk foci in remote Amazonian settings.

Population at risk. The PAR in the Americas was relatively low given the geographic limits of *P. vivax* due to low population densities in areas of high transmission. It was estimated that there were 137.45 million people living at risk of *P. vivax* in the Americas in 2010. The majority of the PAR was in areas of unstable transmission (64%; 87.66 million) and 36% (49.79 million) were exposed to stable *P. vivax* transmission. Brazil had the largest PAR with 45.22 million people living at any risk, which is 26% of its total population and represents 33% of the region's PAR). The largest number of people living at stable risk (13.02 million) was in Brazil, followed by Venezuela with 4.43 million at stable risk and 22.87 million total at risk (83% of the national population and 17% of the regional PAR). Colombia was estimated to have 7.53 million at stable risk and 19.15 million at risk (56% of its total population and 14% of the region's total). El Salvador, with a PAR 3.17 million, did not have any individuals living at stable risk. French Guiana (PAR 158,000) and Suriname (PAR 30,000; the lowest PAR in the region) did not have any populations living at unstable risk. The portion of the PAR at stable transmission in Belize was also nearly 100%; only 30 individuals out of 220,000 at risk were in unstable transmission areas. Of the 1.5 billion people living at unstable and 964 million at stable transmission around the globe, 5.8% and 5.2% of those were found in the Americas, respectively.

Vectors. Of the 20 potential vectors (Table 1.2) of *P. vivax* in the Americas, nine are considered DVS. These include *Anopheles albimanus*, *An. albitarsis* complex, *An. aquasalis*, *An. darlingi*, *An. freeborni*, *An. marajoara*, *An. nuneztovari* complex, *An. pseudopunctipennis* and *An. quadrimaculatus* (Sinka et al., 2010b). The predicted distribution of these species is shown in Fig. 1.9D. There were 4141 georeferenced spatiotemporally unique occurrence points obtained for *Anopheles* DVS from 25 countries in the Americas. Twenty-two percent of the occurrence records (n = 926) referred to *An. darlingi* and 20% (n = 851) to *An. albimanus*. The remaining species had occurrence data of between 63 (*An. freeborni*) and 572 (*An. quadrimaculatus s.l.*) records. Of the total number of point locations identified (n = 1509), the greatest number came from the United States (n = 377), followed by Brazil (n = 304) and Belize (n = 228). The *Pv*MECs with the fewest point records (<10) were French Guiana (n = 7), El Salvador (n = 3), Paraguay (n = 2), Guyana (n = 1), Honduras (n = 1) and Nicaragua (n = 1).

Distribution maps indicate a relatively straightforward vector profile across the *Pv*MECs of the Americas. Of the nine DVS in this region, *An. darlingi* and *An. albitarsis s.l.* met the criteria for further description because these species are predicted to have a wide distribution in the region and conclusive supporting evidence indicating transmission of *P. vivax* in the wild, indeed, the distribution of *An. darlingi* is remarkably similar to that of *P. vivax* in the region (1.3C2 and 1.9D). Although *An. darlingi* is considered a riverine species inhabiting forest locations, it has been observed to take advantage of deforestation, exploiting areas with reduced canopy cover compared to those more densely forested locations. The larval habitats of this species are typically clear and natural water bodies such as slow-flowing, clear rivers and streams. Adult female *An. darlingi* are both exo- and endophagic, though they tend to rest outdoors regardless of where their blood meal was taken. Host preference also varies within this species as does biting activity (time of biting), which may adapt to correspond to human behaviour (i.e. late night human activity influences or causes late night biting by *An. darlingi*).

The Albitarsis Complex is composed of five sibling species including the known vectors, *An. albitarsis*, *An. deaneorum* and *An. marajoara*. Members of this complex exhibit some behaviours that are similar to *An. darlingi*, for example, the adult females tend to rest indoors but will bite indiscriminately without much host preference both indoors and outdoors. The larval habitats tend to be clear, sunlit, freshwater.

The Americas summary. National surveillance data have shown a widespread decrease in morbidity and mortality from both major *Plasmodium* species across the American continent since 2000 (WHO/PAHO, 2008). This is due largely to the successful implementation of integrated vector management (Roberts et al., 1997; Butler and Roberts, 2000; Rojas et al., 2001; Killeen et al., 2002; Roberts et al., 2002; Shiff, 2002; WHO/PAHO, 2006) and has led eight of the 21 *Pv*MECs in the Americas (Argentina, Belize, Costa Rica, El Salvador, Mexico, Nicaragua, Panama and Paraguay) to target malaria elimination (The Global Health Group and the Malaria Atlas Project, 2011). However, the prevalence estimates of *P. vivax* in the Americas are heterogeneous with isolated areas of intense transmission (Fig. 1.3C2). Vector occurrence data coverage was uniformly low in the American *Pv*MECs. Improved vector research in the Americas will advance the maps of species-specific distributions, which is essential for the continued success of vector management to curb malaria morbidity and mortality. Pockets of high transmission (*Pv*PR > 7%) were observed in large areas of the region. However, it is important to note that these were generally in areas with low population estimates. Sparse prevalence data resulted in high uncertainty estimates in the region that were greatly reduced when the density of the population in endemic areas was taken into consideration (Fig. 1.4C). *Anopheles darlingi* appears to be the primary vector in this region; however, eight of the potential vectors of *P. vivax* in this region lack conclusive evidence regarding their potential to transmit the parasite. Therefore, further research is needed to decisively incriminate the vectors of *P. vivax* in the Americas.

Plasmodium vivax makes up the vast majority of malaria transmission that occurs in the Americas (Arevalo-Herrera et al., 2010). Survey data from this region were relatively sparse and records represent a picture of heterogeneous transmission levels. The Americas contributed a small fraction (5.5%) of global PAR of *P. vivax*, but comprised nearly a quarter (22%) of the global area at risk. Areas of high transmission in highly dispersed populations may present unique challenges to malaria control. To advance elimination efforts, high-resolution mapping will be needed to accurately illustrate the degree of heterogeneity in this region and areas that require the greatest resources. This will demand more data from areas with high predicted prevalence and uncertainty, such as Honduras and Nicaragua in Central America and northwest Brazil in South America. Although Brazil is a large and populous country, its areas of stable transmission occur in sparsely populated regions of the Amazon basin. In 2010, Brazil had the seventh largest PAR of *P. vivax* globally (45 million), the largest area at risk (4.90 million km^2) as well as the largest area at stable risk (4.40 million km^2) and, therefore, has an important role to play in the region's future of malaria control and elimination.

4.4. Africa+

The estimates for endemicity and populations at risk of *P. vivax* malaria in Africa+ (defined here as Africa, Saudi Arabia and Yemen) are mitigated by the high prevalence of Duffy negativity in these populations, and the vector situation in this region is relatively straightforward.

Defining the limits of transmission. High *P. falciparum* endemicity in Africa, coupled with high prevalences of Duffy negativity, has meant that collection of *P. vivax*–specific data has not been a priority in the past. The data available from which to estimate the limits of *P. vivax* transmission is, therefore, limited. *Plasmodium vivax* annual parasite incidence (*Pv*API) data were only available from six Africa+ countries (13%). The last year of reporting available was 2009 for four countries (Dijbouti, Namibia, South Africa and Swaziland); whilst for the remaining two countries (Saudi Arabia and Yemen) the last available reports were from 2006. Based on the limited *Pv*API and *Pv*PR data available for Africa+, and the presence of suitable vectors and climatic conditions, forty-six countries are assumed to be *P. vivax* endemic in this region.

The area at risk was estimated to span over 22 million km^2 of Africa, Yemen and Saudi Arabia. Despite the historical misconception that *P. vivax* is absent from the African continent, 84% of the 22.46 million km^2 of land in the region was at some risk of *P. vivax* transmission (Fig. 1.3D2). However, the vast majority (92%; 20.60 million km^2) of total area at risk was estimated to house unstable transmission.

Estimating endemicity. There were 1640 records of prevalence data from 97 sources obtained for Africa+, 79% of which reported an absence of *P. vivax*. Approximately 16% of the global *Pv*PR data records used in the modelling were from surveys conducted in Africa, Yemen and Saudi Arabia. Ethiopia, Zambia and Sudan were the most data rich countries, contributing 50% (n = 826), 18% (n = 295) and 18% (n = 290) of the total data, respectively.

The predicted prevalence estimates were uniformly low for areas at stable risk in Africa, Yemen and Saudi Arabia as shown in Fig. 1.3D2. The point estimates of predicted $Pv\text{PR}_{1-99}$ rarely exceeded 2%. The uncertainty of the predictions in this region was also very low (Fig. 1.4D). The incorporation of information regarding the proportion of Duffy negative individuals strengthened the predictions in this area and compensated for the sparse parasite rate data available from this region (Fig. 1.3D1, Fig. 1.4D and Fig. 1.7D). Areas with higher $Pv\text{PR}_{1-99}$ (Ethiopia, South Sudan and Madagascar) predictions were also found to have the highest uncertainty, likely because these areas were outside the range of high Duffy negativity and could not borrow from the certainty conferred by that restriction. As with the Americas, the uncertainty in Africa+ was reduced in the population-weighted predictions because of the low population density found in parts of the continent, particularly Madagascar (Fig. 1.4D2). Uncertainty remained relatively high in the highlands of Ethiopia, which are more densely populated and are therefore important targets for control.

Population at risk. The PAR in Africa+ was low given the large geographic coverage of the region because of the high prevalence of Duffy negativity (Fig. 1.3D1 and Fig. 1.7D). There were an estimated 86 million people living at risk of *P. vivax* in Africa, Yemen and Saudi Arabia in 2010, which was 3.5% of the global total. More than half (56%), or 48.72 million, of the PAR were those living in unstable transmission and the remaining 37.66 million (44%) lived in areas at the level of stable transmission. Nine countries had an estimated PAR of zero, all of which are located in West Africa: Côte d'Ivoire, Ghana, Guinea, The Gambia, Guinea-Bissau, Senegal, Sierra Leone and São Tomé and Príncipe. Togo, Equatorial Guinea, Gabon and Burkina Faso also had a total PAR of less than one thousand individuals. Thirty-four countries in the Africa+ region had zero individuals at risk of stable transmission. Of the 12 countries with stable transmission, Madagascar was the only country where the entire PAR (5.23 million) experienced stable transmission. Madagascar has the fourth highest PAR in the region; the highest were Ethiopia, Sudan and Yemen with 35.19 million (41% of the PAR in the Africa+ region), 14.78 million (17%) and 13.06 million (15%) people at risk in each, respectively.

Although the high proportion of Duffy-negative individuals in Africa is thought to result in an absence of the parasite across the continent, evidence of transmission in Africa+ (Guerra et al., 2010; Ménard et al., 2010) supported the decision to assume stable transmission in all areas of Africa+ that were not excluded by *Pv*API data or biological masks. Areas were then reclassified following $Pv\text{PR}_{1-99}$ predictions such that locations with endemicity levels below 1% were reclassified as unstable; this was the majority of the region. High proportions of Duffy negativity are especially common among the population of West Africa. Ghana, for example, had a PAR of 24 million individuals when accounting for environmental exclusions and before the inclusion of Duffy negativity, which subsequently reduced the PAR of Ghana to zero. The incorporation of the Duffy negativity layer reduced the PAR of the Africa+ region from 840 million to 86 million, which is 3.5% of the global PAR of *P. vivax*.

Vectors. The high impact of malaria on the African continent is largely due to the efficiency of the African DVS in transmitting *P. falciparum*. However, *P. vivax* may also play an important role. As more countries on the continent move towards elimination, there may be in shift in the endemicity of *P. vivax* as levels of *P. falciparum* decline. It is essential to know which vector species are capable of transmitting *P. vivax*.

Africa is the home to two of the most efficient vector species of human malaria: *An. gambiae* and *An. funestus* (Gillies and de Meillon, 1968; Coluzzi, 1999). Of the 10 potential vectors of *P. vivax* in Africa+ (Table 1.2), eight (*Anopheles arabiensis*, *An. funestus* complex, *An. gambiae* complex, *An. Labranchiae*, *An. melas*, *An. merus*, *An. moucheti* and *An. nili* complex (Sinka et al., 2010a)) are considered to be DVS of human malaria, but their importance in transmitting *P. vivax* is still uncertain. For example, it is unclear whether the forest vector *An. moucheti* is a poor/non-vector of *P. vivax* or that this parasite simply does not exist in areas where this species occurs (Table 1.2). The distributions of *An. arabiensis*, *An. funestus s.l.* and *An. gambiae s.l.* and the DVS that occur in Saudi Arabia and Yemen are shown in Fig. 1.9E. Only the three primary DVS are shown in Africa because of the lack of evidence of the ability for the other African DVS to transmit *P. vivax* (Table 1.2). For the African DVS there were 8338 spatially and temporally unique occurrence records. The majority of these occurrence records were obtained for *An. funestus s.l.* (n = 2692), *An. arabiensis* (n = 2301) and *An. gambiae s.l.* (n = 2291). Point data were obtained from 44 countries. The largest number of point records per country were from Kenya (*n* = 757), followed by Tanzania and Cameroon (*n* = 383 for both nations). For the most part, the other countries on the continent had less than 100 point records, with the exception of Burkina Faso (*n* = 310), Equatorial Guinea (*n* = 113), Ghana (*n* = 106), Madagascar (*n* = 198), Mali (*n* = 166), Nigeria (*n* = 190), Senegal (*n* = 209), Sudan (*n* = 125), the Gambia (*n* = 192) and Uganda (*n* = 135). African countries classified as *P. vivax* endemic that had very few (≤5) vector occurrence point records were Central African Republic (*n* = 3), Congo (*n* = 2), Liberia (*n* = 4), Namibia (*n* = 5) and Togo (*n* = 1).

The compiled species distribution maps (Fig. 1.9E) illustrate a relatively straightforward picture of the distribution of the region's DVS: *Anopheles arabiensis*, *An. funestus s.l.*, and *An. gambiae s.l.* which all have broad distributions in the region and are confirmed vectors of *P. vivax*. These three DVS dominate in heterogeneous ranges of different pairs and combinations (of one another) in such a way that the species are present on their own in only focused locations. The co-dominant range of the *An. funestus* complex and *An. gambiae* in Central Africa is surrounded by an 'envelope' that houses all three primary DVS, that is further surrounded by *An. arabiensis* and the Funestus Complex, and then only *An. arabiensis* on the periphery. *Anopheles arabiensis* tolerates drier environments and is therefore absent from the forested areas of western Central Africa. The *An. funestus* complex distribution indicates a presence throughout all of sub-Saharan Africa, including Madagascar, but excluding much of southern Africa. *Anopheles gambiae* has a more complex distribution across a band from East (including Madagascar) to West Africa.

The bionomics of all the African DVS of malaria are summarised in full elsewhere (Sinka et al., 2010a). Here, we briefly describe the behaviours of those three DVS identified as potential *P. vivax* vectors in the region. *Anopheles arabiensis*, *An. funestus s.l.*, and *An. gambiae s.l.* are known to be primary vectors of *P. falciparum*, but have also been incriminated as vectors of *P. vivax* through the detection *P. vivax* circumsporozoite proteins in wild-caught specimens (Table 1.2). *Anopheles arabiensis* is often described as zoophilic, exophagic and exophilic, yet its behaviour appears to be quite variable, depending on location. For example, *An. arabiensis* found in West Africa are generally more anthropophilic and endophagic than those in the East. Such behavioural variability may enhance this species' ability to transmit *P. vivax* (or any human malaria) allowing it to adapt to avoid control methods such as IRS. Moreover, with peak biting times ranging from evening (1900) to early morning (0300), *An. arabiensis* may also avoid control via ITNs.

Anopheles arabiensis tends to be found in drier climates within Africa and larval habitats are typically small, temporary, clear, sunlit freshwater pools similar to those of *An. gambiae s.l.* However, *An. arabiensis* larvae have also been sampled from large or small man-made water bodies, including rice fields, as well as flowing or even brackish waters.

There is less variation observed in the bionomics of *An. funestus s.l.* relative to *An. arabiensis*. This species is known to be highly anthropophilic and endophilic and have a late biting time (after 2200), making IRS or ITNs highly effective interventions (although pyrethroid resistance has now been reported (Hargreaves et al., 2000; Coetzee and Fontenille, 2004)). The larval habitats of the Funestus Complex tend to be large, permanent (or semi-permanent) bodies of freshwater such as a pond, lake edge or rice field, where the larvae use emergent plants as protection against predation. There is some variation in the behaviour of members of the complex and further investigations using molecular identification methods are needed to determine the distribution of the different subtypes and the bionomics they exhibit.

Anopheles gambiae s.l. is perhaps the most well-known vector of human malaria and the most studied. It occupies a wide geographic range and is considered to be a highly efficient vector because of its highly anthropophilic biting behaviour and relatively long adult life stage. Females generally feed late at night indoors and also rest indoors, again making IRS and ITNs successful intervention strategies. However, in studies comparing feeding and resting location preference, *An. gambiae s.s.* exhibited both indoor and outdoor feeding and resting habits (Sinka et al., 2010a). This is likely ascribed to different chromosomal and molecular forms now identified within the species (Bockarie et al., 1993). The larval habitats of *An. gambiae s.l.* were long thought to be restricted to clear sunlit, temporary pools such as puddles or hoof prints, however, larvae have been reported from turbid and even polluted water and from large semi-permanent water sources such as rice fields. The variation of larval habitats is again attributed to divergences of the chromosomal or molecular forms.

The vector species of *P. vivax* in Africa occupy wide geographic ranges across the region and exhibit variation in behaviours both among and within individual species or species complexes. This may present challenges for control, yet, the tendency for these species to bite and rest indoors at night makes control methods such as ITNs effective control strategies. Vivax malaria is not currently the main focus of malaria research in this region and incrimination studies on wild populations have yet to be performed on half of the potential DVS in this region (Table 1.2). Further research identifying which anopheline species are the most efficient vectors of this parasite will be beneficial to the Africa+ region as goals of control and elimination of *P. falciparum* are realised and the focus moves to address *P. vivax*.

Africa+ summary. Estimates of the PAR of *P. vivax* in Africa, which is climatically well suited for malaria transmission, were very low, with the exception of countries around the Horn of Africa. This was because of the assumption that Duffy negative individuals, found in high frequencies on the continent (Howes et al., 2011), are refractory to *P. vivax* infection. This supposition was incorporated into the model, despite observations of *P. vivax* malaria infections in Duffy-negative individuals in Madagascar (Menard et al., 2010) and on the mainland of the continent (Ryan et al., 2006; Mendes et al., 2011; Wurtz et al., 2011). While this information contradicts our working assumption of complete protection, there is insufficient evidence to determine if these are more than just rare occurrences on the continent that would have a significant effect on the epidemiology of *P. vivax* in Africa.

Prevalence estimates for Africa are characterised by low predicted values and high levels of uncertainty (Fig. 1.4D). The PR data from Africa that served as the input data for the endemicity predictions were sparse, with the exception of a few regions (Fig. 1.3D1). This is largely because *P. vivax* is not the main priority for Africa. There were 753 million people at risk of stable *P. falciparum* in Africa in 2010 (Gething et al., 2011a), compared to the 38 million at stable risk *P. vivax*. However, 30% (228 million) of the 753 million were living in regions of low stable transmission (*Pf*$PR_{2-10} \leq 5\%$); the prevalence of *P. falciparum* in these areas falls to a point where elimination is viable, the situation of *P. vivax* will increase relatively, a consequence of its tendency to be the last parasite standing during elimination efforts (Garnham, 1951; Yekutiel, 1960; Pampana, 1969; Wernsdorfer et al., 2009; Tatem et al., 2010). This reinforces the need for increased vector surveillance and incrimination of species specifically for *P. vivax*. The endemicity map (Fig. 1.3D2) presented indicates that while *P. vivax* is present at very low levels in Africa, it is circulating. Consideration for how those prevalence estimates may be affected by decreasing *P. falciparum* levels must be considered.

4.5. Areas Where Lack of Geographical Data is Acute

Asia. Although there was thorough coverage of annual clinical incidence data and a large number of prevalence data records, given the large area of *Pv*MECs in Central Asia (20.5 million km^2), there were still large regions with a dearth of data (Fig. 1.3A1). These regions are highlighted by the uncertainty maps (Fig. 1.4A). India, the nation with the largest PAR of *P. vivax* globally, had disproportionately little prevalence data available. Given that parts of India are predicted to experience intense transmission, high-resolution mapping is needed to identify foci of transmission and this will only be achieved through improved surveillance coverage or the release of data that have not been previously shared. The uncertainty map also highlights lack of data in Myanmar. Regions of high uncertainty, which were also present in the population-weighted estimates, correspond with areas plotted to be highly endemic ($PvPR_{1-99} > 7\%$). Improved certainty in predictions in this country would be beneficial as neighbouring Thailand and China work towards malaria elimination. There would also be utility in the provision of contemporary data as progress towards elimination is made (e.g. in China) and transmission dynamics are therefore altered.

Asia-Pacific. Coverage of annual clinical incidence and prevalence survey data was relatively strong for this region (Fig. 1.3B1). However, the vast majority of the prevalence data originated from Indonesia. Sites on the island of New Guinea and parts of Borneo had very sparse prevalence data and were therefore predicted with greater uncertainty (Fig. 1.4B1 and B2). These regions have some of the most intense *P. vivax* transmission settings in the world. Improved surveillance would help localise hotspots and generate prime targets for control. The Philippines should also be noted as being a region with high uncertainty that was retained even after accounting for population density in the predictions. The smaller islands of this region would benefit from improved vector data collection since many had very little vector data available. This would improve the fidelity of the species distribution models in this region to better inform control decisions.

The Americas. Relative to other regions with stable *P. vivax* transmission, there were very little prevalence data available for this region (Fig. 1.3C1). Improved PR data from this part of the world, particularly those countries with intense transmission (Brazil, Honduras and Nicaragua), would benefit *P. vivax* mapping efforts. High-resolution mapping is needed to accurately map these highly focalized transmission settings, and this will require broad coverage survey data. There was good vector data coverage in this region, which is likely linked to the research and implementation of successful vector control efforts in the Americas. However, data collection on vector occurrence could be further improved. Some of the countries with the highest transmission (Honduras and Nicaragua) had the fewest occurrence records available.

Africa+. The *P. vivax* data available from Africa+ were also sparse, a reflection of the perception that the parasite is largely absent from the continent

(Fig. 1.3D1). In this region, the importance of *P. vivax* is overshadowed by *P. falciparum* and as such is not recorded in many countries. However, *P. vivax* infection has been observed in Duffy-positive and -negative individuals in the region, hence, it would be prudent for countries to monitor the prevalence of the parasite. While prevalence data were lacking from the region as a whole, the areas that would benefit most from improved surveillance are those with higher intensity transmission: Ethiopia, Madagascar and Somalia. This would reduce the uncertainty of the predictions and allow for better monitoring of control efforts. Lastly, the vector data available for the DVS in the region was relatively poor. Although the vector profile of Africa+ is straightforward, increased occurrence records may capture intricacies in distribution patterns and help distinguish vector and non-vector species of species complexes.

5. DISCUSSION

Plasmodium vivax malaria imposes serious public health burdens and is the most widespread of all the human malarias, particularly in women and children in poorly resourced communities (Poespoprodjo et al., 2008, 2009). Robust evidence demonstrates that *P. vivax*, despite long-held convention, is not a 'benign' infection (Baird, 2007; Price et al., 2007b). Moreover, studies show that 60-year-old drugs used to treat vivax malaria are failing throughout much of the *P. vivax*-endemic world (Baird, 2009; Price et al., 2009). The hypnozoitocidal component of that treatment, primaquine, is a deeply flawed therapy due to the threat it imposes to G6PD deficient patients (Baird, 2007). The reader is also referred to the chapter by Howes et al. that appears elsewhere in this thematic issue of *Advances in Parasitology* (Chapter 4, Volume 81). These issues are brought into focus as international targets, such as the Millennium Development Goals (http://www.mdgmonitor.org/), are developed to halt or mitigate malaria incidence and further goals are set for the elimination of the disease (WHO, 2007; Feachem and Sabot, 2008; The Global Health Group and the Malaria Atlas Project, 2011). Neglect of the impact and study of *P. vivax* is incompatible with these expressed goals.

The *P. vivax* PR estimates reviewed here are, across the American, African and Asian regions, uniformly low in comparison with PR estimates derived for *P. falciparum* (Gething et al., 2011a). The PR spectrum for *P. vivax* ranges from 0% to 7%, whereas for *P. falciparum*, it is 0 to 70%. This seems to imply *P. falciparum* transmission is an order-of-magnitude greater, but this is potentially misleading. The absolute prevalence of *P. vivax* in heavily endemic zones may also reach or exceed 70%. Relatively low parasite densities in blood (arising from its strict preference for reticulocytes) may lead to high rates of false-negative diagnoses by microscopy or rapid diagnostic tests (RDTs) (Mueller et al., 2009a). Microscopic diagnoses very often underestimate the true prevalence of *P. vivax* in blood in both high- and low-transmission settings (Mueller et al., 2009b; da Silva et al., 2010; Harris et al., 2010; Katsuragawa et al., 2010; Steenkeste et al., 2010). Further details regarding the diagnosis of *P. vivax* may be found in a review elsewhere in this thematic volume of *Advances in Parasitology* (Chapter 4, Volume 80). Missed diagnosis is important in mixed species infections where *P. vivax* may be a minor contributor to parasitaemia and often overlooked (Mayxay et al., 2004). There is no proven evidence that this is due to cross-species immunity, rather that plasmodia species are mutually suppressive in mixed infections (Richie, 1988). While *P. falciparum* tends to dominate *P. vivax*, infection with *P. vivax* reduces the intensity of falciparum infections (Snounou and White, 2004). Mixed infections are complex and often under-diagnosed (Mayxay et al., 2004) and it is difficult to interpret the effect they may have on prevalence estimates. The reader is referred to a review of acquired immunity to *P. vivax* provided in this volume (Chapter 3, Volume 81).

The difference in PR spectrums between *P. vivax* and *P. falciparum* could also be a reflection of the age range used as the input data for the endemicity model. To model endemicity, *P. vivax* predictions were standardised across the 1–99 age range, rather than the 2–10 years of age range, which was used for *P. falciparum* (Gething et al., 2011a). Malaria transmission peaks in 2–10-year olds and, therefore, the use of the 1–99-year age range 'dilutes' the prevalence estimates. Regardless, it is evident that the association between the risk of disease and parasite prevalence is markedly different for *P. vivax* than *P. falciparum*, with significant risk of *P. vivax* disease at lower parasite densities.

The maps reviewed here represent progress towards an improved understanding of the epidemiology of this unique malaria parasite. Included in this work is the first ever *P. vivax*-specific global endemicity map and updated limits of transmission. These maps are intended to aid control strategy formulation and operational decision-making. Stratification by transmission intensity has been supported by mathematical modelling in the context of *P. falciparum* management (Smith et al., 2006, 2008; Okell et al., 2008; Smith and Hay, 2009; Chitnis et al., 2010a, 2010b; Griffin et al., 2010; Ross et al., 2011), but *P. vivax* is rarely differentiated by endemicity level.

Consensus has yet to be reached in defining suitable control-oriented strata for *P. vivax*. Without such stratified control, there has been little impetus to fill that knowledge gap and generate reliable stratified risk maps. These *P. vivax* mapping efforts may also be unified with those for *P. falciparum* (Hay et al., 2009; Gething et al., 2011a) to help rectify this and elucidate where control efforts of the two parasites can be amalgamated. An example of such is the potential use of artemisinin-combination therapy (ACT) as presumptive treatment for diagnosed malaria in co-endemic areas discussed in the review of the use of anti-malarial drugs to reduce *P. vivax* transmission that is provided elsewhere in this volume (Chapter 5).

Primaquine, the only drug currently licensed to treat the liver stage of the parasite is contraindicated in individuals with G6PD deficiency. The map of *P. vivax* endemicity is an important complement to the recently developed map of G6PD deficiency (Howes et al., 2012) and will help to identify areas with high prevalence of both *P. vivax* and G6PD deficiency. Overlaying these is essential for estimating the potential risk of adverse outcomes that could occur from treatment with primaquine in areas where G6PD deficiency testing cannot be guaranteed (Ruwende and Hill, 1998; Cappellini and Fiorelli, 2008). The reader is again referred to a review of G6PD deficiency that details the geographic distribution, genetic variants and the implication of primaquine therapy provided elsewhere in this thematic issue of *Advances in Parasitology* (Chapter 4, Volume 81).

Plasmodium vivax endemicity maps provide bench-marks for progress in control and elimination. This information is increasingly needed by international organizations and groups that are once again assessing the prospect of eradication of all species of human malaria (Lines et al., 2007; Feachem and Sabot, 2008; Greenwood, 2008; RBMP, 2008; Mendis et al., 2009; Malaria Eradication Research Agenda, 2011b). At present, these maps are of particular importance outside of Africa, where *P. vivax* is the primary threat. Thirty of the 95 *Pv*MECs are in Asia and their populations comprise 91% of the global PAR of *P. vivax*. Knowledge of the global distribution and impact of *P. vivax* is also important to estimate market size and investment priorities for those developing targets for drug (Malaria Eradication Research Agenda, 2011a) and vaccine (Brown et al., 2009; Malaria Eradication Research Agenda, 2011c) research and development. The maps also facilitate national priority setting and advocacy.

The maps of dominant vector species of human malaria presented here (Fig. 1.9) and information regarding the potential vector species of *P. vivax* (Table 1.2) highlight a further knowledge gap in our understanding of *P.*

vivax compared to *P. falciparum*. Classification of anopheline species as vectors of malaria has typically meant vectors of *P. falciparum* malaria. This is most evident in the Africa+ region, but conclusive information is also missing from the Asian and American regions, where *P. vivax* is the dominant parasite. In many countries, the status of vector species is determined by absence of evidence rather than decisive evidence of absence of the parasite. The research into transmission of *P. vivax* is lagging behind that of *P. falciparum*. Improved vector surveillance would benefit all regions, but not without incrimination studies that focus specifically on *P. vivax*.

Currently, the cartographic foundations for estimating the public health burden of *P. vivax* do not exist and this is perhaps the highest priority for moving forward. The link between *P. vivax* prevalence and clinical burden must be established for regions where the disease is monoendemic as well as where it is sympatric with *P. falciparum*. National estimates of *P. vivax* rely on routine case reporting sources that vary in their fidelity and are often crudely distinguished from *P. falciparum* (WHO, 2011). Cartographic estimates of *P. vivax* burden would provide valuable information regarding the impact of the disease independent of inherent biases that may accompany health system data (Gupta et al., 2009; Rowe et al., 2009; Hay et al., 2010a, 2010b; Malaria Eradication Research Agenda, 2011b; Mueller et al., 2011). There is also a need to define the burden of *P. vivax* in clinically vulnerable groups such as pregnant women (Nosten et al., 1999; Mueller et al., 2009a) and, most notably, children. To estimate the burden of *P. vivax*, it will first be necessary to gain a better understanding of the impact of relapsing infections on prevalence. The degree to which relapses contribute to clinical disease is known to vary geographically, but the cause and pattern of the variation is still poorly understood (Battle et al., 2011; Betuela et al., 2011; White, 2011) and calls for further investigation. The extent to which *P. falciparum* affects the burden of *P. vivax* remains unknown, and now emerges as an important question in elimination strategy (Maitland et al., 1996, 1997; Snounou and White, 2004; Genton et al., 2008).

This chapter described the global public health significance of *P. vivax* in light of new cartographic technology and evidence, along with the fundamentally important emerging understanding of the infection as threatens to life. The parasite occurs across a broader geographic range, in more diverse habitats, in more anopheline vector species, and threatens more people than *P. falciparum*. Vivax malaria is an overwhelmingly Asian and Asia-Pacific problem, home to 91% of the 2.49 billion PAR, dozens of mosquito vector species, and endemic habitats as distinct as the temperate Korean peninsula and tropical New Guinea. This is the same region where treatments are failing due to drug resistance or reluctance to use threatening and impractical treatments like primaquine. Vivax malaria today threatens many people with very serious illness and is especially difficult to prevent and treat. The evidence presented in this chapter and others in this thematic issue disclaim the long-held presumption of inconsequence with infection by *P. vivax*.

6. METHODS

The full methods that would allow the reader to reproduce these maps are given in Gething et al. (2012) and Sinka et al. (2010b). Here, we bring together the full suite of methodologies and present a set of summaries that highlight the data used and the assumptions made to allow the value of the resulting estimates to be assessed.

6.1. Defining the Limits of *P. vivax* Transmission

To adequately assess the global burden of *P. vivax*, the limits of the infection and the global distribution of risk must first be identified. Knowledge of the spatial distribution, and the clinical incidence within those limits, provide a foundation on which control efforts and measures of progress can be based. MAP began these efforts through defining the spatial limits of infection and the PAR, informed by ecological variables and the distribution of Duffy negative individuals, and then applied those limits to mapping the transmission and varying levels of *P. vivax* endemicity globally.

6.1.1. International Limits of P. vivax

The global spatial limits of *P. vivax* malaria were first defined for 2009 (Guerra et al., 2010) and the methods and results have since been updated for 2010 (Gething et al., 2012). A list of 95 *P. vivax* malaria endemic countries (*Pv*MECs), illustrated in Fig. 1.10A1–D1, was identified using previous methods based on international health and travel guidelines (Centers for Disease Control and Prevention, 2009; Guerra et al., 2010; WHO, 2010b; Gething et al., 2011a). The *Pv*MECs were grouped into three regions: the Americas; Africa, Saudi Arabia and Yemen (Africa+); and Central and South East Asia (CSE Asia), which was further divided into Asia and Asia-Pacific in order to resolve PAR estimates (referred to here as Asia and Asia-Pacific) (Fig. 1.2). The borders of these countries, along with national survey information and relevant, published sources and personal communications, defined the first version of the *P. vivax* spatial limits map (Guerra et al., 2010).

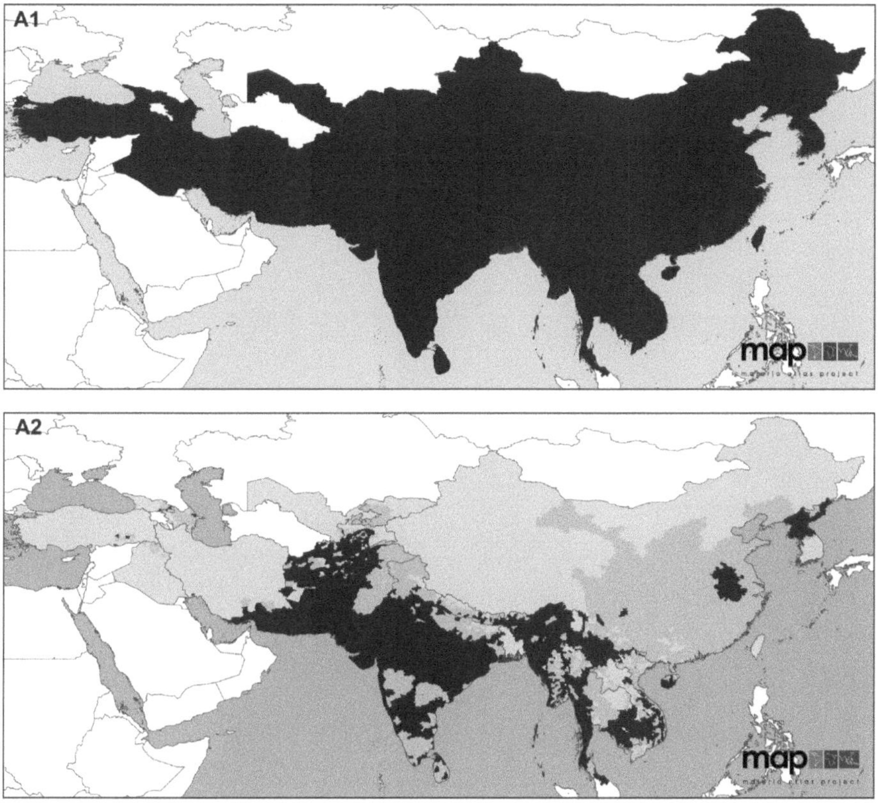

Figure 1.10 Map sequence illustrating the different exclusion layers applied by region. Maps are shown by region: Asia (A), Asia-Pacific (B), the Americas (C) and Africa+. Panel 1 = all regions of the regional *P. vivax* endemic countries; 2 = downgrading or exclusion of risk informed by annual parasite incidence data (*Pv*API); 3 = additional exclusion of risk informed by the biological temperature mask; 4 = additional downgrading or exclusion of risk informed by the aridity mask; 5 = additional downgrading or exclusion of risk informed by medical intelligence and international travel and health guidelines; 6 = the final limits definition after additionally downgrading risk in stable areas predicted to have very low prevalence by the model-based geostatistics (MBG) model. Stable transmission is shown in red, unstable transmission in pink, *P. vivax* malaria free areas in grey and countries non-endemic for *P. vivax* or outside of the region in white. For interpretation of the references to colour in this figure legend, the reader is referred to the online version of this book.

*Pv*API data, which report the number of confirmed *P. vivax* malaria cases per administrative unit per 1000 people per annum (p.a.) from *Pv*MECs, were used to further constrain the spatial limits of infection (Fig. 1.10A2–D2). *Pv*API data were obtained from a variety of sources which are provided in detail elsewhere (Gething et al., 2012). Data were unavailable for 41 of

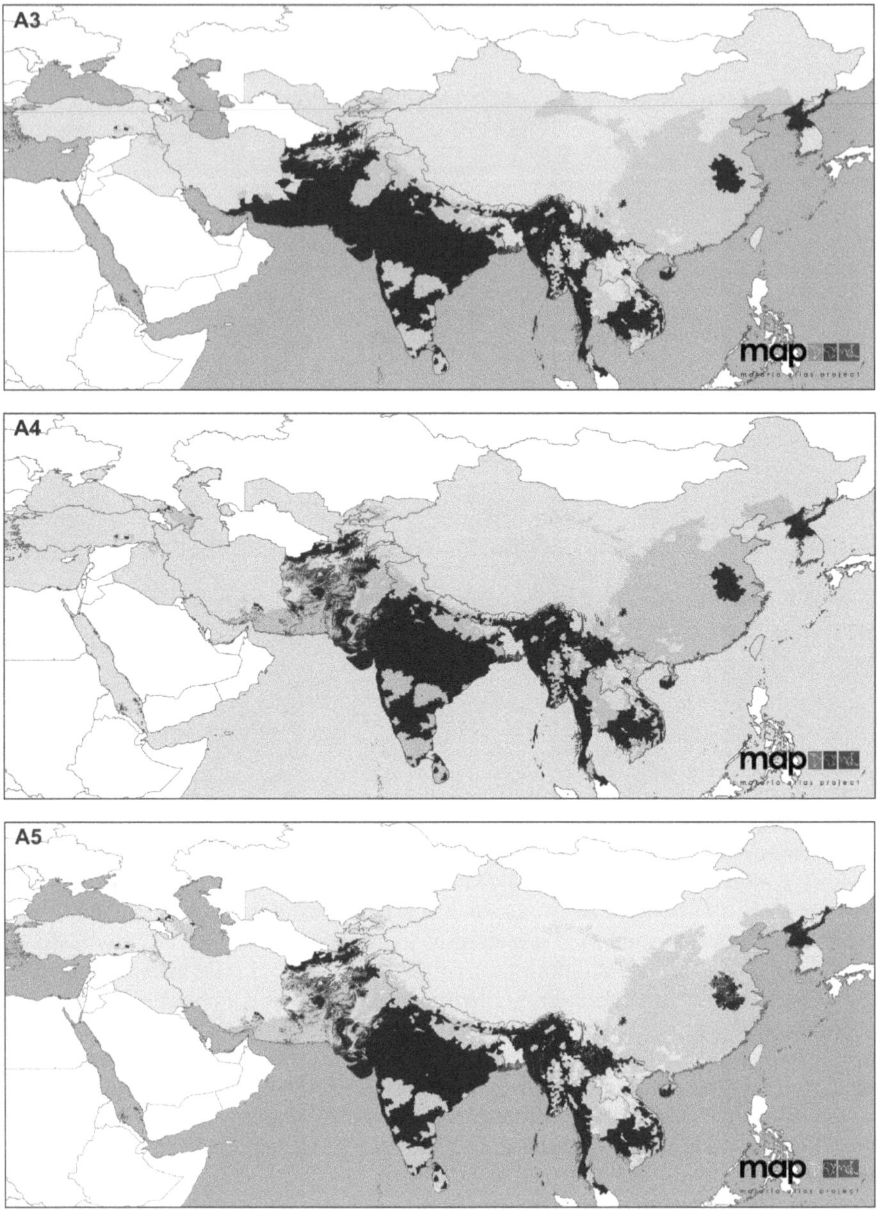

Figure 1.10, cont'd

the 95 *Pv*MECs, which were all within the Africa+ region, with the exception of Uzbekistan. Ideally, the *Pv*API data were available per administrative unit per year with each record containing information regarding the number of people in each administrative unit and the number of *P. falciparum*

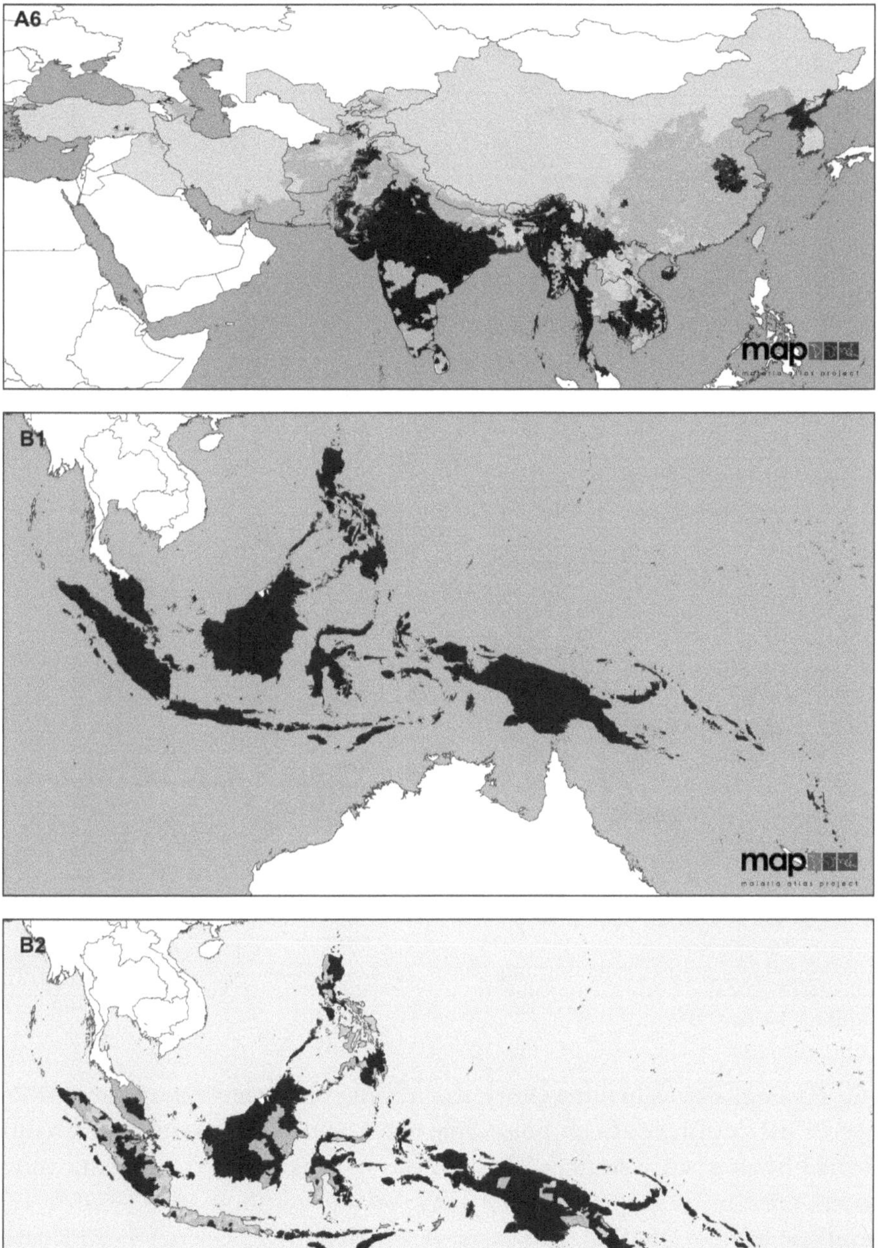

Figure 1.10, cont'd

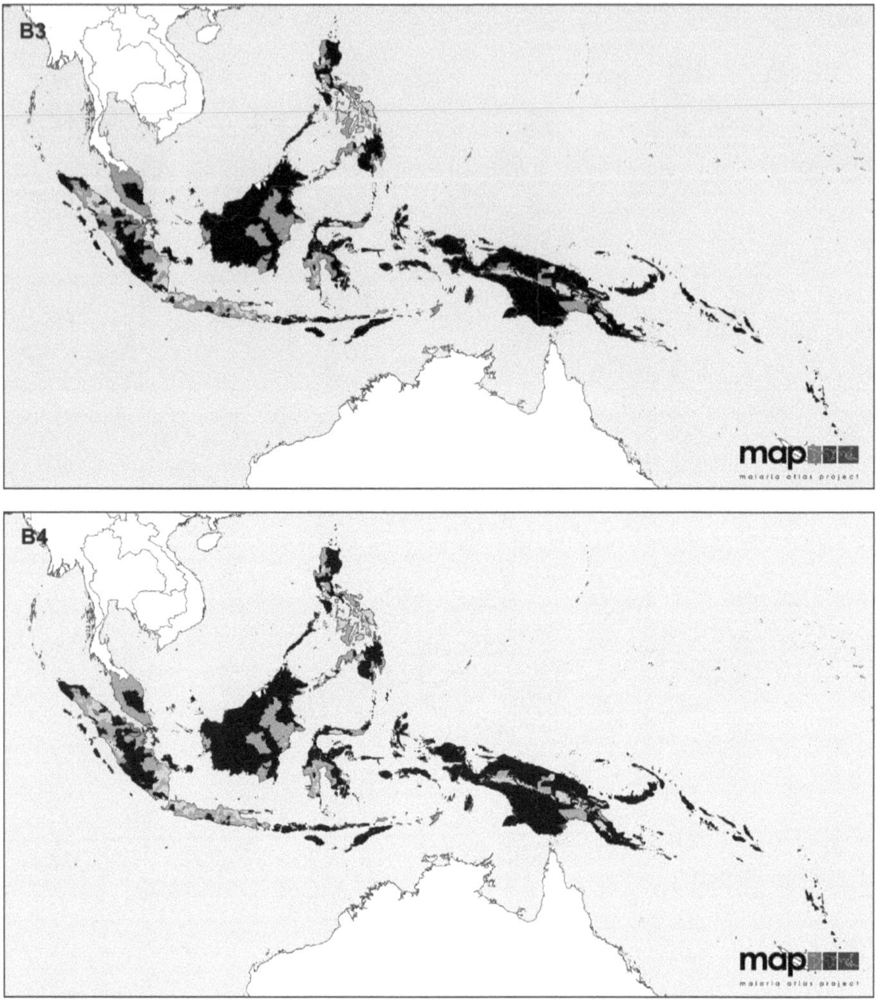

Figure 1.10, cont'd

and *P. vivax* cases. When necessary, the missing data were extrapolated from *Pv*API data estimates from preceding years or parasite species ratios confirmed by alternative sources. The aim was to obtain data for the four most recent previous years (up to 2010) at the second administrative (ADMIN2) level (or third, ADMIN3, if data were available). To map the *Pv*API data, the information was reconciled to the digital administrative boundaries available from the 2009 Global Administrative Unit Layers (GAUL) from the Food and Agriculture Association of the United Nations (FAO) within the Food Security for Action Programme (FAO, 2008). *Pv*API data were

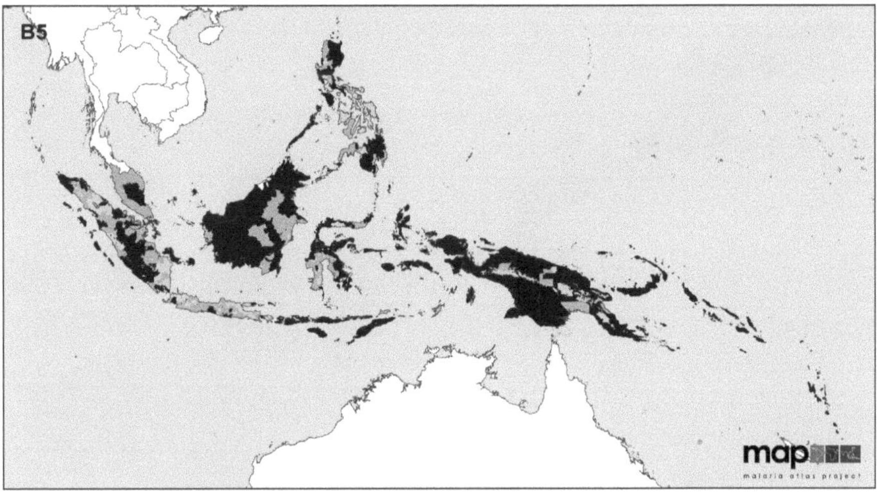

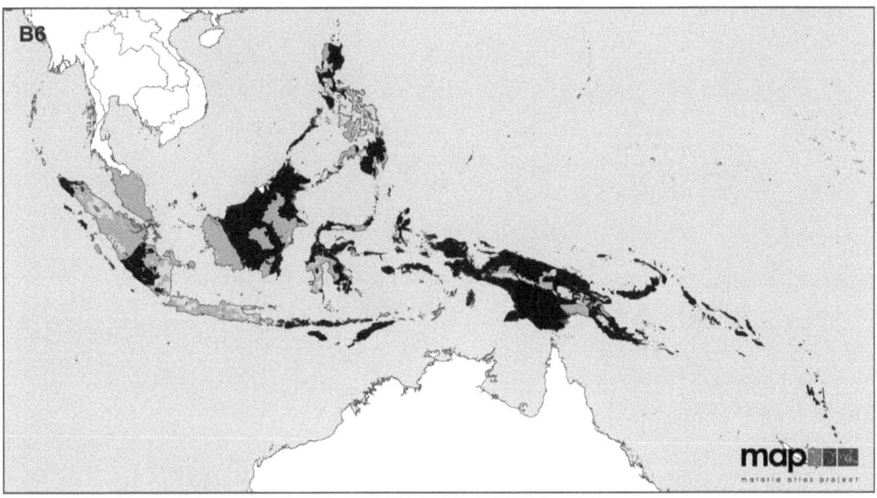

Figure 1.10, cont'd

arranged to classify areas at risk into three categories: malaria free, unstable (<0.1 cases per 1000 p.a.) or stable (≥0.1 cases per 1000 p.a.) transmission. These values were based on the Global Malaria Eradication Programme classifications; API values of <0.1‰ were considered a reliable indication to cease IRS and to shift to the consolidation phase of eradication (Pampana, 1969; Yekutiel, 1980; Guerra et al., 2008; Hay et al., 2008). This followed a transition from a cut off of 0.5 cases per 1000 p.a., which was deemed less reliable because malaria would at times return after the cessation for spraying at an API of 0.5‰ (Guerra et al., 2008). This was likely because

Figure 1.10, cont'd

surveillance, whether passive or active, was not wholly reliable or reflective of true endemicity levels. The more conservative categorization that is also applied here, of unstable transmission equating to less than 0.1 cases per 1000 p.a. allowed for less confidence in the fidelity of the prevalence source information and accounted for inaccuracies in district or provincial level reporting (Snow et al., 2005; Erhart et al., 2007; Sharma, 2007).

Biological masks were applied to further constrain the risk in the *P. vivax* endemic regions. Environmental conditions can suppress malaria transmission by limiting various components of the transmission cycle. Temperature has been shown to affect vector survival, emergence and feeding rates (Ross, 1911; Detinova, 1962; Mahmood and Reisen, 1981; Ahumada et al., 2004). The most limiting effect of temperature on malaria transmission is the

Figure 1.10, cont'd

interaction between vector lifespan and the length of the sporogonic cycle; as temperature varies throughout the year, as does the length of the extrinsic incubation period during which the parasite matures to the sporozoite stage within the gut of the mosquito (Nikolaev, 1935). For transmission to occur, the anopheline population must survive long enough for sporogony of the parasite to occur. A model that assesses the effects of temperature on *P. vivax* over time (Gething et al., 2011b) was used to generate a grid of temperature suitability proportional to vectorial capacity, a measure of transmission potential (Garrett-Jones, 1964; Smith and McKenzie, 2004). Daily vector survival was calculated as a function of local temperature and vector lifespan was assumed to be 1 month (Kiszewski et al., 2004), with exceptions being made for areas with longer living species (*Anopheles sergentii* and

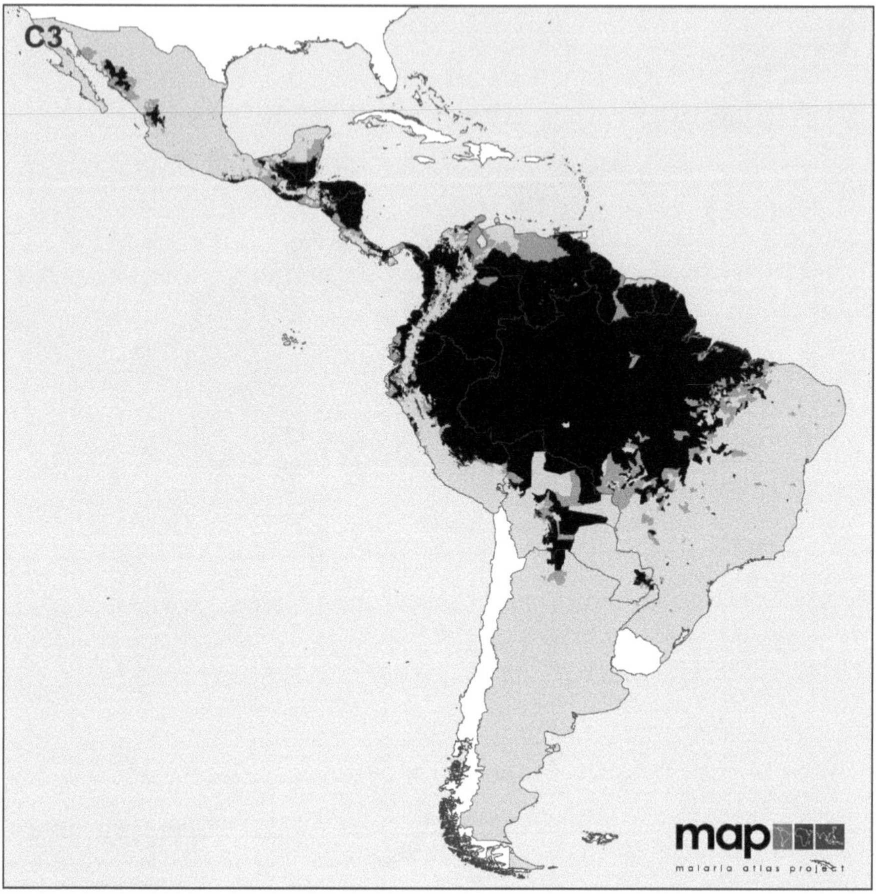

Figure 1.10, cont'd

An. superpictus) (Guerra et al., 2008; Sinka et al., 2010a). The interaction between vector lifespan and the length of the sporogonic cycle was modelled for each pixel of the temperature suitability grid (Fig. 1.5), so that pixels in which there was no time during the year that sporogony could be completed were classified as being at no risk for *P. vivax* transmission (Fig. 1.10A3–D3).

A second environmental driver of suitability for *P. vivax* transmission is availability of sufficient moisture. A mask of aridity was used to exclude areas where arid conditions would preclude anopheline survival at all life stages (Shililu et al., 2004; Gray and Bradley, 2005). Arid areas were identified using the global GlobCover Land Cover product (ESA/ESA GlobCover Project, led by MEDIAS-France/POSTEL) (Bicheron et al., 2008). The aridity mask (Fig. 1.6) was treated differently from the temperature mask

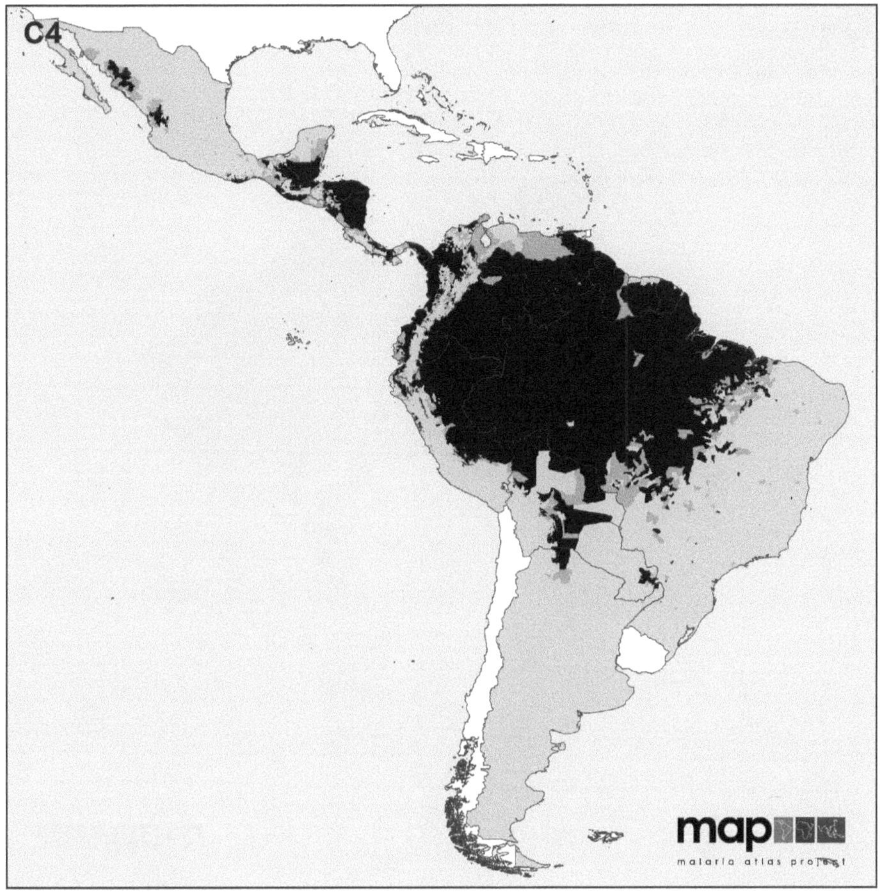

Figure 1.10, cont'd

to accommodate for the possibility of human and vector species adapting to arid conditions so that risk classes were downgraded depending on the land cover. GlobCover bare areas classified as stable risk according to the *Pv*API were downgraded to unstable, and unstable risk was stepped down to malaria free (Fig. 1.10A4–D4).

Medical intelligence was also used to refine the limits of *P. vivax* risk (Fig. 1.10A5–D5). International travel guidelines (Centers for Disease Control and Prevention, 2009; WHO, 2010b) and data obtained through personal communications were applied to identify urban and sub-national malaria free-areas. Urban areas have less malaria transmission than rural areas because of the distinct ecological conditions that result from a man-made environment (Hay et al., 2005; Tatem et al., 2008). Different species

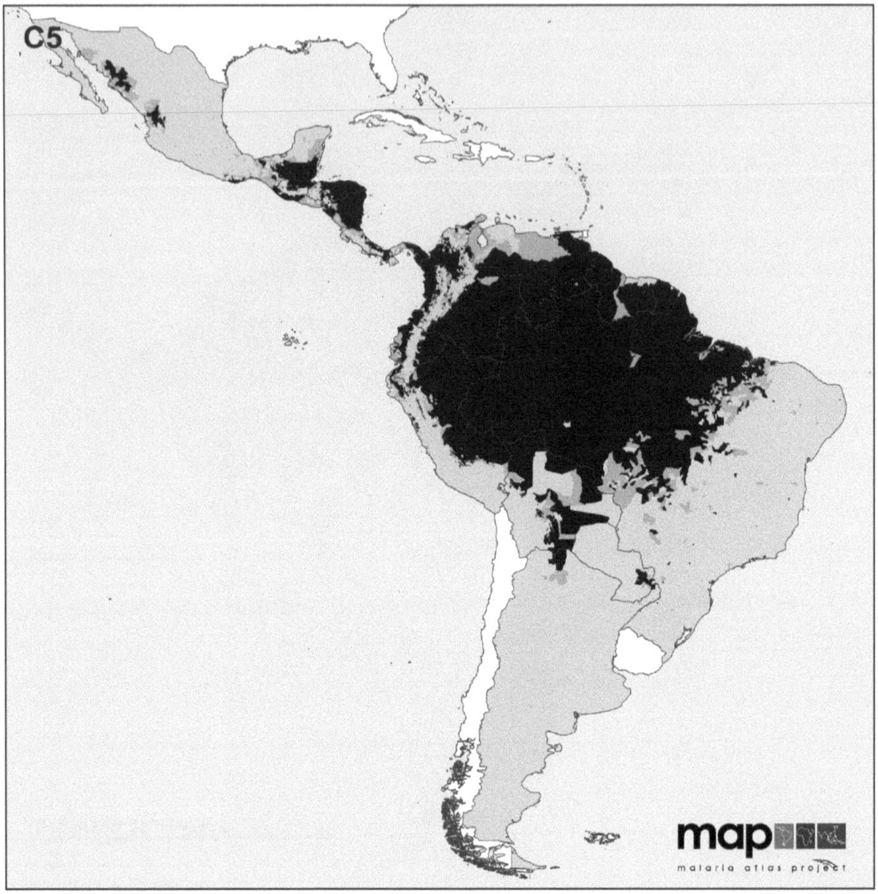

Figure 1.10, cont'd

of *Anopheles* mosquitoes respond at varying degrees, but urbanisation has been shown to reduce malaria transmission across Africa and in parts of the Americas (Hay et al., 2005; de Castro et al., 2006). In Asia, a principal vector, *An. stephensi*, has adapted to urban environments by breeding in artificial water collections (Sharma et al., 1993; Sinka et al., 2011) to the extent that stable transmission has been observed in the majority of cities (70 out of 86 cities examined) in India reporting annual parasite index data (Akhtar et al., 2009). *Anopheles culicifacies s.l.* has also been shown to transmit malaria in urban settings, but population densities and sporozoite rates indicate that this species is more affected by the environmental changes of urban areas (Nalin et al., 1985; Sharma et al., 1993; Sharma, 1995; Sinka et al., 2011). Therefore, it was assumed that urban transmission was maintained by only

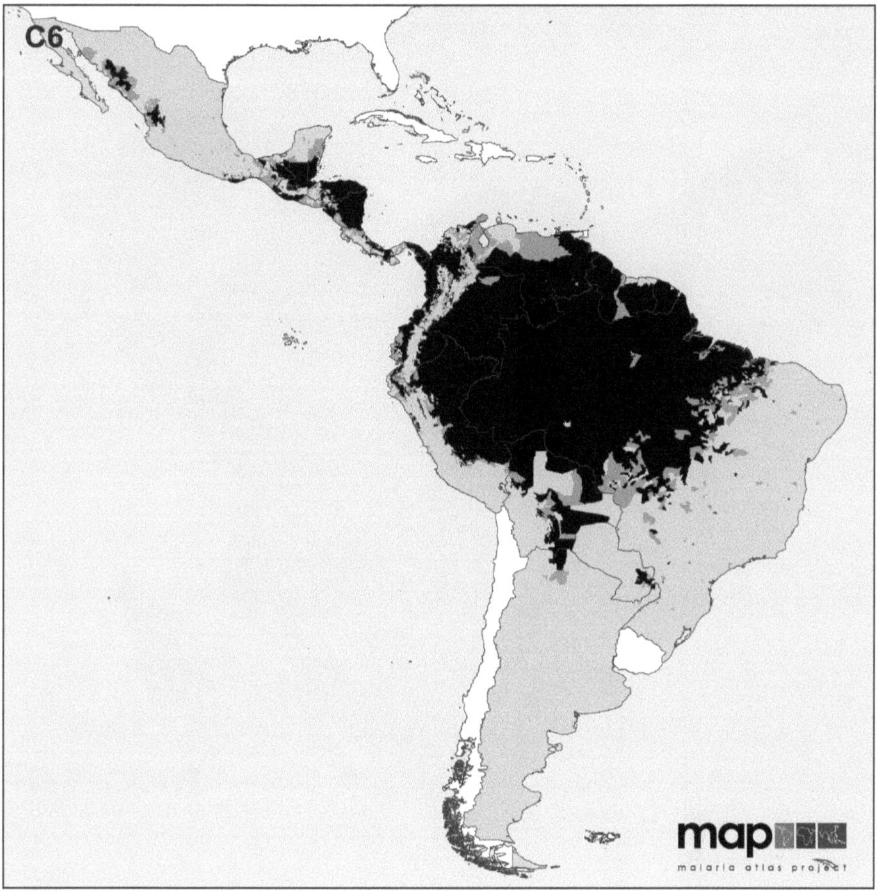

Figure 1.10, cont'd

An. stephensi. Cities specified as being malaria free by the international travel and health guidelines (Centers for Disease Control and Prevention, 2009; WHO, 2010b) or other sources were mapped using the GRUMP urban extent layer (Balk et al., 2006). Urban areas outside of the range of *An. stephensi* were classified as being at no risk of *P. vivax* transmission and areas within the vector species' range were down-graded by one risk level (stable to unstable; unstable to malaria free) accordingly. In addition to urban areas, some administrative areas, such as islands, were also declared malaria free by the international travel and health guidelines and a few specific personal communications. These areas were also classified as no risk if they were not already specified as such by the *Pv*API layer. The final limits for each region are shown in Fig. 1.10A6–D6.

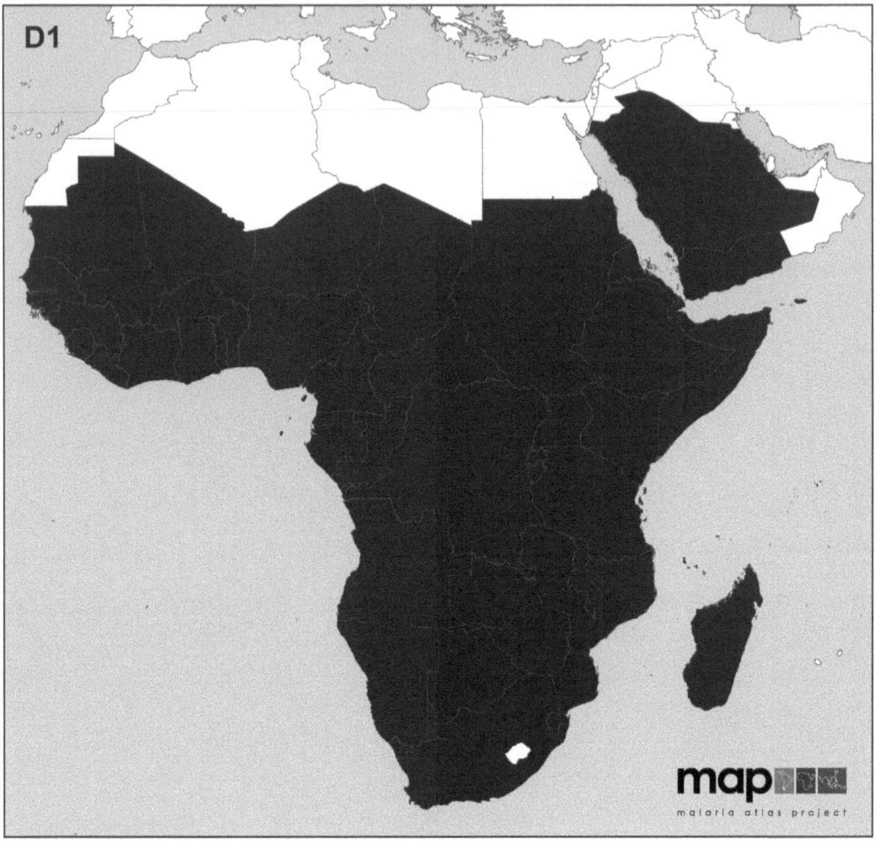

Figure 1.10, cont'd

6.1.2. The Availability of PvAPI Data by Region

Plasmodium vivax API data were available from 53 countries at a variety of administrative levels from years ranging between 2002 and 2010.

Asia. Annual parasite index data were aggregated at a variety of administrative levels. Azerbaijan, Bhutan, Cambodia, Kyrgyzstan and Thailand provided data at the first administrative (ADMIN1) level. China reported data from areas known to have little to no transmission regions at ADMIN1, and at ADMIN3 in the remaining areas. Myanmar and Nepal reported all data from ADMIN3 units. The remaining countries (Afghanistan, Bangladesh, Georgia, India, Iran, Iraq, Lao PDR, Pakistan, Republic of Korea, Sri Lanka, Tajikistan, Turkey and Viet Nam) collected *Pv*API data from ADMIN2 units. *Pv*API data were available from 4443 risk units in Asia.

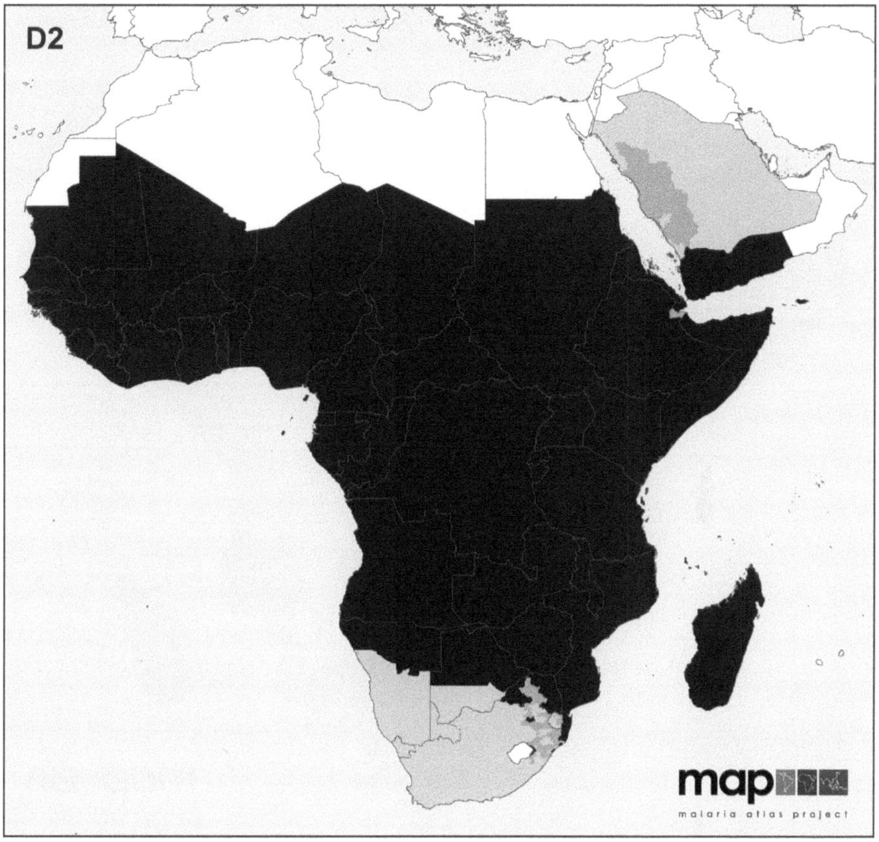

Figure 1.10, cont'd

Asia-Pacific. Annual parasite index data were provided for all of the *P. vivax* endemic countries in Asia-Pacific. Only Malaysia provided data up until 2010. Indonesia and Timor-Leste had data until 2008 available and 2007 was the last reporting year for the remaining countries. Malaysia, the Solomon Islands, Timor-Leste and Vanuatu provided data at the ADMIN1 level and Indonesia, Papua New Guinea and the Philippines reported data from ADMIN2 units. *Pv*API data were obtained from a total of 559 risk units in Asia-Pacific.

*The Americas. Pv*API data were available from all *Pv*MECs in the Americas. For the years reported, which varied by country, six countries (Belize, El Salvador, Guatemala, Guyana, Nicaragua and Suriname) reported data at the first administrative (ADMIN1) level and Venezuela reported data at the ADMIN1 and ADMIN2 level. Eleven countries (Argentina, Bolivia, Brazil, Colombia, Costa Rica, Ecuador, French Guiana, Honduras, Mexico,

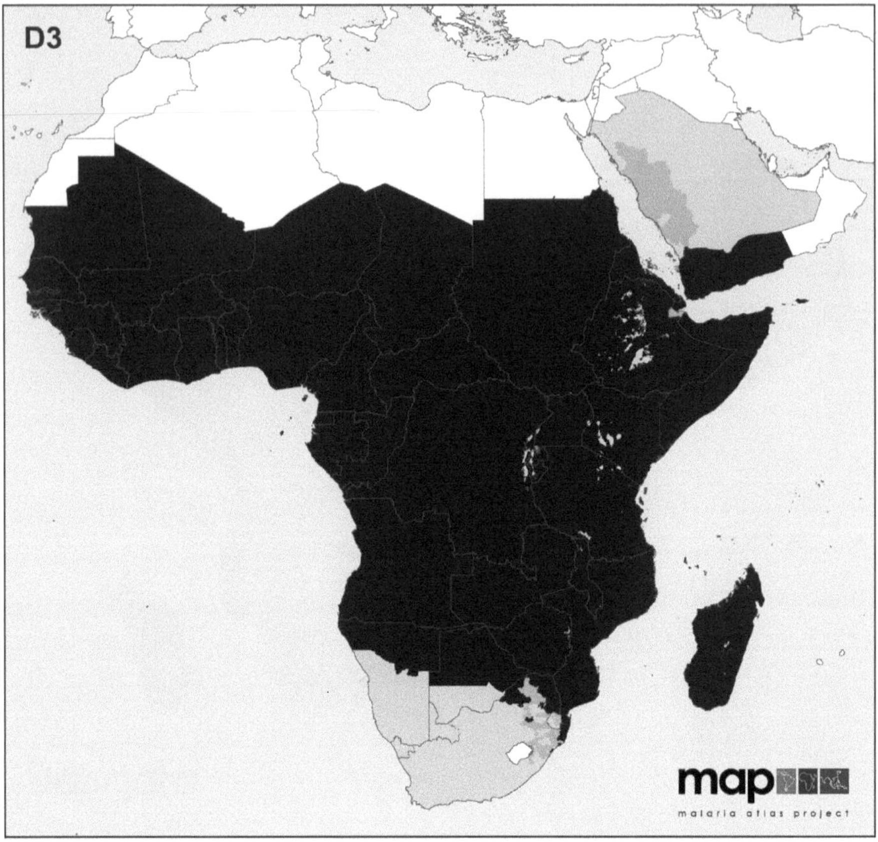

Figure 1.10, cont'd

Panama and Paraguay) reported ADMIN2 data and Peru reported data from ADMIN3 units. Data were available for a total of 12,514 administrative (or risk) units in the Americas.

Africa+. *Pv*API data were aggregated at a variety of levels by the six (out of 46) *Pv*MECs that reported data in the region. For the years of available data, Dijbouti, Saudi Arabia and Yemen reported ADMIN1 level data, Namibia reported from a mix of ADMIN1 and ADMIN2 units, and South Africa and Swaziland had data available at the ADMIN2 level. Data were provided for a total of 377 risk units for the Africa+ region.

6.1.3. The Global Distribution of the Duffy Blood Group

The distribution of Duffy-negative populations was used as another exclusion layer to constrain the limits of *P. vivax* transmission (Fig. 1.7). The reader

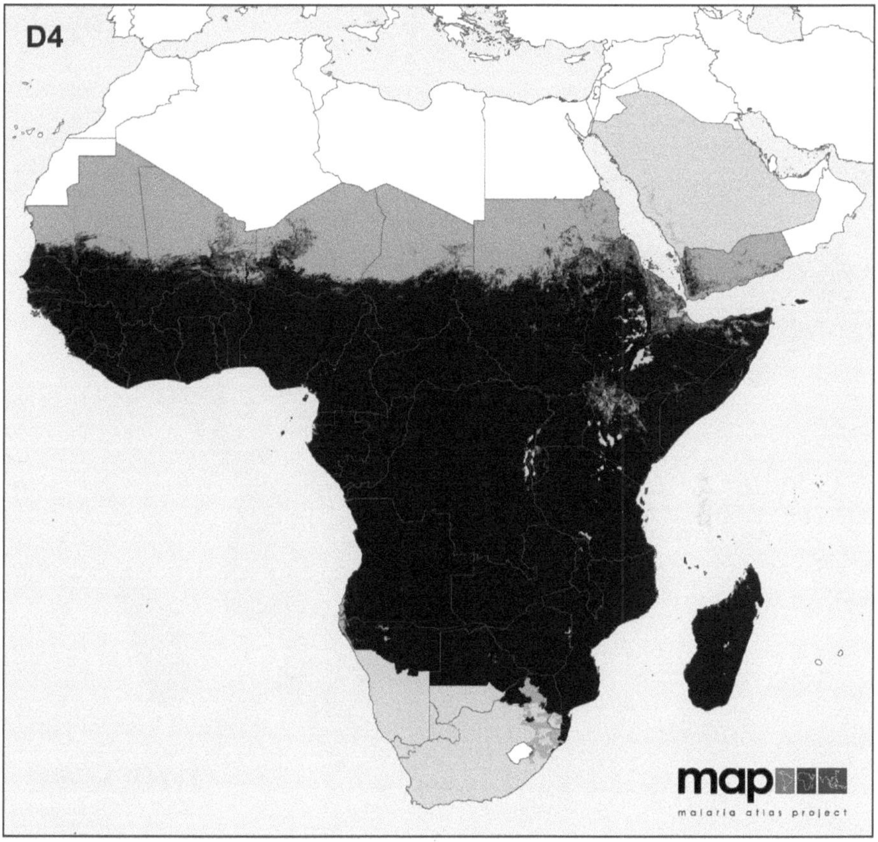

Figure 1.10, cont'd

is again referred to a detailed review of the effect of the lacking Duffy antigen on resistance to *P. vivax* provided in this thematic issue of *Advances in Parasitology* (Chapter 2, Volume 81). The Duffy antigen refers to a receptor expressed on the surface of red blood cells, which *P. vivax* has been shown to be dependent upon for erythrocytic invasion (Miller et al., 1976; Barnwell et al., 1989; Wertheimer and Barnwell, 1989). Duffy-negative individuals, who lack the antigen, are therefore largely refractory to *P. vivax* infection and high frequencies of the phenotype are presumed to suppress *P. vivax* endemicity in areas that would otherwise be well suited for transmission. A continuous map of the Duffy-negative phenotype, described in detail elsewhere (Howes et al., 2011) and briefly here, was used as the exclusion surface.

To model the global distribution of the Duffy-negative phenotype, a database of Duffy blood group surveys was assembled. Surveys of Duffy-variant

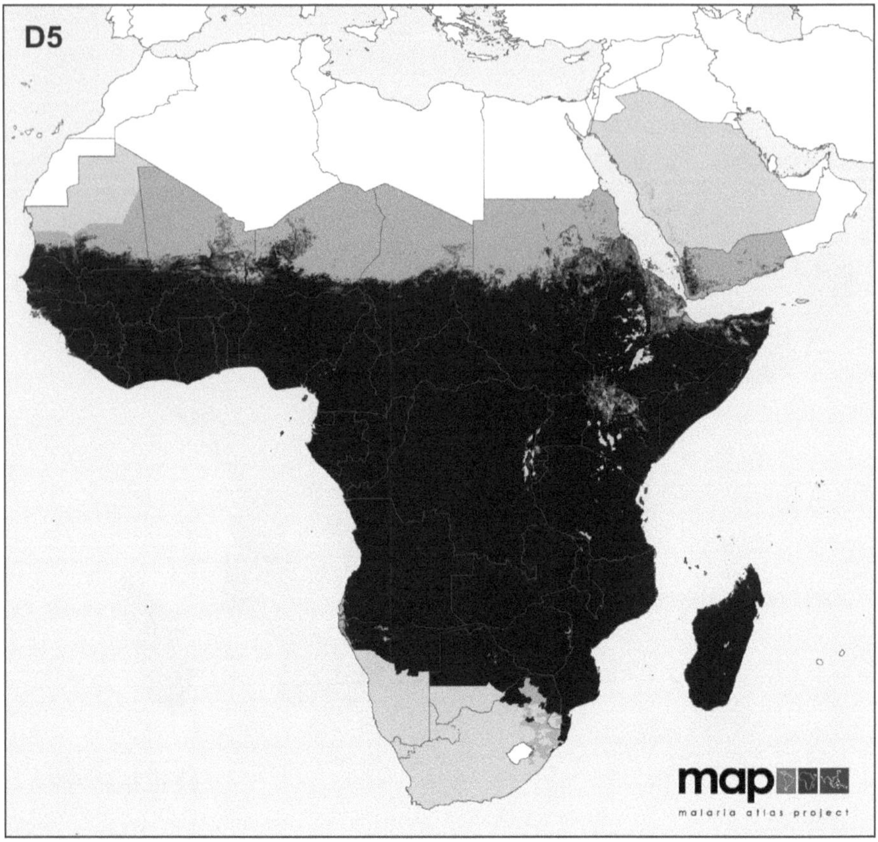

Figure 1.10, cont'd

frequencies were compiled from systematic searches of published literature, personal communications and sources obtained through previously published databases (Mourant et al., 1976; Cavalli-Sforza et al., 1994). Surveys dating back to 1950, when the blood group was first described (Cutbush and Mollison, 1950), were included. Results were refined so that only population-based surveys were used and potentially biased samples were removed. Survey data were geopositioned following guidelines described previously by MAP (Guerra et al., 2007).

In addition to the frequency of the blood group variants, the diagnostic method used in each survey was recorded to classify the type of information provided to ultimately inform the model. The Duffy antigen has two main variant forms, Fy^a and Fy^b. These differ by a single amino acid substitution (Gly42Asp), which is encoded in two alleles, *FY*A* and

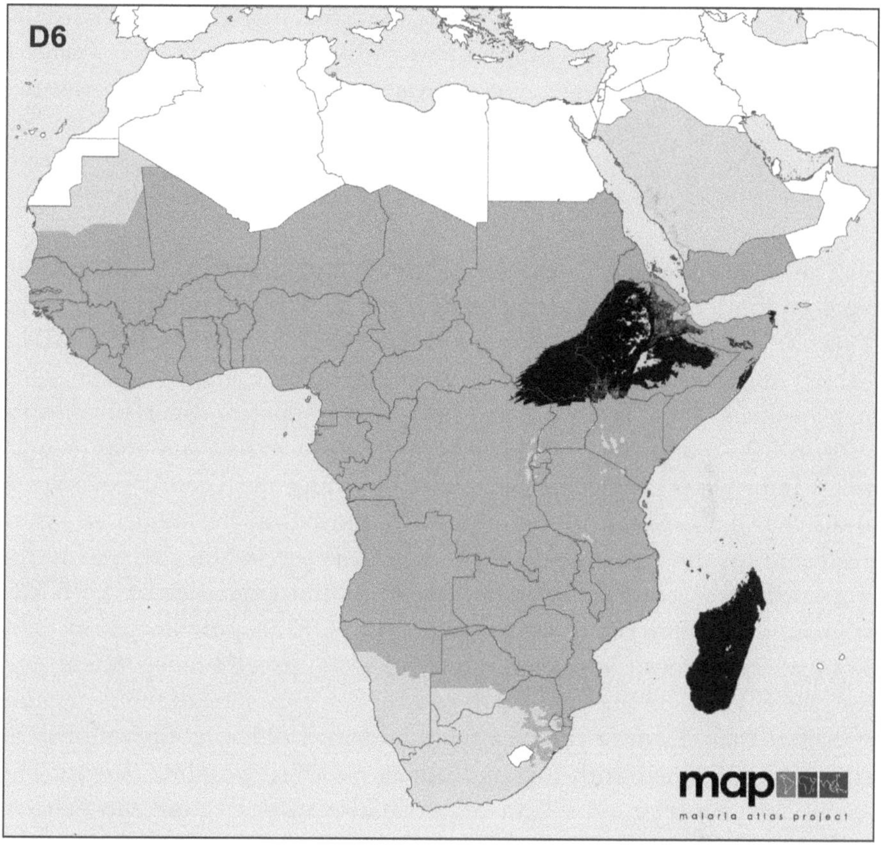

Figure 1.10, cont'd

*FY**B*, that vary based on a single base substitution (G125A) (Zimmerman, 2004; Langhi and Bordin, 2006). A separate single base substitution in the gene's promoter region (T-33C) may block the expression of the gene and result in a null 'erythrocyte silent' (ES) phenotype. This mutation is most commonly associated with the $FY^{\star}B$ ($FY^{\star}B^{ES}$) allele (the $FY^{\star}A^{ES}$ allele is extremely rare) (Langhi and Bordin, 2006; Sellami et al., 2008). These four alleles ($FY^{\star}A$, $FY^{\star}B$, $FY^{\star}B^{ES}$, $FY^{\star}A^{ES}$) result in 10 possible genotypes and four possible phenotypes: Fy(a+b+), Fy(a+b−), Fy(a−b−) and Fy(a−b−) (Howes et al., 2011). Methods to diagnose Duffy types can focus on different aspects of the system: the individual polymorphic sites, the genotype, the presence or absence of the antigen, and which particular type of antigen is being expressed. The assembled data were therefore grouped into five types according to which diagnostic approach was used:

i) *Genotype*, where full genotypes were reported; ii) *Phenotype*, where full serological diagnoses (anti-Fy^a and anti-Fy^b sera) were used; iii) *Promoter*, if results only stated antigen expression or non-expression (no distinction between Fy^a and Fy^b); iv) *Phenotype-a*, if only Fy^a was tested for (no distinction between Fy^b and the negative phenotype); and v) *Phenotype-b*, if only Fy^b was tested for (no distinction between Fy^a and the negative phenotype).

This dataset formed the evidence-base for a geostatistical model used to predict a continuous map of the prevalence of the Duffy-negative phenotype [Fy(a−b−)], the phenotype encoded by the $FY{\star}B^{ES}/FY{\star}B^{ES}$ genotype. The $FY{\star}A^{ES}$ is highly infrequent and so was modelled as a small constant rather than a spatially variable allele. Although not all of the five data types directly informed the prevalence of Duffy negativity, the model was able to infer useful information directly from each (by ruling out certain genotypes), so all were included as model inputs. To allow this, the model structure was based on the frequencies of the two loci involved in the system: the promoter type (determining expression *versus* non-expression at the T-33C site) and the coding region (determining the G125A polymorphism). The Bayesian framework considered the two loci as spatially independent random fields, and used the survey data to map the frequency of the expression of each variant. Each of the data types informed different aspects of these loci: some informed both loci and others excluded possible variants. The model also incorporated a land cover variable to distinguish sub-Saharan African populations from others on the continent. This was to inform the model of the high probability of association between silencing mutation of the Fy^b variant and the $FY{\star}B^{ES}$ allele, which was observed in high frequencies across sub-Saharan Africa.

The Bayesian model-based geostatistical framework predicted Duffy group expression frequencies in all geographic locations across a 5 × 5 km grid to generate a continuous global surface of each variant. The Duffy negativity phenotype was expressed by the squared frequency of the $FY{\star}B^{ES}$ allele. The map reveals that the homozygous null phenotype is highly constrained to sub-Saharan African populations, with localized high-frequency areas in the Americas. Historical perceptions have supported the assumption that *P. vivax* is absent from much of the African continent (Rosenberg, 2007). However, evidence of autochthonous transmission within Africa indicated that areas of the continent should not be excluded a priori. This map, therefore, provided an evidence-based exclusion layer for Duffy-negative populations resistant to *P. vivax* infection in Africa.

6.2. Global Endemicity of *P. vivax*

6.2.1. P. vivax *Parasite Rate Data*

To map *P. vivax* malaria endemicity within the limits of stable transmission, in addition to the updated geographic boundaries and corresponding populations at risk of *P. vivax*, an updated georeferenced database of *P. vivax* parasite rate (*Pv*PR) survey data were compiled. PR was used here because it is the most ubiquitous of the malariometric measures of risk (Hay et al., 2009). *Pv*PR data represent the proportion of a randomly sampled population to have detectable *P. vivax* parasitaemia when screened via microscopy or RDTs and is the most consistently measured index of malaria endemicity. The final database consisted of unique surveys obtained from published and unpublished literature sources spanning the period 1985–2010. The database included information on the survey origin, how the location was determined (georeferencing method), time period, age group, sample size and diagnostic method used.

The *Pv*PR database was made up of 9970 spatiotemporally unique records from 432 different sources. Data were available from 53 countries, 12 of which were in the Americas, 19 in the Africa+ region, 15 in Asia and 7 in Asia-Pacific. There were 44 *Pv*MECs not represented in the database, most of which were in Africa, with the exception of Argentina, Azerbaijan, Belize, Bhutan, Korea DPR, El Salvador, Georgia, Guyana, Iran, Kyrgyzstan, Panama, Paraguay, Republic of Korea and Uzbekistan. Details of the PR data that were input into the model from each region are given below.

6.2.2. *Modelling* P. vivax *Endemicity*

To generate a continuous surface of *P. vivax* endemicity using PR data, a flexible modelling framework based on model-based geostatistics (MBG) (Diggle et al., 1998; Diggle and Ribeiro, 2007) was used. With areas of stable transmission converted into a 5 × 5 km grid, MBG models allow for endemicity values to be predicted at each pixel as a function of the geographically varying mean of survey values and a weighted average of neighbouring data values. MBG models are well suited for predicting endemicity values, in this case PRs, for a number of reasons. First, the mean PR values may be defined as a function of multiple environmental covariates that influence malaria transmission. Second, a covariance function may be employed to define the spatial heterogeneity of the PR data and, in turn, define the appropriate weight for each data point when generating a prediction. Third, uncertainty can be based on the nature and density of data surrounding a pixel. Fitting MBG models with Bayesian inference and a Markov chain Monte Carlo (MCMC) algorithm produces uncertainty metrics around the final predictions as well as the model inputs in the form of predictive posterior distributions (Patil et al., 2011). Areas with the least uncertainty are those with a large number of recent surveys with relatively homogenous results, whereas greater uncertainty would be found in places with sparse or old surveys with ranges of different observed PRs.

To model a global endemicity surface of *P. vivax* malaria, an MBG framework that had been successfully employed for falciparum malaria (Hay et al., 2009; Gething et al., 2011a) was used following modifications made to accommodate biological features unique to *P. vivax*. The modelling methods incorporate surveys from a wide time period such that older surveys are given less weight than recent ones. The environmental covariates included were those that had an a priori expectation to affect malaria transmission intensity. These were urban areas defined by the GRUMP urban extent product (Balk et al., 2006; CIESIN/IFPRI/WB/CIAT, 2007), a long-term average vegetation index product used as a proxy for available moisture for vector reproduction and survival (Hay et al., 2006; Scharlemann et al., 2008), and the temperature suitability index derived from the model described above, which identifies areas suitable for transmission based on the requirements of vector survival and sporogony (Gething et al., 2011b).

PR data were standardised by age because of variation in infection rates observed in different age groups. It is often observed that malaria prevalence rises rapidly in infancy before reaching a plateau in early childhood and declining through adolescence and adulthood. This phenomenon was modelled using a previously described framework (Smith et al., 2007) to standardise for the prevalence variability among age groups. The model was informed with finely age-stratified *Pv*PR surveys to represent vivax-specific age profiles (Mueller et al., 2009b; Lin et al., 2010) and was used to convert all the observed survey prevalence values to standardised age-independent values for use in the MBG modelling. Predictions were made for all-age prevalence estimates for individuals aged one to 99 years ($PvPR_{1-99}$). Children aged <1 year were not included because of the potential confounding effect of maternal antibodies, but all other ages were included. This deviated from the method of using the 2- to 10-year cohort for falciparum malaria (Guerra et al., 2007; Gething et al., 2011a), because all age ranges are typically sampled for *P. vivax*, which is found at relatively lower prevalence rates.

To determine the endemicity of *P. vivax*, it was also important to incorporate Duffy negativity into the modelling framework because of the refractory nature of the phenotype to the parasite. The map of Duffy negativity described above informed the model of the fraction of Duffy-negative individuals in the population at each pixel of the predicted surface. The parasite endemicity prediction could be made, therefore, from the vivax-susceptible or Duffy positive portion of the population. This meant that the proportion of *Pv*PR could not exceed the percentage of the population who were Duffy positive and that predictions in data-sparse portions of Africa could borrow strength from the Duffy negativity surface because estimates were limited to a more restricted range of potential outcomes.

The endemicity surface that results from the MBG modelling framework described is a 5 × 5 km grid of predictions for $PvPR_{1-99}$ within the limits of stable *P. vivax* transmission. For practical reasons, and because areas endemic for *P. vivax* have distinct ecological, entomological and epidemiological characteristics, the *P. vivax* endemic world was divided into four regions: the Americas, Africa+, Asia and Asia-Pacific. Separate models were fitted to each region so that a $PvPR_{1-99}$ estimate averaged across the 12 months of 2010 was found. The endemicity map was created by using the mean of each posterior distribution as a point estimate and uncertainty was shown as the ratio of the posterior distribution IQR to its mean. The IQR was found to express the precision with which the $PvPR_{1-99}$ values were predicted. Calculation of the ratio of the IQR of each posterior distribution to its mean generated an index that demonstrated how the model performance varied with data density in different locations. This index was also weighted by population density to generate a map to show where high levels of uncertainty may be operationally significant.

6.3. The Refined Population at Risk of *P. vivax*

The various data sources described above were combined to produce the final spatial limits map (Fig. 1.10A6–D6). After the implementation of MBG modelling, some of the regions that were estimated as being within the limits of stable transmission were downgraded to unstable transmission. If the model outputs of *Pv*PR were extremely low due to a large abundance of surveys reporting zero infections in that area or, in Africa, because of high Duffy negativity, a decision rule was applied such that pixels that were predicted with high certainty (probability >0.9) to be less than 1% *Pv*PR were reassigned as unstable.

The PAR of *P. vivax* malaria was estimated using the constrained limits of infection and population values for 2010 projected from the year 2000 GRUMP *beta* version population counts. The result was a 1 × 1 km spatial grid of population surface of the number of people living at stable or unstable risk in each country as well as surface area of the regions at risk. The population surface, along with uncertainty maps, was also used to calculate a population-weighted index of uncertainty. The initial PAR based on the *Pv*MECs was 5.36 billion individuals in a land area of 69 million km^2. After applying the *Pv*API data (1.3 billion PAR excluded), temperature suitability (sporogony duration; 61 million PAR excluded), aridity (32 million PAR excluded), medical intelligence (713 million PAR excluded) and the Duffy negativity (768 million PAR excluded) layers, the remaining PAR was 2.49 billion in an area of 44 million km^2. This indicates that *P. vivax* was endemic across approximately a third of the world's surface. Half of that area was found in Africa (51%) and a quarter was in both America (22%) and Asia (27%). However, high population density in parts of Asia and the large proportion of the protective Duffy-negative phenotype found in African populations, meant that 82% of the 2.49 billion people at risk were in Asia with the remaining 17% being spread across Asia-Pacific (9%), the Americas (6%) and Africa (3%). Well over half of the PAR lived in areas of unstable transmission (62%; 1.52 billion) where transmission was very low and unlikely to exceed one case per 10,000 people per annum. A total of 965 million people were estimated to be living at risk of stable transmission. The endemicity maps (Fig. 1.3A2–D2) demonstrate that the transmission potential in and among stable transmission areas may vary greatly, even within relatively small geographic areas.

6.4. Mapping the Range of Dominant Vector Species

Vector distribution maps have long been used as a tool to aid malaria control globally. Examples of these maps date back to the 1950s and include vector species maps (May, 1951) and the ecological zones of malaria epidemiology determined by Macdonald based on climatology and known vector species ranges (Macdonald, 1957). More recent ecological (Mouchet et al., 2004) and global vector distribution (Kiszewski et al., 2004) maps have been widely adopted by the malaria research community. A recent series of publications by MAP has attempted to update malaria vector distribution maps with a comprehensive and extensive evidence base (Hay et al., 2010c; Sinka et al., 2010a, 2010b, 2011, 2012). There are 465 formally described species of *Anopheles* mosquitoes and more than 50 unnamed species and species complexes (Harbach, 2011). Approximately 70 species and species complexes have been incriminated to transmit malaria parasites (Service and Townson, 2002) and of those, 41 have been identified as DVS (Hay et al., 2010c). Determination of vector dominance is generally based on factors that increase overall vectorial capacity (Takken and Lindsay, 2003), including abundance, propensity for feeding on humans and the mean adult longevity (to determine if the species lives long enough to transmit the parasite) (Hay et al., 2010c). In the process of determining vector dominance, it was noted which anophelines may have the potential to transmit *P. vivax*. A literature search of 'vivax' and the *Anopheles* species name was performed to identify evidence supporting wild transmission of the vivax parasite. The list of the *Anopheles* species, species complexes and groups as well as the evidence gathered regarding the potential for transmission of *P. vivax* are shown in Table 1.2.

To predict the geographic range of the 41 DVS of malaria, known occurrence points, expert opinion maps, ecological data and modelling techniques were applied. To begin, 15,837 occurrence records from 4800 sources were acquired from systematic searches of formal and informal literature sources and compiled into a comprehensive database (Hay et al., 2010c; Sinka et al., 2012). Expert opinion (EO) maps were then digitized from exhaustive searches of published maps, which are referenced in detail elsewhere (Sinka et al., 2010a, 2010b, 2011), and refined by consultation with a TAG. A suite of environmental and climatic variables known to shape vector distribution landscapes (such as elevation, land surface temperature and precipitation) were also included in the database (Sinka et al., 2010b). BRT modelling methodology (Elith et al., 2008) was applied to generate a predicted distribution map for each DVS. Distributions were generated for nine species/species complexes in the Americas; 13 for Africa, Europe and the Middle East; and 19 in Asia, 14 of which were in Asia (five were only in Asia), and 16 in Asia-Pacific (three were only in Asia-Pacific). Information was also gathered regarding the bionomics of the DVS, which greatly affects the potential impacts of common malaria interventions such as ITNs and IRS. Behaviours that were searched for and catalogued by species included larval site and habitat types, and adult resting and biting behaviours.

The predicted ranges of the DVS varied greatly across the regions, with relatively straightforward vector profiles in the Americas and Africa and very complex vector distributions across Asia. The distribution maps illustrate a probability of occurrence, but do not indicate the predicted prevalence. A positive (coloured) pixel does indicates that the probability of occurrence is >0.5 (>0.5 and ≤1.0) and that a negative pixel (not coloured/grey) represents a probability of occurrence <0.5 (0–0.5). The regional maps shown here are generated from an amalgamation of individual DVS distribution maps. The TAG identified the top three DVS per country (if the country had more than three DVS) and ranked the DVS by their importance. Where there was great variation within countries (e.g. Indonesia), more detailed spatial information regarding the DVS was gathered. The rankings were then used to determine the order in which the species-specific distribution maps were layered to generate global-scale maps for the *Anopheles* vectors of greatest public health importance.

LIST OF ABBREVIATIONS

ACT Artemisinin-based combination therapy
ADMIN Administrative unit
Africa+ Africa, Yemen and Saudi Arabia
Asia Mainland Asia excluding the Malaysian Peninsula
Asia-Pacific The Malaysian Peninsula and the islands of Asia and the Pacific
CSE Asia Central and South East Asia
DVS Dominant vector species or species complexes
ELISA Enzyme-linked immunosorbent assay
G6PD Glucose-6-phosphate dehydrogenase
GRUMP Global Rural-Urban Mapping Project
IQR Inter-quartile range
Lao PDR Lao People's Democratic Republic
Korea DPR Democratic People's Republic of Korea
MAP Malaria Atlas Project
MBG Model-based geostatistics
MCMC Markov chain Monte Carlo
PAR Populations at risk
***Pv*API** *P. vivax* annual parasite incidence
***Pv*MEC** *P. vivax* malaria endemic country
***Pv*PR** *P. vivax* parasite rate
***Pv*PR$_{1-99}$** *P. vivax* parasite rate in 1–99-year olds
RDT Rapid diagnostic test
s.l. *sensu lato*
s.s. *sensu stricto*
TAG Technical advisory group

ACKNOWLEDGEMENTS

The large global assembly of parasite prevalence data was critically dependent on the generous contributions of data made by a large number of people in the malaria research and control communities and these individuals are listed on the MAP website (http://www.map.ac.uk/acknowledgements). We thank David Pigott and Jennie Charlton for proofreading the manuscript. SIH is funded by a Senior Research Fellowship from the Wellcome Trust (#095066), which also supports PWG, CAG and KEB. IRFE is funded by grants from the University of Oxford—Li Ka Shing Foundation Global Health Program and the Oxford Tropical Network. CLM is funded by a Biomedical Resources Grant from the Wellcome

Trust (#091835). The initial vector database and species distribution maps were created while MES was funded by a project grant from the Wellcome Trust (#083534) to SIH. The compilation of the global distributions was conducted with support and funding from the Bill and Melinda Gates Foundation via the VECNet consortium (http://www.vecnet.org/). REH is funded by a Biomedical Resources Grant from the Wellcome Trust (#085406). RNP is a Wellcome Trust Senior Fellow in Clinical Science (#091625). JKB is supported by a Wellcome Trust grant (#B9RJIXO). PWG and SIH also acknowledge support from the RAPIDD program of the Science & Technology Directorate, Department of Homeland Security, and the Fogarty International Center, National Institutes of Health (http://www.fic.nih.gov). This work reviewed here forms part of the output of the Malaria Atlas Project (MAP, http://www.map.ox.ac.uk), principally funded by the Wellcome Trust, UK (http://www.wellcome.ac.uk). MAP also acknowledges the support of the Global Fund to Fight AIDS, Tuberculosis, and Malaria (http://www.theglobalfund.org). The funders had no role in data collection and analysis, decision to publish, or preparation of the manuscript.

REFERENCES

Achee, N.L., Korves, C.T., Bangs, M.J., Rejmankova, E., Lege, M., Curtin, D., Lenares, H., Alonzo, Y., Andre, R.G., Roberts, D.R., 2000. *Plasmodium vivax* polymorphs and *Plasmodium falciparum* circumsporozoite proteins in *Anopheles* (Diptera: Culicidae) from Belize, Central America. J. Vector Ecol. 25, 203–211.

Adak, T., Kaur, S., Singh, O.P., 1999. Comparative susceptibility of different members of the *Anopheles culicifacies* complex to *Plasmodium vivax*. Trans. R. Soc. Trop. Med. Hyg. 93, 573–577.

Adak, T., Sharma, V.P., Orlov, V.S., 1998. Studies on the *Plasmodium vivax* relapse pattern in Delhi, India. Am. J. Trop. Med. Hyg. 59, 175–179.

Adak, T., Singh, O.P., Das, M.K., Wattal, S., Nanda, N., 2005. Comparative susceptibility of three important malaria vectors *Anopheles stephensi, Anopheles fluviatilis*, and *Anopheles sundaicus* to *Plasmodium vivax*. J. Parasitol. 91, 79–82.

Adak, T., Singh, O.P., Nanda, N., Sharma, V.P., Subbarao, S.K., 2006. Isolation of a *Plasmodium vivax* refractory *Anopheles culicifacies* strain from India. Trop. Med. Int. Health 11, 197–203.

Ahumada, J.A., Lapointe, D., Samuel, M.D., 2004. Modeling the population dynamics of *Culex quinquefasciatus* (Diptera: Culicidae), along an elevational gradient in Hawaii. J. Med. Entomol. 41, 1157–1170.

Akhtar, R., Dutt, A., Wadhwa, V., 2009. Malaria resurgence in urban India: lessons from health planning strategies. In: Akhtar, R., Dutt, A., Wadhwa, V. (Eds.), Malaria in South Asia: Eradication and Resurgence During the Second Half of the Twentieth Century (Advances in Asian Human-Environmental Research), Spring, London, pp. 141–155.

Alam, M.S., Khan, M.G., Chaudhury, N., Deloer, S., Nazib, F., Bangali, A.M., Haque, R., 2010. Prevalence of anopheline species and their *Plasmodium* infection status in epidemic-prone border areas of Bangladesh. Malar. J. 9, 15.

Alonso, P.L., Brown, G., Arevalo-Herrera, M., Binka, F., Chitnis, C., Collins, F., Doumbo, O.K., Greenwood, B., Hall, B.F., Levine, M.M., Mendis, K., Newman, R.D., Plowe, C.V., Rodriguez, M.H., Sinden, R., Slutsker, L., Tanner, M., 2011. A research agenda to underpin malaria eradication. PLoS Med. 8, e1000406.

Am, N.T., Riec, L.Q., Huyen, V.T., Lan, N.B., 1993. Etudes entomo-épidémiologiques du paludisme dans la zone côtière de Ho Chi Minh-Ville, 1990–1992. Cahiers Santé 3, 464–473.

Amerasinghe, F.P., Amerasinghe, P.H., Peiris, J.S., Wirtz, R.A., 1991a. Anopheline ecology and malaria infection during the irrigation development of an area of the Mahaweli Project, Sri Lanka. Am. J. Trop. Med. Hyg. 45, 226–235.

Amerasinghe, F.P., Amerasinghe, P.H., Peiris, J.S., Wirtz, R.A., 1991b. Anopheline ecology and malaria infection during the irrigation development of an area of the Mahaweli Project, Sri Lanka. Am. J. Trop. Med. Hyg. 45, 226–235.

Amerasinghe, F.P., Indrajith, N.G., 1994. Postirrigation breeding patterns of surface water mosquitoes in the Mahaweli Project, Sri Lanka, and comparisons with preceding developmental phases. J. Med. Entomol. 31, 516–523.

Anonymous, 2009. Kenya Malaria Monitoring and Evaluation Plan 2009–2017.

Anstey, N.M., Russell, B., Yeo, T.W., Price, R.N., 2009. The pathophysiology of vivax malaria. Trends Parasitol. 25, 220–227.

Antonio-Nkondjio, C., Simard, F., Awono-Ambene, P., Ngassam, P., Toto, J.C., Tchuinkam, T., Fontenille, D., 2005. Malaria vectors and urbanization in the equatorial forest region of south Cameroon. Trans. R. Soc. Trop. Med. Hyg. 99, 347–354.

Arevalo-Herrera, M., Chitnis, C., Herrera, S., 2010. Current status of *Plasmodium vivax* vaccine. Hum. Vaccin. 6, 124–132.

Attenborough, R.D., Burkot, T.R., Gardner, D.S., 1997. Altitude and the risk of bites from mosquitoes infected with malaria and filariasis among the Mianmin people of Papua New Guinea. Trans. R. Soc. Trop. Med. Hyg. 91, 8–10.

Awab, G.R., Pukrittayakamee, S., Imwong, M., Dondorp, A.M., Woodrow, C.J., Lee, S.J., Day, N.P., Singhasivanon, P., White, N.J., Kaker, F., 2010. Dihydroartemisinin-piperaquine versus chloroquine to treat vivax malaria in Afghanistan: an open randomized, non-inferiority, trial. Malar. J. 9, 105.

Baimai, V., Harbach, R.E., Sukowati, S., 1988. Cytogenetic evidence for two species within the current concept of the malaria vector *Anopheles leucosphyrus* in Southeast Asia. J. Am. Mosq. Control Assoc. 4, 44–50.

Baird, J.K., 2004. Chloroquine resistance in *Plasmodium vivax*. Antimicrob. Agents Chemother. 48, 4075–4083.

Baird, J.K., 2007. Neglect of *Plasmodium vivax* malaria. Trends Parasitol. 23, 533–539.

Baird, J.K., 2009. Resistance to therapies for infection by *Plasmodium vivax*. Clin. Microbiol. Rev. 22, 508–534.

Baird, J.K., 2010. Eliminating malaria – all of them. Lancet 376, 1883–1885.

Baird, J.K., Hoffman, S.L., 2004. Primaquine therapy for malaria. Clin. Infect. Dis. 39, 1336–1345.

Baird, J.K., Rieckmann, K.H., 2003. Can primaquine therapy for vivax malaria be improved?. Trends Parasitol. 19, 115–120.

Baird, J.K., Schwartz, E., Hoffman, S.L., 2007. Prevention and treatment of vivax malaria. Curr. Infect. Dis. Rep. 9, 39–46.

Baker, E.Z., Beier, J.C., Meek, S.R., Wirtz, R.A., 1987. Detection and quantification of *Plasmodium falciparum* and *P. vivax* infections in Thai-Kampuchean Anopheles (Diptera: Culicidae) by enzyme-linked immunosorbent assay. J. Med. Entomol. 24, 536–541.

Balk, D.L., Deichmann, U., Yetman, G., Pozzi, F., Hay, S.I., Nelson, A., 2006. Determining global population distribution: methods, applications and data. Adv. Parasitol. 62, 119–156.

Barcus, M.J., Laihad, F., Sururi, M., Sismadi, P., Marwoto, H., Bangs, M.J., Baird, J.K., 2002. Epidemic malaria in the Menoreh Hills of Central Java. Am. J. Trop. Med. Hyg. 66, 287–292.

Barik, T.K., Sahu, B., Swain, V., 2009. A review on *Anopheles culicifacies*: from bionomics to control with special reference to Indian subcontinent. Acta Trop. 109, 87–97.

Barnwell, J.W., Nichols, M.E., Rubinstein, P., 1989. *In vitro* evaluation of the role of the Duffy blood group in erythrocyte invasion by *Plasmodium vivax*. J. Exp. Med. 169, 1795–1802.

Basseri, H.R., Doosti, S., Akbarzadeh, K., Nategpour, M., Whitten, M.M., Ladoni, H., 2008. Competency of *Anopheles stephensi mysorensis* strain for *Plasmodium vivax* and the role of inhibitory carbohydrates to block its sporogonic cycle. Malar. J. 7, 131.

Battle, K.E., Van Boeckel, T., Gething, P.W., Baird, J.K., Hay, S.I., 2011. A review of the geographical variations in *Plasmodium vivax* relapse rate. Am. J. Trop. Med. Hyg. 85 (Suppl. 451-A-516), 470.

Betuela, I., Rosanas, A., Kiniboro, B., Stanisic, D., Samol, L., Lazzari, E., Bassat, Q., del Portillo, H., Siba, P., Alonso, P., Muller, I., 2011. Relapses are contributing significantly to risk of *P. vivax* infection and disease in Papua New Guinean children 1-5 years of age. Trop. Med. Int. Health 16, 60–61.

Beutler, E., 1994. G6PD deficiency. Blood 84, 3613–3636.

Bicheron, P., Defourny, P., Brockmann, C., Schouten, L., Vancutsem, C., Huc, M., Bontemps, S., Leroy, M., Achard, F., Herold, M., Ranera, F. and Arino, O. (2008). GLOBCOVER: Products Description and Validation Report.

Bigoga, J.D., Manga, L., Titanji, V.P., Coetzee, M., Leke, R.G., 2007. Malaria vectors and transmission dynamics in coastal south-western Cameroon. Malar. J. 6, 5.

Black, R.E., Cousens, S., Johnson, H.L., Lawn, J.E., Rudan, I., Bassani, D.G., Jha, P., Campbell, H., Walker, C.F., Cibulskis, R., Eisele, T., Liu, L., Mathers, C., 2010. Global, regional, and national causes of child mortality in 2008: a systematic analysis. Lancet 375, 1969–1987.

Bockarie, M.J., Dagoro, H., 2006. Are insecticide-treated bednets more protective against *Plasmodium falciparum* than *Plasmodium vivax*-infected mosquitoes?. Malar. J. 5, 15.

Bockarie, M.J., Service, M.W., Toure, Y.T., Traore, S., Barnish, G., Greenwood, B.M., 1993. The ecology and behaviour of the forest form of Anopheles gambiae s.s. Parassitologia 35 (Suppl.), 5–8.

Bousema, T., Drakeley, C., 2011. Epidemiology and infectivity of *Plasmodium falciparum* and *Plasmodium vivax* gametocytes in relation to malaria control and elimination. Clin. Microbiol. Rev. 24, 377–410.

Boyd, M.F., Kitchen, S.F., 1937. On the infectiousness of patients infected with *Plasmodium vivax* and *Plasmodium falciparum*. Am. J. Trop. Med. Hyg. s1–17, 253–262.

Branquinho, M.S., Marrelli, M.T., Curado, I., Natal, D., Barata, J.M., Tubaki, R., Carreri-Bruno, G.C., de Menezes, R.T., Kloetzel, J.K., 1997. Infection of *Anopheles (Kerteszia) cruzii* by *Plasmodium vivax* and *Plasmodium vivax* variant VK247 in the municipalities of Sao Vicente and Juquitiba, Sao Paulo. Rev. Panam. Salud Publica 2, 189–193.

Brooker, S., Leslie, T., Kolaczinski, K., Mohsen, E., Mehboob, N., Saleheen, S., Khudonazarov, J., Freeman, T., Clements, A., Rowland, M., Kolaczinski, J., 2006. Spatial epidemiology of *Plasmodium vivax*, Afghanistan. Emerg. Infect. Dis. 12, 1600–1602.

Brown, G.V., Moorthy, V.S., Reed, Z., Mendis, K., Arevalo-Herrera, M., Alonso, P., 2009. Priorities in research and development of vaccines against *Plasmodium vivax* malaria. Vaccine 27, 7228–7235.

Burgess, R.W., Young, M.D., 1950. The comparative susceptibility of *Anopheles quadrimaculatus* and *Anopheles freeborni* to infection by *Plasmodium vivax* (St. Elizabeth strain). J. Natl. Malar. Soc. 9, 218–221.

Burkot, T.R., Graves, P.M., Paru, R., Wirtz, R.A., Heywood, P.F., 1988. Human malaria transmission studies in the *Anopheles punctulatus* complex in Papua New Guinea: sporozoite rates, inoculation rates, and sporozoite densities. Am. J. Trop. Med. Hyg. 39, 135–144.

Butler, W.P., Roberts, D.R., 2000. Malaria in the Americas: a model of reemergence. Mil. Med. 165, 897–902.

Cappellini, M.D., Fiorelli, G., 2008. Glucose-6-phosphate dehydrogenase deficiency. Lancet 371, 64–74.

Cavalli-Sforza, L.L., Menozzi, P., Piazza, A., 1994. The History and Geography of Human Genes. Princeton University Press, Princeton.

Centers for Disease Control and Prevention, 2009. CDC Health Information for International Travel 2010.

Chang, M.S., Hii, J., Buttner, P., Mansoor, F., 1997. Changes in abundance and behaviour of vector mosquitoes induced by land use during the development of an oil palm plantation in Sarawak. Trans. R. Soc. Trop. Med. Hyg. 91, 382–386.

Chen, N., Auliff, A., Rieckmann, K., Gatton, M., Cheng, Q., 2007. Relapses of *Plasmodium vivax* infection result from clonal hypnozoites activated at predetermined intervals. J. Infect. Dis. 195, 934–941.
Chitnis, N., Schapira, A., Smith, D.L., Hay, S.I., Smith, T., Steketee, R.W., Global Partnership to Roll Back Malaria, 2010a. Mathematical modeling to support malaria control and elimination. Roll Back Malaria Progress and Impact Series 5, 48.
Chitnis, N., Schapira, A., Smith, T., Steketee, R., 2010b. Comparing the effectiveness of malaria vector-control interventions through a mathematical model. Am. J. Trop. Med. Hyg. 83, 230–240.
Chow, C.Y., 1970. Bionomics of malaria vectors in the Western Pacific Region. Southeast Asian J. Trop. Med. Public Health 1, 40–57.
CIESIN/IFPRI/WB/CIAT, 2007. Global Rural Urban Mapping Project (GRUMP) Alpha: Gridded Population of the World, Version 2, with Urban Reallocation (GPW-UR). URL: http://sedac.ciesin.columbia.edu/gpw.
Coatney, G.R., 1976. Relapse in malaria – an enigma. J. Parasitol. 62, 3–9.
Coatney, G.R., Cooper, W.C., 1948. Recrudescence and relapse in vivax malaria. In: Proceedings of the 4th International Congress on Tropical Medicine.
Coetzee, M., Fontenille, D., 2004. Advances in the study of *Anopheles funestus*, a major vector of malaria in Africa. Insect Biochem. Mol. Biol. 34, 599–605.
Cogswell, F.B., 1992. The hypnozoite and relapse in primate malaria. Clin. Microbiol. Rev. 5, 26–35.
Coleman, R.E., Kumpitak, C., Ponlawat, A., Maneechai, N., Phunkitchar, V., Rachapaew, N., Zollner, G., Sattabongkot, J., 2004. Infectivity of asymptomatic *Plasmodium*-infected human populations to *Anopheles dirus* mosquitoes in western Thailand. J. Med. Entomol. 41, 201–208.
Coleman, R.E., Sithiprasasna, R., Kankaew, P., Kiaattiut, C., Ratanawong, S., Khuntirat, B., Sattabongkot, J., 2002. Naturally occurring mixed infection of *Plasmodium vivax* VK210 and *P. vivax* VK247 in *Anopheles* mosquitoes (Diptera: Culicidae) in western Thailand. J. Med. Entomol. 39, 556–559.
Collins, W.E., Jeffery, G.M., 1996. Primaquine resistance in *Plasmodium vivax*. Am. J. Trop. Med. Hyg. 55, 243–249.
Collins, W.E., Sullivan, J.S., Nace, D., Williams, T., Sullivan, J.J., Galland, G.G., Grady, K.K., Bounngaseng, A., 2002. Experimental infection of *Anopheles farauti* with different species of Plasmodium. J. Parasitol. 88, 295–298.
Collins, W.E., Sullivan, J.S., Strobert, E., Galland, G.G., Williams, A., Nace, D., Williams, T., Barnwell, J.W., 2009. Studies on the Salvador I strain of *Plasmodium vivax* in non-human primates and anopheline mosquitoes. Am. J. Trop. Med. Hyg. 80, 228–235.
Collins, W.E., Warren, M., Contacos, P.G., Skinner, J.C., Richardson, B.B., Kearse, T.S., 1980. The Chesson strain of *Plasmodium vivax* in *Aotus* monkeys and anopheline mosquitoes. J. Parasitol. 66, 488–497.
Collins, W.E., Warren, M., Skinner, J.C., Sutton, B.B., 1985. Infectivity of two strains of *Plasmodium vivax* to *Anopheles albitarsis* mosquitoes from Colombia. J. Parasitol. 71, 771–773.
Coluzzi, M., 1999. The clay feet of the malaria giant and its African roots: hypotheses and inferences about origin, spread and control of *Plasmodium falciparum*. Parassitologia 41, 277–283.
Conn, J.E., Wilkerson, R.C., Segura, M.N.O., De Souza, R.T.L., Schlichting, C.D., Wirtz, R.A., Povoa, M.M., 2002. Emergence of a new neotropical malaria vector facilitated by human migration and changes in land use. Am. J. Trop. Med. Hyg. 66, 18–22.
Cutbush, M., Mollison, P.L., 1950. The Duffy blood group system. Heredity (Edinb) 4, 383–389.
da Silva-Vasconcelos, A., Kato, M.Y., Mourao, E.N., de Souza, R.T., Lacerda, R.N., Sibajev, A., Tsouris, P., Povoa, M.M., Momen, H., Rosa-Freitas, M.G., 2002. Biting indices, host-seeking activity and natural infection rates of anopheline species in Boa Vista, Roraima, Brazil from 1996 to 1998. Mem. Inst. Oswaldo Cruz 97, 151–161.

da Silva, A.N., Santos, C.C., Lacerda, R.N., Machado, R.L., Povoa, M.M., 2006. Susceptibility of *Anopheles aquasalis* and *An. darlingi* to *Plasmodium vivax* VK210 and VK247. Mem. Inst. Oswaldo Cruz 101, 547–550.
da Silva, N.S., da Silva-Nunes, M., Malafronte, R.S., Menezes, M.J., D'Arcadia, R.R., Komatsu, N.T., Scopel, K.K., Braga, E.M., Cavasini, C.E., Cordeiro, J.A., Ferreira, M.U., 2010. Epidemiology and control of frontier malaria in Brazil: lessons from community-based studies in rural Amazonia. Trans. R. Soc. Trop. Med. Hyg. 104, 343–350.
Daskova, N.G., Rasnicyn, S.P., 1982. Review of data on susceptibility of mosquitoes in the USSR to imported strains of malaria parasites. Bull. World Health Organ. 60, 893–897.
de Arruda, M., Carvalho, M.B., Nussenzweig, R.S., Maracic, M., Ferreira, A.W., Cochrane, A.H., 1986. Potential vectors of malaria and their different susceptibility to *Plasmodium falciparum* and *Plasmodium vivax* in northern Brazil identified by immunoassay. Am. J. Trop. Med. Hyg. 35, 873–881.
de Castro, M.C., Monte-Mor, R.L., Sawyer, D.O., Singer, B.H., 2006. Malaria risk on the Amazon frontier. Proc. Natl. Acad. Sci. U.S.A. 103, 2452–2457.
de Zulueta, J., Ramsdale, C.D., Coluzzi, M., 1975. Receptivity to malaria in Europe. Bull. World Health Organ. 52, 109–111.
Deane, L.M., 1986. Malaria vectors in Brazil. Mem. Inst. Oswaldo Cruz 81, 5–14.
Detinova, T.S., 1962. Age-grouping methods in Diptera of medical importance, with special reference to some vectors of malaria.
DFID, 2010. Breaking the Cycle: Saving Lives and Protecting the Future. vol. 65, The UK's Framework for Results for Malaria in the Developing World.
Diggle, P.J., Ribeiro, P.J., 2007. Model-based Geostatistics. Springer, New York.
Diggle, P.J., Tawn, J.A., Moyeed, R.A., 1998. Model-based geostatistics. J. R. Stat. Soc. Ser. C Appl. Stat. 47, 299–326.
Douglas, N.M., Anstey, N.M., Angus, B.J., Nosten, F., Price, R.N., 2010. Artemisinin combination therapy for vivax malaria. Lancet Infect. Dis. 10, 405–416.
El Said, S.E., El Sawaf, B., Gebril, M., Kenawy, M.A., Azzab, A., 1983. Bionomics of anopheline mosquitoes in the Faiyoum Governate, Egypt, in relation to transmission and control of malaria. J. Egypt Public Health Assoc. 58, 189–241.
Elith, J., Leathwick, J.R., Hastie, T., 2008. A working guide to boosted regression trees. J. Anim. Ecol. 77, 802–813.
Erhart, A., Thang, N.D., Xa, N.X., Thieu, N.Q., Hung, L.X., Hung, N.Q., Nam, N.V., Toi, L.V., Tung, N.M., Bien, T.H., Tuy, T.Q., Cong, L.D., Thuan, L.K., Coosemans, M., D'Alessandro, U., 2007. Accuracy of the health information system on malaria surveillance in Vietnam. Trans. R. Soc. Trop. Med. Hyg. 101, 216–225.
Eshghi, N., Motabar, M., Javadian, E., Manoutcheri, A.V., 1976. Biological features of *Anopheles fluviatilis* and its role in the transmission of malaria in Iran. Trop. Geogr. Med. 28, 41–44.
FAO, 2008. The Global Administrative Unit Layers (GAUL): Technical Aspects.
Faulde, M.K., Hoffmann, R., Fazilat, K.M., Hoerauf, A., 2007. Malaria reemergence in northern Afghanistan. Emerg. Infect. Dis. 13, 1402–1404.
Feachem, R., Sabot, O., 2008. A new global malaria eradication strategy. Lancet 371, 1633–1635.
Feachem, R.G.A., Phillips, A.A., Targett, G.A., Malaria Elimination Group (Eds.), 2009. Shrinking the Malaria Map: A Prospectus on Malaria Elimination, The Global Health Group, University of California - Santa Cruz Global Health Sciences, San Francisco, U.S.A.
Flores-Mendoza, C., Fernandez, R., Escobedo-Vargas, K.S., Vela-Perez, Q., Schoeler, G.B., 2004. Natural *Plasmodium* infections in *Anopheles darlingi* and *Anopheles benarrochi* (Diptera: Culicidae) from eastern Peru. J. Med. Entomol. 41, 489–494.
Foley, D.H., Torres, E.P., Mueller, I., Bryan, J.H., Bell, D., 2003. Host-dependent *Anopheles flavirostris* larval distribution reinforces the risk of malaria near water. Trans. R. Soc. Trop. Med. Hyg. 97, 283–287.

Fontenille, D., Lepers, J.P., Campbell, G.H., Coluzzi, M., Rakotoarivony, I., Coulanges, P., 1990. Malaria transmission and vector biology in Manarintsoa, high plateaux of Madagascar. Am. J. Trop. Med. Hyg. 43, 107–115.
Frances, S.P., Klein, T.A., Wirtz, R.A., Eamsila, C., Pilakasiri, C., Linthicum, K.J., 1996. *Plasmodium falciparum* and *P. vivax* circumsporozoite proteins in anophelines (Diptera: Culicidae) collected in eastern Thailand. J. Med. Entomol. 33, 990–991.
Fryauff, D.J., Tuti, S., Mardi, A., Masbar, S., Patipelohi, R., Leksana, B., Kain, K.C., Bangs, M.J., Richie, T.L., Baird, J.K., 1998. Chloroquine-resistant *Plasmodium vivax* in transmigration settlements of West Kalimantan, Indonesia. Am. J. Trop. Med. Hyg. 59, 513–518.
Galappaththy, G.N., Omari, A.A., Tharyan, P., 2007. Primaquine for preventing relapses in people with *Plasmodium vivax* malaria. Cochrane Database Syst. Rev. (Online), CD004389.
Garnham, P.C., 1951. Some effects on the community of malaria eradication with special reference to the relapse phenomenon. East Afr. Med. J. 28, 6–10.
Garnham, P.C., 1989. Malaria Parasites of Man: Life-cycles and Morphology Excluding Ultrastructure). Malaria: Principles and Practice of Malariology. Churchill Livingstone, Edinburgh, pp. 61–96.
Garnham, P.C., Bray, R.S., Bruce-Chwatt, L.J., Draper, C.C., Killick-Kendrick, R., Sergiev, P.G., Tiburskaja, N.A., Shute, P.G., Maryon, M., 1975. A strain of *Plasmodium vivax* characterized by prolonged incubation: morphological and biological characteristics. Bull. World Health Organ. 52, 21–32.
Garrett-Jones, C., 1964. Prognosis for interruption of malaria transmission through assessment of the mosquito's vectorial capacity. Nature 204, 1173–1175.
Genton, B., D'Acremont, V., Rare, L., Baea, K., Reeder, J.C., Alpers, M.P., Muller, I., 2008. *Plasmodium vivax* and mixed infections are associated with severe malaria in children: a prospective cohort study from Papua New Guinea. PLoS Med. 5, e127.
Gething, P.W., Elyazar, I.R.F., Moyes, C.M., Smith, D.L., Battle, K.E., Guerra, C.A., Patil, A.P., Tatem, A.J., Howes, R.E., Myers, M.F., George, D.B., Horby, P., Wertheim, H.F.L., Price, R.N., Mueller, I., Baird, J.K., Hay, S.I., 2012. A long neglected world malaria map: *Plasmodium vivax* endemicity in 2010. PLoS Negl. Trop. Dis. 6 e1814.
Gething, P.W., Patil, A.P., Hay, S.I., 2010a. Quantifying aggregated uncertainty in *Plasmodium falciparum* malaria prevalence and populations at risk via efficient space-time geostatistical joint simulation. PLoS Comput. Biol. 6 e1000724.
Gething, P.W., Patil, A.P., Smith, D.L., Guerra, C.A., Elyazar, I.R., Johnston, G.L., Tatem, A.J., Hay, S.I., 2011a. A new world malaria map: *Plasmodium falciparum* endemicity in 2010. Malar. J. 10, 378.
Gething, P.W., Smith, D.L., Patil, A.P., Tatem, A.J., Snow, R.W., Hay, S.I., 2010b. Climate change and the global malaria recession. Nature 465, 342–345.
Gething, P.W., Van Boeckel, T.P., Smith, D.L., Guerra, C.A., Patil, A.P., Snow, R.W., Hay, S.I., 2011b. Modelling the global constraints of temperature on transmission of *Plasmodium falciparum* and *P. vivax*. Parasit. Vectors 4, 92.
Gillies, M.T., de Meillon, B. (Eds.), 1968. The Anophelinae of Africa South of the Sahara (Ethiopian Zoogeographical Region), The South African Institute for Medical Research, Johannesburg.
Ginsberg, M.M., 1991. Transmission of malaria in San Diego County, California. West J. Med. 154, 465–466.
Girod, R., Gaborit, P., Carinci, R., Issaly, J., Fouque, F., 2008. *Anopheles darlingi* bionomics and transmission of *Plasmodium falciparum, Plasmodium vivax* and *Plasmodium malariae* in Amerindian villages of the Upper-Maroni Amazonian forest, French Guiana. Mem. Inst. Oswaldo Cruz 103, 702–710.
Global Partnership to Roll Back Malaria, Johansson, E.W., Cibulskis, R.E., Steketee, R.W, 2010. Malaria funding and resource utilization: the first decade of Roll Back Malaria. 95.

Gonzalez-Ceron, L., Rodriguez, M.H., Chavez-Munguia, B., Santillan, F., Nettel, J.A., Hernandez-Avila, J.E., 2007. *Plasmodium vivax*: impaired escape ofVk210 phenotype ookinetes from the midgut blood bolus of *Anopheles pseudopunctipennis*. Exp. Parasitol. 115, 59–67.

Gray, E.M., Bradley, T.J., 2005. Physiology of desiccation resistance in *Anopheles gambiae* and *Anopheles arabiensis*. Am. J. Trop. Med. Hyg. 73, 553–559.

Greenwood, B.M., 2008. Control to elimination: implications for malaria research. Trends Parasitol. 24, 449–454.

Griffin, J.T., Hollingsworth, T.D., Okell, L.C., Churcher, T.S., White, M., Hinsley, W., Bousema, T., Drakeley, C.J., Ferguson, N.M., Basanez, M.G., Ghani, A.C., 2010. Reducing *Plasmodium falciparum* malaria transmission in Africa: a model-based evaluation of intervention strategies. PLoS Med. 7, e1000324.

Guerra, C.A., Gikandi, P.W., Tatem, A.J., Noor, A.M., Smith, D.L., Hay, S.I., Snow, R.W., 2008. The limits and intensity of *Plasmodium falciparum* transmission: implications for malaria control and elimination worldwide. PLoS Med. 5, e38.

Guerra, C.A., Hay, S.I., Lucioparedes, L.S., Gikandi, P.W., Tatem, A.J., Noor, A.M., Snow, R.W., 2007. Assembling a global database of malaria parasite prevalence for the Malaria Atlas Project. Malar. J. 6, 17.

Guerra, C.A., Howes, R.E., Patil, A.P., Gething, P.W., Van Boeckel, T.P., Temperely, W.H., Kabaria, C.W., Tatem, A.J., Manh, B.H., Elyazar, I., Baird, J.K., Snow, R.W., Hay, S.I., 2010. The international limits and population at risk of *Plasmodium vivax* transmission in 2009. PLoS Negl. Trop. Dis. 4 e774.

Guerra, C.A., Snow, R.W., Hay, S.I., 2006a. Defining the global spatial limits of malaria transmission in 2005. Adv. Parasitol. 62, 157–179.

Guerra, C.A., Snow, R.W., Hay, S.I., 2006b. Mapping the global extent of malaria in 2005. Trends Parasitol. 22, 353–358.

Gupta, S., Gunter, J.T., Novak, R.J., Regens, J.L., 2009. Patterns of *Plasmodium vivax* and *Plasmodium falciparum* malaria underscore importance of data collection from private health care facilities in India. Malar. J. 8, 227.

Harbach, R.E., 2004. The classification of genus *Anopheles* (Diptera: Culicidae): a working hypothesis of phylogenetic relationships. Bull. Entomol. Res. 94, 537–553.

Harbach, R.E., 2011. Genus *ANOPHELES* Meigen, 1818. Mosquito Taxonomic Inventory. URL: http://mosquito-taxonomic-inventory.info/genus-anopheles-meigen-1818.

Hardin, S., Santa Maria, M.E. and Liaw, C.F., 1973. Experimental Infection of *Anopheles donaldi* Reid with *Plasmodium falciparum* and *Plasmodium vivax*.

Hargreaves, K., Koekemoer, L.L., Brooke, B.D., Hunt, R.H., Mthembu, J., Coetzee, M., 2000. *Anopheles funestus* resistant to pyrethroid insecticides in South Africa. Med. Vet. Entomol. 14, 181–189.

Harris, I., Sharrock, W.W., Bain, L.M., Gray, K.A., Bobogare, A., Boaz, L., Lilley, K., Krause, D., Vallely, A., Johnson, M.L., Gatton, M.L., Shanks, G.D., Cheng, Q., 2010. A large proportion of asymptomatic *Plasmodium* infections with low and sub-microscopic parasite densities in the low transmission setting of Temotu Province, Solomon Islands: challenges for malaria diagnostics in an elimination setting. Malar. J. 9, 254.

Harrison, B.A., Scanlon, J.E., 1975. Medical entomology studies II: the subgenus *Anopheles* in Thailand. Contrib. Am. Entomol. Inst. 12, 1–308.

Hay, S.I., Gething, P.W., Snow, R.W., 2010a. India's invisible malaria burden. Lancet 376, 1716–1717.

Hay, S.I., Guerra, C.A., Gething, P.W., Patil, A.P., Tatem, A.J., Noor, A.M., Kabaria, C.W., Manh, B.H., Elyazar, I.R., Brooker, S., Smith, D.L., Moyeed, R.A., Snow, R.W., 2009. A world malaria map: *Plasmodium falciparum* endemicity in 2007. PLoS Med. 6, e1000048.

Hay, S.I., Guerra, C.A., Tatem, A.J., Atkinson, P.M., Snow, R.W., 2005. Urbanization, malaria transmission and disease burden in Africa. Nat. Rev. Microbiol. 3, 81–90.

Hay, S.I., Guerra, C.A., Tatem, A.J., Noor, A.M., Snow, R.W., 2004. The global distribution and population at risk of malaria: past, present, and future. Lancet Infect. Dis. 4, 327–336.
Hay, S.I., Okiro, E.A., Gething, P.W., Patil, A.P., Tatem, A.J., Guerra, C.A., Snow, R.W., 2010b. Estimating the global clinical burden of *Plasmodium falciparum* malaria in 2007. PLoS Med. 7, e1000290.
Hay, S.I., Sinka, M.E., Okara, R.M., Kabaria, C.W., Mbithi, P.M., Tago, C.C., Benz, D., Gething, P.W., Howes, R.E., Patil, A.P., Temperley, W.H., Bangs, M.J., Chareonviriyaphap, T., Elyazar, I.R., Harbach, R.E., Hemingway, J., Manguin, S., Mbogo, C.M., Rubio-Palis, Y., Godfray, H.C., 2010c. Developing global maps of the dominant *Anopheles* vectors of human malaria. PLoS Med. 7, e1000209.
Hay, S.I., Smith, D.L., Snow, R.W., 2008. Measuring malaria endemicity from intense to interrupted transmission. Lancet Infect. Dis. 8, 369–378.
Hay, S.I., Snow, R.W., 2006. The Malaria Atlas Project: developing global maps of malaria risk. PLoS Med. 3, e473.
Hay, S.I., Tatem, A.J., Graham, A.J., Goetz, S.J., Rogers, D.J., 2006. Global environmental data for mapping infectious disease distribution. Adv. Parasitol. 62, 37–77.
Hii, J., Kan, S., Pereira, M., Parmar, S.S., Campos, R.L., Chan, M.K., 1985. Bancroftian filariasis and malaria in island and hinterland populations in Sabah, Malaysia. Trop. Geogr. Med. 37, 93–101.
Howes, R.E., Patil, A.P., Piel, F.B., Nyangiri, O.A., Kabaria, C.W., Gething, P.W., Zimmerman, P.A., Barnadas, C., Beall, C.M., Gebremedhin, A., Menard, D., Williams, T.N., Weatherall, D.J., Hay, S.I., 2011. The global distribution of the Duffy blood group. Nat. Commun. 2, 266.
Howes, R.E., Piel, F.B., Patil, A.P., Nyangiri, O.A., Gething, P.W., Dewi, M., Hogg, M.M., Battle, K.E., Padilla, C.D., Baird, J.K., Hay, S.I., 2012. G6PD deficiency prevalence and estimates of affected populations in malaria endemic countries: geostatistical model-based map. PLoS Med., 9, e1001339
Hulden, L., 2009. The decline of malaria in Finland–the impact of the vector and social variables. Malar. J. 8, 94.
Hulden, L., Hulden, L., Heliovaara, K., 2008. Natural relapses in vivax malaria induced by Anopheles mosquitoes. Malar. J. 7, 64.
Imwong, M., Snounou, G., Pukrittayakamee, S., Tanomsing, N., Kim, J.R., Nandy, A., Guthmann, J.P., Nosten, F., Carlton, J., Looareesuwan, S., Nair, S., Sudimack, D., Day, N.P., Anderson, T.J., White, N.J., 2007. Relapses of *Plasmodium vivax* infection usually result from activation of heterologous hypnozoites. J. Infect. Dis. 195, 927–933.
Iyengar, M.O., 1954. Vector of malaria in Kabul, Afghanistan. Trans. R. Soc. Trop. Med. Hyg. 48, 319–324.
James, S.P., 1931. Some general results of a study of induced malaria in England. Trans. R. Soc. Trop. Med. Hyg. 24, 478–525.
Jensen, T., Cockburn, A.F., Kaiser, P.E., Barnard, D.R., 1996. Human blood-feeding rates among sympatric sibling species of *Anopheles quadrimaculatus* mosquitoes in northkern Florida. Am. J. Trop. Med. Hyg. 54, 523–525.
Joshi, D., Choochote, W., Park, M.H., Kim, J.Y., Kim, T.S., Suwonkerd, W., Min, G.S., 2009. The susceptibility of *Anopheles lesteri* to infection with Korean strain of *Plasmodium vivax*. Malar. J. 8, 42.
Junkum, A., Jitpakdi, A., Jariyapan, N., Komalamisra, N., Somboon, P., Suwonkerd, W., Saejeng, A., Bates, P.A., Choochote, W., 2005. Susceptibility of two karyotypic forms of *Anopheles aconitus* (Diptera: Culicidae) to *Plasmodium falciparum* and *P. vivax*. Rev. Inst. Med. Trop. Sao Paulo 47, 333–338.
Kaneko, A., Taleo, G., Kalkoa, M., Yaviong, J., Reeve, P.A., Ganczakowski, M., Shirakawa, C., Palmer, K., Kobayakawa, T., Bjorkman, A., 1998. Malaria epidemiology, glucose 6-phosphate dehydrogenase deficiency and human settlement in the Vanuatu Archipelago. Acta Trop. 70, 285–302.

Kasap, H., 1990. Comparison of experimental infectivity and development of *Plasmodium vivax* in *Anopheles sacharovi* and *An. superpictus* in Turkey. Am. J. Trop. Med. Hyg. 42, 111–117.

Katsuragawa, T.H., Gil, L.H., Tada, M.S., de Almeida e Silva, A., Costa, J.D., Araujo Mda, S., Escobar, A.L., da Silva, L.H., 2010. The dynamics of transmission and spatial distribution of malaria in riverside areas of Porto Velho, Rondonia, in the Amazon region of Brazil. PLoS One 5, e9245.

Kenawy, M.A., Beier, J.C., Asiago, C.M., el Said, S.E., Roberts, C.R., 1990. Interpretation of low-level *Plasmodium* infection rates determined by ELISA for anophelines (Diptera: Culicidae) from Egyptian oases. J. Med. Entomol. 27, 681–685.

Killeen, G.F., Fillinger, U., Kiche, I., Gouagna, L.C., Knols, B.G., 2002. Eradication of *Anopheles gambiae* from Brazil: lessons for malaria control in Africa?. Lancet Infect. Dis. 2, 618–627.

Kiszewski, A., Mellinger, A., Spielman, A., Malaney, P., Sachs, S.E., Sachs, J., 2004. A global index representing the stability of malaria transmission. Am. J. Trop. Med. Hyg. 70, 486–498.

Klein, T.A., Lima, J.B., Tada, M.S., Miller, R., 1991. Comparative susceptibility of anopheline mosquitoes in Rondonia, Brazil to infection by *Plasmodium vivax*. Am. J. Trop. Med. Hyg. 45, 463–470.

Kochar, D.K., Das, A., Kochar, S.K., Saxena, V., Sirohi, P., Garg, S., Kochar, A., Khatri, M.P., Gupta, V., 2009. Severe *Plasmodium vivax* malaria: a report on serial cases from Bikaner in northwestern India. Am. J. Trop. Med. Hyg. 80, 194–198.

Langhi Jr., D.M., Bordin, J.O., 2006. Duffy blood group and malaria. Hematology 11, 389–398.

Lardeux, F., Loayza, P., Bouchite, B., Chavez, T., 2007. Host choice and human blood index of *Anopheles pseudopunctipennis* in a village of the Andean valleys of Bolivia. Malar. J. 6, 8.

Lee, D.K., 1998. Effect of two rice culture methods on the seasonal occurrence of mosquito larvae and other aquatic animals in rice fields of southwestern Korea. J. Vector Ecol. 23, 161–170.

Lee, H.I., Lee, J.S., Shin, E.H., Lee, W.J., Kim, Y.Y., Lee, K.R., 2001. Malaria transmission potential by *Anopheles sinensis* in the Republic of Korea. Korean J. Parasitol. 39, 185–192.

Lee, H.W., Shin, E.H., Cho, S.H., Lee, H.I., Kim, C.L., Lee, W.G., Moon, S.U., Lee, J.S., Lee, W.J., Kim, T.S., (2002). Detection of vivax malaria sporozoites naturally infected in anopheline mosquitoes from endemic areas of northern parts of Gyeonggi-do (Province) in Korea. Korean J. Parasitol. 40, 75–81.

Li, J., Collins, W.E., Wirtz, R.A., Rathore, D., Lal, A., McCutchan, T.F., 2001. Geographic subdivision of the range of the malaria parasite *Plasmodium vivax*. Emerg. Infect. Dis. 7, 35–42.

Lien, J.C., Kawengian, B.A., Partono, F., Lami, B., Cross, J.H., 1977. A brief survey of the mosquitoes of South Sulawesi, Indonesia, with special reference to the identity of *Anopheles barbirostris* (Diptera: Culicidae) from the Margolembo area. J. Med. Entomol. 13, 719–727.

Lin, E., Kiniboro, B., Gray, L., Dobbie, S., Robinson, L., Laumaea, A., Schopflin, S., Stanisic, D., Betuela, I., Blood-Zikursh, M., Siba, P., Felger, I., Schofield, L., Zimmerman, P., Mueller, I., 2010. Differential patterns of infection and disease with *P. falciparum* and *P. vivax* in young Papua New Guinean children. PLoS One 5, e9047.

Lindsay, S., Kirby, M., Baris, E., Bos, R., 2004. Environmental management for malaria control in the East Asia and Pacific region.

Lindsay, S.W., Thomas, C.J., 2001. Global warming and risk of vivax malaria in Great Britain. EcoHealth 2, 80–84.

Lines, J., Whitty, C.J., Hanson, K., 2007. Prospects for eradication and elimination of malaria: a technical briefing for DFID.

Linthicum, K.J., 1988. A revision of the Argyritarsis section of the subgenus *Nyssorhynchus* of *Anopheles* (Diptera: Culicidae). Mosq. Syst. 20, 98–271.

Livingstone, F.B., 1984. The Duffy blood groups, vivax malaria, and malaria selection in human populations: a review. Hum. Biol. 56, 413–425.

Loaiza, J.R., Bermingham, E., Scott, M.E., Rovira, J.R., Conn, J.E., 2008. Species composition and distribution of adult *Anopheles* (Diptera: Culicidae) in Panama. J. Med. Entomol. 45, 841–851.

Luxemburger, C., Perea, W.A., Delmas, G., Pruja, C., Pecoul, B., Moren, A., 1994. Permethrin-impregnated bed nets for the prevention of malaria in schoolchildren on the Thai-Burmese border. Trans. R. Soc. Trop. Med. Hyg. 88, 155–159.

Lysenko, A.J., Beljaev, A.E., 1969. An analysis of the geographical distribution of *Plasmodium ovale*. Bull. World Health Organ. 40, 383–394.

Lysenko, A.J., Semashko, I.N., 1968. Geography of malaria. A medico-geographic profile of an ancient disease. In: Lebedew, A.W. (Ed.), Itogi Nauki: Medicinskaja Geografija, Academy of Sciences, USSR, Moscow, pp. 25–146.

Macdonald, G., 1957. Local Features of Malaria. The Epidemiology and Control of Malaria. Oxford University Press, London, pp. 63–99.

Mahgoub, H., Gasim, G.I., Musa, I.R., Adam, I., 2012. Severe *Plasmodium vivax* malaria among Sudanese children at New Halfa Hospital, Eastern Sudan. Parasit. Vectors 5, 154.

Mahmood, F., Reisen, W.K., 1981. Duration of the gonotrophic cycle of *Anopheles culicifacies* Giles and *Anopheles stephensi* Liston, with observations on reproductive activity and survivorship during winter in Punjab province, Pakistan. Mosq. News. 41, 41–50.

Maitland, K., Williams, T.N., Bennett, S., Newbold, C.I., Peto, T.E., Viji, J., Timothy, R., Clegg, J.B., Weatherall, D.J., Bowden, D.K., 1996. The interaction between *Plasmodium falciparum* and *P. vivax* in children on Espiritu Santo island, Vanuatu. Trans. R. Soc. Trop. Med. Hyg. 90, 614–620.

Maitland, K., Williams, T.N., Newbold, C.I., 1997. *Plasmodium vivax* and *P. falciparum*: biological interactions and the possibility of cross-species immunity. Parasitol. Today 13, 227–231.

Malaria Eradication Research Agenda, 2011a. A research agenda for malaria eradication: drugs. PLoS Med. 8, e1000402.

Malaria Eradication Research Agenda, 2011b. A research agenda for malaria eradication: monitoring, evaluation, and surveillance. PLoS Med. 8, e1000400.

Malaria Eradication Research Agenda, 2011c. A research agenda for malaria eradication: vaccines. PLoS Med. 8, e1000398.

Maldonado, Y.A., Nahlen, B.L., Roberto, R.R., Ginsberg, M., Orellana, E., Mizrahi, M., McBarron, K., Lobel, H.O., Campbell, C.C., 1990. Transmission of *Plasmodium vivax* malaria in San Diego County, California, 1986. Am. J. Trop. Med. Hyg. 42, 3–9.

Manguin, S., Garros, C., Dusfour, I., Harbach, R.E., Coosemans, M., 2008. Bionomics, taxonomy, and distribution of the major malaria vector taxa of *Anopheles* subgenus *Cellia* in Southeast Asia: an updated review. Infect. Genet. Evol. 8, 489–503.

May, J.M., 1951. Map of the world distribution of malaria vectors. Geogr. Rev. 41, 638–639.

Mayxay, M., Pukrittayakamee, S., Newton, P.N., White, N.J., 2004. Mixed-species malaria infections in humans. Trends Parasitol. 20, 233–240.

McKenzie, F.E., Jeffery, G.M., Collins, W.E., 2002. *Plasmodium vivax* blood-stage dynamics. J. Parasitol. 88, 521–535.

McLaughlin, C., Levy, J., Noonan, K., Rosqueta, K., 2009. Lifting the burden of malaria: an investment guide for impact-driven philanthropy, 79.

Meek, S.R., 1995. Vector control in some countries of Southeast Asia: comparing the vectors and the strategies. Ann. Trop. Med. Parasitol. 89, 135–147.

Ménard, D., Barnadas, C., Bouchier, C., Henry-Halldin, C., Gray, L.R., Ratsimbasoa, A., Thonier, V., Carod, J.F., Domarle, O., Colin, Y., Bertrand, O., Picot, J., King, C.L., Grimberg,

B.T., Mercereau-Puijalon, O., Zimmerman, P.A., 2010. *Plasmodium vivax* clinical malaria is commonly observed in Duffy-negative Malagasy people. Proc. Natl. Acad. Sci. U.S.A. 107, 5967–5971.

Mendes, C., Dias, F., Figueiredo, J., Mora, V.G., Cano, J., de Sousa, B., do Rosario, V.E., Benito, A., Berzosa, P., Arez, A.P., 2011. Duffy negative antigen is no longer a barrier to *Plasmodium vivax* – molecular evidences from the African West Coast (Angola and Equatorial Guinea). PLoS Negl. Trop. Dis. 5, e1192.

Mendis, K., Rietveld, A., Warsame, M., Bosman, A., Greenwood, B., Wernsdorfer, W.H., 2009. From malaria control to eradication: the WHO perspective. Trop. Med. Int. Health 14, 802–809.

Mendis, K., Sina, B.J., Marchesini, P., Carter, R., 2001. The neglected burden of *Plasmodium vivax* malaria. Am. J. Trop. Med. Hyg. 64, 97–106.

Miller, L.H., Mason, S.J., Clyde, D.F., McGinniss, M.H., 1976. The resistance factor to *Plasmodium vivax* in blacks. The Duffy-blood-group genotype, FyFy. N. Engl. J. Med. 295, 302–304.

Moonen, B., Cohen, J.M., Snow, R.W., Slutsker, L., Drakeley, C., Smith, D.L., Abeyasinghe, R.R., Rodriguez, M.H., Maharaj, R., Tanner, M., Targett, G., 2010. Operational strategies to achieve and maintain malaria elimination. Lancet 376, 1592–1603.

Moreno, M., Cano, J., Nzambo, S., Bobuakasi, L., Buatiche, J.N., Ondo, M., Micha, F., Benito, A., 2004. Malaria Panel Assay versus PCR: detection of naturally infected *Anopheles melas* in a coastal village of Equatorial Guinea. Malar. J. 3, 20.

Morsy, T.A., el Kadry, A.A., Salama, M.M., Sabry, A.H., el Sharkawy, I.M., 1995. Studies on the bionomics and vector competence of adult anopheline mosquitoes in El Faiyum Governorate, Egypt. J. Egypt Soc. Parasitol. 25, 213–244.

Mouchet, J., Carnevale, P., Coosemans, M., Julvez, J., Manguin, S., Richard-Lenoble, D., Sircoulon, J., 2004. Biodiversité du paludisme dans le monde. John Libbey Eurotext, Montrouge, France.

Mourant, A.E., Kopec, A.C., Domaniewska-Sobczak, K., 1976. The Distribution of the Human Blood Groups and Other Polymorphisms. Oxford University Press, London.

Mueller, I., Galinski, M.R., Baird, J.K., Carlton, J.M., Kochar, D.K., Alonso, P.L., del Portillo, H.A., 2009a. Key gaps in the knowledge of *Plasmodium vivax*, a neglected human malaria parasite. Lancet Infect. Dis. 9, 555–566.

Mueller, I., Slutsker, L., Tanner, M., 2011. Estimating the burden of malaria: the need for improved surveillance. PLoS Med. 8, e1001144.

Mueller, I., Widmer, S., Michel, D., Maraga, S., McNamara, D.T., Kiniboro, B., Sie, A., Smith, T.A., Zimmerman, P.A., 2009b. High sensitivity detection of *Plasmodium* species reveals positive correlations between infections of different species, shifts in age distribution and reduced local variation in Papua New Guinea. Malar. J. 8, 41.

Nalin, D.R., Mahood, F., Rathor, H., Muttalib, A., Sakai, R., Chowdhry, M.A., Safdar, G., ul Haq, I., Munir, M., Suleiman, M., et al., 1985. A point survey of periurban and urban malaria in Karachi. J. Trop. Med. Hyg. 88, 7–15.

Nguyen, D.M., Tran, D.H., Harbach, R.E., Elphick, J., Linton, Y.M., 2000. A new species of the Hyrcanus Group of *Anopheles,* subgenus *Anopheles*, a secondary vector of malaria in coastal areas of southern Vietnam. J. Am. Mosq. Control Assoc. 16, 189–198.

Nigatu, W., Petros, B., Lulu, M., Adugna, N., Wirtz, R., 1994. Species composition, feeding and resting behaviour of the common anthropophilic anopheline mosquitoes in relation to malaria transmission in Gambella, south west Ethiopia. Insect Sci. Appl. 15, 371–377.

Nikolaev, B.P., 1935. On the influence of temperature on the development of malaria *plasmodia* inside the mosquito.

Nosten, F., McGready, R., Simpson, J.A., Thwai, K.L., Balkan, S., Cho, T., Hkirijaroen, L., Looareesuwan, S., White, N.J., 1999. Effects of *Plasmodium vivax* malaria in pregnancy. Lancet 354, 546–549.

Oberst, R.B., Schultz, G.W., Laughlin, L.W., Sy, N.E., Stantos, M.N., Casimiro, C., 1988. Epidemiological study of malaria in Palawan. Philippine J. Microbiol. Infect. Dis. 17, 41–48.
Obsomer, V., Defourny, P., Coosemans, M., 2007. The *Anopheles dirus* complex: spatial distribution and environmental drivers. Malar. J. 6, 26.
Okell, L.C., Drakeley, C.J., Bousema, T., Whitty, C.J., Ghani, A.C., 2008. Modelling the impact of artemisinin combination therapy and long-acting treatments on malaria transmission intensity. PLoS Med. 5, e226 discussion e226.
Pampana, E., 1969. A Textbook of Malaria Eradication. Oxford University Press, London.
PATH, 2011. Staying the Course? Malaria Research and Development in a Time of Economic Uncertainty.
Patil, A.P., Gething, P.W., Piel, F.B., Hay, S.I., 2011. Bayesian geostatistics in health cartography: the perspective of malaria. Trends Parasitol. 27, 246–253.
Patil, A.P., Okiro, E.A., Gething, P.W., Guerra, C.A., Sharma, S.K., Snow, R.W., Hay, S.I., 2009. Defining the relationship between *Plasmodium falciparum* parasite rate and clinical disease: statistical models for disease burden estimation. Malar. J. 8, 186.
Pigott, D.M., Atun, R., Moyes, C.M., Hay, S.I., Gething, P.W., 2012. Funding for malaria control 2006–2010: a comprehensive global assessment. Malar. J. 11, 246.
Poespoprodjo, J.R., Fobia, W., Kenangalem, E., Lampah, D.A., Hasanuddin, A., Warikar, N., Sugiarto, P., Tjitra, E., Anstey, N.M., Price, R.N., 2009. Vivax malaria: a major cause of morbidity in early infancy. Clin. Infect. Dis. 48, 1704–1712.
Poespoprodjo, J.R., Fobia, W., Kenangalem, E., Lampah, D.A., Warikar, N., Seal, A., McGready, R., Sugiarto, P., Tjitra, E., Anstey, N.M., Price, R.N., 2008. Adverse pregnancy outcomes in an area where multidrug-resistant *Plasmodium vivax* and *Plasmodium falciparum* infections are endemic. Clin. Infect. Dis. 46, 1374–1381.
Povoa, M.M., Conn, J.E., Schlichting, C.D., Amaral, J.C., Segura, M.N., Da Silva, A.N., Dos Santos, C.C., Lacerda, R.N., De Souza, R.T., Galiza, D., Santa Rosa, E.P., Wirtz, R.A., 2003. Malaria vectors, epidemiology, and the re-emergence of *Anopheles darlingi* in Belem, Para, Brazil. J. Med. Entomol. 40, 379–386.
Prakash, A., Bhattacharyya, D.R., Mohapatra, P.K., Mahanta, J., 2001. Estimation of vectorial capacity of *Anopheles dirus* (Diptera: Culicidae) in a forest-fringed village of Assam (India). Vector Borne Zoonotic Dis. 1, 231–237.
Prakash, A., Bhattacharyya, D.R., Mohapatra, P.K., Mahanta, J., 2004. Role of the prevalent *Anopheles* species in the transmission of *Plasmodium falciparum* and *P. vivax* in Assam state, north-eastern India. Ann. Trop. Med. Parasitol. 98, 559–568.
Price, L., Planche, T., Rayner, C., Krishna, S., 2007a. Acute respiratory distress syndrome in *Plasmodium vivax* malaria: case report and review of the literature. Trans. R. Soc. Trop. Med. Hyg. 101, 655–659.
Price, R.N., Douglas, N.M., Anstey, N.M., 2009. New developments in *Plasmodium vivax* malaria: severe disease and the rise of chloroquine resistance. Curr. Opin. Infect. Dis. 22, 430–435.
Price, R.N., Tjitra, E., Guerra, C.A., Yeung, S., White, N.J., Anstey, N.M., 2007b. Vivax malaria: neglected and not benign. Am. J. Trop. Med. Hyg. 77, 79–87.
Prudencio, M., Rodriguez, A., Mota, M.M., 2006. The silent path to thousands of merozoites: the *Plasmodium* liver stage. Nat. Rev. Microbiol. 4, 849–856.
Pukrittayakamee, S., Imwong, M., Looareesuwan, S., White, N.J., 2004. Therapeutic responses to antimalarial and antibacterial drugs in vivax malaria. Acta Trop. 89, 351–356.
Quinones, M.L., Ruiz, F., Calle, D.A., Harbach, R.E., Erazo, H.F., Linton, Y.M., 2006. Incrimination of *Anopheles* (*Nyssorhynchus*) *rangeli* and *An.* (*Nys.*) *oswaldoi* as natural vectors of *Plasmodium vivax* in Southern Colombia. Mem. Inst. Oswaldo Cruz 101, 617–623.
Rahman, W.A., Che'Rus, A., Ahmad, A.H., 1997. Malaria and *Anopheles* mosquitos in Malaysia. Southeast Asian J. Trop. Med. Public Health 28, 599–605.

Ramsey, J.M., Salinas, E., Bown, D.N., Rodriguez, M.H., 1994. Plasmodium vivax sporozoite rates from *Anopheles albimanus* in southern Chiapas, Mexico. J. Parasitol. 80, 489–493.

Ratcliff, A., Siswantoro, H., Kenangalem, E., Maristela, R., Wuwung, R.M., Laihad, F., Ebsworth, E.P., Anstey, N.M., Tjitra, E., Price, R.N., 2007. Two fixed-dose artemisinin combinations for drug-resistant falciparum and vivax malaria in Papua, Indonesia: an open-label randomised comparison. Lancet 369, 757–765.

Rattanarithikul, R., Konishi, E., Linthicum, K.J., 1996. Detection of *Plasmodium vivax* and *Plasmodium falciparum* circumsporozoite antigen in anopheline mosquitoes collected in southern Thailand. Am. J. Trop. Med. Hyg. 54, 114–121.

RBMP, 2008. The global malaria action plan for a malaria free world.

Richie, T.L., 1988. Interactions between malaria parasites infecting the same vertebrate host. Parasitology 96 (Pt 3), 607–639.

Roberts, D., 1996. Mosquitoes (Diptera:Culicidae) breeding in brackish water: female ovipositional preferences or larval survival?. J. Med. Entomol. 33, 525–530.

Roberts, D.R., Laughlin, L.L., Hsheih, P., Legters, L.J., 1997. DDT, global strategies, and a malaria control crisis in South America. Emerg. Infect. Dis. 3, 295–302.

Roberts, D.R., Masuoka, P., Yu, A.Y., 2002. Determinants of malaria in the Americas. In: Casman, E.A., Dowlatabadi, H. (Eds.), The Contextual Determinants of Malaria, Resources for the Future Press, Washington, D.C., pp. 35–58.

Rodriguez, M.H., Gonzalez-Ceron, L., Hernandez, J.E., Nettel, J.A., Villarreal, C., Kain, K.C., Wirtz, R.A., 2000. Different prevalences of *Plasmodium vivax* phenotypes VK210 and VK247 associated with the distribution of *Anopheles albimanus* and *Anopheles pseudopunctipennis* in Mexico. Am. J. Trop. Med. Hyg. 62, 122–127.

Rojas, W., Botero, S., Garcia, H.I., 2001. An integrated malaria control program with community participation on the Pacific Coast of Colombia. Cad. Saude Publica 17 (Suppl), 103–113.

Romi, R., Pierdominici, G., Severini, C., Tamburro, A., Cocchi, M., Menichetti, D., Pili, E., Marchi, A., 1997. Status of malaria vectors in Italy. J. Med. Entomol. 34, 263–271.

Romi, R., Sabatinelli, G., Majori, G., 2001. Could malaria reappear in Italy? Emerg. Infect. Dis. 7, 915–919.

Rosenberg, R., 2007. *Plasmodium vivax* in Africa: hidden in plain sight? Trends Parasitol. 23, 193–196.

Ross, A., Maire, N., Sicuri, E., Smith, T., Conteh, L., 2011 Determinants of the cost-effectiveness of intermittent preventive treatment for malaria in infants and children. PLoS One 6, e18391.

Ross, R., 1911. The Prevention of Malaria. John Murray, London.

Rowe, A.K., Kachur, S.P., Yoon, S.S., Lynch, M., Slutsker, L., Steketee, R.W., 2009. Caution is required when using health facility-based data to evaluate the health impact of malaria control efforts in Africa. Malar. J. 8, 209.

Rowland, M., Mohammed, N., Rehman, H., Hewitt, S., Mendis, C., Ahmad, M., Kamal, M., Wirtz, R., 2002. Anopheline vectors and malaria transmission in eastern Afghanistan. Trans. R. Soc. Trop. Med. Hyg. 96, 620–626.

Rozeboom, L.E., Laird, R.L., 1942. *Anopheles* (Kerteszia) *bellator* Dyar and Knab as vector of malaria in Trinidad, British West Indies. Am. J. Trop. Med. Hyg. 22, 83–91.

Ruwende, C., Hill, A., 1998. Glucose-6-phosphate dehydrogenase deficiency and malaria. J. Mol. Med. (Berl.) 76, 581–588.

Ryan, J.R., Stoute, J.A., Amon, J., Dunton, R.F., Mtalib, R., Koros, J., Owour, B., Luckhart, S., Wirtz, R.A., Barnwell, J.W., Rosenberg, R., 2006. Evidence for transmission of *Plasmodium vivax* among a Duffy antigen negative population in Western Kenya. Am. J. Trop. Med. Hyg. 75, 575–581.

Sachs, J., Malaney, P., 2002. The economic and social burden of malaria. Nature 415, 680–685.

Sallum, M.A.M., Peyton, E.L., Wilkerson, R.C., 2005. Six new species of the *Anopheles leucosphyrus* group, reinterpretation of *An. elegans* and vector implications. Med.Vet. Entomol. 19, 158–199.
Sattabongkot, J., Tsuboi, T., Zollner, G.E., Sirichaisinthop, J., Cui, L., 2004. *Plasmodium vivax* transmission: chances for control?. Trends Parasitol. 20, 192–198.
Scharlemann, J.P., Benz, D., Hay, S.I., Purse, B.V., Tatem, A.J., Wint, G.R., Rogers, D.J., 2008. Global data for ecology and epidemiology: a novel algorithm for temporal Fourier processing MODIS data. PLoS One 3, e1408.
Schultz, G.W., 1992. Biting activity of mosquitos (Diptera: Culicidae) at a malarious site in Palawan, Republic of The Philippines. Southeast Asian J. Trop. Med. Public Health 23, 464–469.
Sellami, M.H., Kaabi, H., Midouni, B., Dridi, A., Mojaat, N., Boukef, M.K., Hmida, S., 2008. Duffy blood group system genotyping in an urban Tunisian population. Ann. Hum. Biol. 35, 406–415.
Seng, C.M., Matusop, A., Sen, F.K., 1999. Differences in *Anopheles* composition and malaria transmission in the village settlements and cultivated farming zone in Sarawak, Malaysia. Southeast Asian J. Trop. Med. Public Health 30, 454–459.
Senise, L., Motoki, M.T., Sallum, M.A.M., 2006. Notes on the holotype of *Anopheles marajoara* Galvao & Damasceno (Diptera, Culicidae). Rev. Bras. Entomol. 50, 453–457.
Service, M.W., Townson, H., 2002. The *Anopheles* vector. In: Gilles, H.M. (Ed.), Essential Malariology, Fourth ed.. Arnold, London, pp. 59–84.
Sharma, R.S., 1995. Urban malaria and its vectors *Anopheles stephensi* and *Anopheles culicifacies* (Diptera: Culicidae) in Gurgaon, India. Southeast Asian J. Trop. Med. Public Health 26, 172–176.
Sharma, S.N., Subbarao, S.K., Choudhury, D.S., Pandey, K.C., 1993. Role of *An. culicifacies* and *An. stephensi* in malaria transmission in urban Delhi. Indian J. Malariol. 30, 155–168.
Sharma, V.P., 2007. Battling the malaria iceberg with chloroquine in India. Malar. J. 6, 105.
Shiff, C., 2002. Integrated approach to malaria control. Clin. Microbiol. Rev. 15, 278–293.
Shililu, J.I., Grueber, W.B., Mbogo, C.M., Githure, J.I., Riddiford, L.M., Beier, J.C., 2004. Development and survival of *Anopheles gambiae* eggs in drying soil: influence of the rate of drying, egg age, and soil type. J. Am. Mosq. Control Assoc. 20, 243–247.
Shin, E.H., Kim, T.S., Lee, H.W., Lee, J.S., Lee, W.J., 2002. Vector competence of *Anopheles lesteri* Baisas and Hu (Diptera: Culicidae) to *Plasmodium vivax* in Korea. Korean J. Parasitol. 40, 41–44.
Shute, P.G., Lupascu, G., Branzei, P., Maryon, M., Constantinescu, P., Bruce-Chwatt, L.J., Draper, C.C., Killick-Kendrick, R., Garnham, P.C., 1976. A strain of *Plasmodium vivax* characterized by prolonged incubation: the effect of numbers of sporozoites on the length of the prepatent period. Trans. R. Soc. Trop. Med. Hyg. 70, 474–481.
Simmons, J.S., 1937. Observations on the importance of *Anopheles punctimacula* as a malaria vector in Panama, and report of experimental infections in *An. neomaculipalpis, An. apicimacula,* and *An. eiseni.* Am. J. Trop. Med. Hyg. 17, 191–212.
Simsek, F.M., Kayans, S., Ozensoy Toz, S., Ozbel, Y., Alten, B., S.T. Chan, A., 2010. Evaluation of the VecTest™ malaria antigen panel assay using *Anopheles sacharovi* specimens in an endemic area, Sanliurfa Province, Turkey. Kafkas. Univ. Vet. Fak. Derg. 16, S231–S234.
Singh, N., Mishra, A.K., 2000. Anopheline ecology and malaria transmission at a new irrigation project area (Bargi Dam) in Jabalpur (Central India). J. Am. Mosq. Control Assoc. 16, 279–287.
Sinka, M.E., Bangs, M.J., Manguin, S., Chareonviriyaphap, T., Patil, A.P., Temperley, W.H., Gething, P.W., Elyazar, I.R., Kabaria, C.W., Harbach, R.E., Hay, S.I., 2011. The dominant *Anopheles* vectors of human malaria in the Asia-Pacific region: occurrence data, distribution maps and bionomic precis. Parasit. Vectors 4, 89.
Sinka, M.E., Bangs, M.J., Manguin, S., Coetzee, M., Mbogo, C.M., Hemingway, J., Patil, A.P., Temperley, W.H., Gething, P.W., Kabaria, C.W., Okara, R.M., Van Boeckel, T., Godfray,

H.C., Harbach, R.E., Hay, S.I., 2010a. The dominant *Anopheles* vectors of human malaria in Africa, Europe and the Middle East: occurrence data, distribution maps and bionomic precis. Parasit. Vectors 3, 117.
Sinka, M.E., Bangs, M.J., Manguin, S., Rubio-Palis, Y., Chareonviriyaphap, T., Coetzee, M., Mbogo, C.M., Hemmingway, J., Patil, A.P., Temperley, W.H., Gething, P.W., Kabaria, C.W., Burkot, T.R., Harbach, R.E., Hay, S.I., 2012. A global map of dominant malaria vectors. Parasit. Vectors 5, 69.
Sinka, M.E., Rubio-Palis, Y., Manguin, S., Patil, A.P., Temperley, W.H., Gething, P.W., Van Boeckel, T., Kabaria, C.W., Harbach, R.E., Hay, S.I., 2010b. The dominant *Anopheles* vectors of human malaria in the Americas: occurrence data, distribution maps and bionomic precis. Parasit. Vectors 3, 72.
Smith, D.L., Guerra, C.A., Snow, R.W., Hay, S.I., 2007. Standardizing estimates of the *Plasmodium falciparum* parasite rate. Malar. J. 6, 131.
Smith, D.L., Hay, S.I., 2009. Endemicity response timelines for *Plasmodium falciparum* elimination. Malar. J. 8, 87.
Smith, D.L., McKenzie, F.E., 2004. Statics and dynamics of malaria infection in *Anopheles* mosquitoes. Malar. J. 3, 13.
Smith, T., Killeen, G.F., Maire, N., Ross, A., Molineaux, L., Tediosi, F., Hutton, G., Utzinger, J., Dietz, K., Tanner, M., 2006. Mathematical modeling of the impact of malaria vaccines on the clinical epidemiology and natural history of *Plasmodium falciparum* malaria: Overview. Am. J. Trop. Med. Hyg. 75, 1–10.
Smith, T., Maire, N., Ross, A., Penny, M., Chitnis, N., Schapira, A., Studer, A., Genton, B., Lengeler, C., Tediosi, F., de Savigny, D., Tanner, M., 2008. Towards a comprehensive simulation model of malaria epidemiology and control. Parasitology 135, 1507–1516.
Snounou, G., White, N.J., 2004. The co-existence of *Plasmodium*: sidelights from falciparum and vivax malaria in Thailand. Trends Parasitol. 20, 333–339.
Snow, R.W., Guerra, C.A., Noor, A.M., Myint, H.Y., Hay, S.I., 2005. The global distribution of clinical episodes of *Plasmodium falciparum* malaria. Nature 434, 214–217.
Sokolova, M.I., Snow, K.R., 2002. Malaria vectors in European Russia. J. Eur. Mosq. Control Assoc. 12, 1–6.
Steenkeste, N., Rogers, W.O., Okell, L., Jeanne, I., Incardona, S., Duval, L., Chy, S., Hewitt, S., Chou, M., Socheat, D., Babin, F.X., Ariey, F., Rogier, C., 2010. Sub-microscopic malaria cases and mixed malaria infection in a remote area of high malaria endemicity in Rattanakiri province, Cambodia: implication for malaria elimination. Malar. J. 9, 108.
Subbarao, S.K., Adak, T., Vasantha, K., Joshi, H., Raghvendra, K., Cochrane, A.H., Nussenzweig, R.S., Sharma, V.P., 1988. Susceptibility of *Anopheles culicifacies* species A and B to *Plasmodium vivax* and *Plasmodium falciparum* as determined by immunoradiometric assay. Trans. R. Soc. Trop. Med. Hyg. 82, 394–397.
Subbarao, S.K., Vasantha, K., Joshi, H., Raghavendra, K., Usha Devi, C., Sathyanarayan, T.S., Cochrane, A.H., Nussenzweig, R.S., Sharma, V.P., 1992. Role of *Anopheles culicifacies* sibling species in malaria transmission in Madhya Pradesh state, India. Trans. R. Soc. Trop. Med. Hyg. 86, 613–614.
Takken, W., Lindsay, S.W., 2003. Factors affecting the vectorial competence of *Anopheles gambiae*: a question of scale. In: Takken, W., Scott, T.W. (Eds.), Ecological Aspects for Application of Genetically Modified Mosquitoes, Kluwer Academic Publishers, Dordrecht, pp. 75–90.
Tanner, M., Hommel, M., 2010. Towards malaria elimination–a new thematic series. Malar. J. 9, 24.
Tatem, A.J., Guerra, C.A., Kabaria, C.W., Noor, A.M., Hay, S.I., 2008. Human population, urban settlement patterns and their impact on *Plasmodium falciparum* malaria endemicity. Malar. J. 7, 218.
Tatem, A.J., Smith, D.L., 2010. International population movements and regional *Plasmodium falciparum* malaria elimination strategies. Proc. Natl. Acad. Sci. U.S.A. 107, 12222–12227.

Tatem, A.J., Smith, D.L., Gething, P.W., Kabaria, C.W., Snow, R.W., Hay, S.I., 2010. Ranking of elimination feasibility between malaria-endemic countries. Lancet 376, 1579–1591.

Taye, A., Hadis, M., Adugna, N., Tilahun, D., Wirtz, R.A., 2006. Biting behavior and *Plasmodium* infection rates of *Anopheles arabiensis* from Sille, Ethiopia. Acta Trop. 97, 50–54.

Temu, E.A., Minjas, J.N., Coetzee, M., Hunt, R.H., Shift, C.J., 1998. The role of four anopheline species (Diptera: Culicidae) in malaria transmission in coastal Tanzania. Trans. R. Soc. Trop. Med. Hyg. 92, 152–158.

The Global Health Group and the Malaria Atlas Project, 2011. Atlas of Malaria-eliminating Countries, 2011. The Global Health Group, Global Health Sciences, University of California, San Francisco, U.S.A.

Tjitra, E., Anstey, N.M., Sugiarto, P., Warikar, N., Kenangalem, E., Karyana, M., Lampah, D.A., Price, R.N., 2008. Multidrug-resistant *Plasmodium vivax* associated with severe and fatal malaria: a prospective study in Papua, Indonesia. PLoS Med. 5, e128.

Ulloa, A., Gonzalez-Ceron, L., Rodriguez, M.H., 2006. Host selection and gonotrophic cycle length of *Anopheles punctimacula* in southern Mexico. J. Am. Mosq. Control Assoc. 22, 648–653.

Vijay, S., Rawat, M., Adak, T., Dixit, R., Nanda, N., Srivastava, H., Sharma, J.K., Prasad, G.B., Sharma, A., 2011. Parasite killing in malaria non-vector mosquito *Anopheles culicifacies* species B: implication of nitric oxide synthase upregulation. PLoS One 6, e18400.

Vittor, A.Y., Gilman, R.H., Tielsch, J., Glass, G., Shields, T., Lozano, W.S., Pinedo-Cancino, V., Patz, J.A., 2006. The effect of deforestation on the human-biting rate of *Anopheles darlingi*, the primary vector of falciparum malaria in the Peruvian Amazon. Am. J. Trop. Med. Hyg. 74, 3–11.

Vythilingam, I., Chan, S.T., Shanmugratnam, C., Tanrang, H., Chooi, K.H., 2005. The impact of development and malaria control activities on its vectors in the Kinabatangan area of Sabah, East Malaysia. Acta Trop. 96, 24–30.

Warren, M., Collins, W.E., Jeffery, G.M., Richardson, B.B., 1980. *Anopheles pseudopunctipennis*: laboratory maintenance and malaria susceptibility of a strain from El Salvador. Am. J. Trop. Med. Hyg. 29, 503–506.

Wells, T.N., Burrows, J.N., Baird, J.K., 2010. Targeting the hypnozoite reservoir of *Plasmodium vivax*: the hidden obstacle to malaria elimination. Trends Parasitol. 26, 145–151.

Wernsdorfer, W., Hay, S.I., Shanks, G.D., 2009. Learning from History. Shrinking the Malaria Map: a Prospectus on Malaria Elimination, pp. 95–107.

Wertheimer, S.P., Barnwell, J.W., 1989. *Plasmodium vivax* interaction with the human Duffy blood group glycoprotein: identification of a parasite receptor-like protein. Exp. Parasitol. 69, 340–350.

White, N.J., 2011. Determinants of relapse periodicity in *Plasmodium vivax* malaria. Malar. J. 10, 297.

WHO, 1986. Abstracts of Recent Chinese Publications on Malaria, 17, World Health Organization, Geneva, Switzerland.

WHO, 2007. Malaria Elimination: a Field Manual for Low and Moderate Endemic Countries, 85, World Health Organization, Geneva, Switzerland.

WHO, 2008. The Global Burden of Disease 2004, 215, World Health Organization, Geneva, Switzerland.

WHO, 2010a. Guidlines for the Treatment of Malaria, second ed.. World Health Organization, Geneva, Switzerland, 194.

WHO, 2010b. International Travel and Health: Situation as on 1 January 2010, World Health Organization, Geneva, Switzerland.

WHO, 2011. World Malaria Report 2011, 246, World Health Organization, Geneva, Switzerland.

WHO/PAHO, 2006. Regional Strategic Plan for Malaria in the Americas 2006–2010, 71, Pan American Health Organization, Office for the Americas, Washington, DC.

WHO/PAHO, 2008. Malaria in the Americas: Time Series Epidemiological Data from 2000 to 2007, Pan American Health Organization, Office for the Americas, Washington, DC.
Winckel, C.W.F., 1955. Long latency in *Plasmodium vivax* infections in a temperate zone. Doc. Med. Geogr. Trop. 7.
Wirtz, R.A., Burkot, T.R., Andre, R.G., Rosenberg, R., Collins, W.E., Roberts, D.R., 1985. Identification of *Plasmodium vivax* sporozoites in mosquitoes using an enzyme-linked immunosorbent assay. Am. J. Trop. Med. Hyg. 34, 1048–1054.
Wooster, M.T., Rivera, D., 1985. Breeding point and larval association of anopheline mosquitoes of northwest Mindoro, Philippines. Southeast Asian J. Trop. Med. Public Health 16, 59–65.
World Bank, 2009. World Development Report 2009: Reshaping Economic Geography.
Wurtz, N., Mint Lekweiry, K., Bogreau, H., Pradines, B., Rogier, C., Ould Mohamed Salem Boukhary, A., Hafid, J.E., Ould Ahmedou Salem, M.S., Trape, J.F., Basco, L.K., Briolant, S., 2011. Vivax malaria in Mauritania includes infection of a Duffy-negative individual. Malar. J. 10, 336.
Yekutiel, P., 1960. Problems of epidemiology in malaria eradication. Bull. World Health Organ. 22, 669–683.
Yekutiel, P., 1980. III The Global Malaria Eradication Campaign. In: Klingberg, M.A. (Ed.), Eradication of Infectious Diseases: a Critical Study, Karger, Basel, pp. 34–88.
Zanzibar Malaria Control Program, 2009. Malaria Elimination in Zanzibar: a Feasibility Assessment.
Zimmerman, P.A., 2004. The enigma of *Plasmodium vivax* malaria and erythrocyte Duffy negativity. In: Dronamraju, K.R. (Ed.), Infectious Disease and Host-pathogen Evolution, Cambridge University Press, Cambridge, pp. 141–172.

CHAPTER TWO

Relapse

Nicholas J. White*[,1], Mallika Imwong†

*Mahidol Oxford Research Unit, Faculty of Tropical Medicine, Mahidol University, Bangkok, Thailand
†Department of Molecular Tropical Medicine and Genetics, Faculty of Tropical Medicine, Mahidol University, Bangkok, Thailand
[1]Corresponding author: nickw@tropmedres.ac

Contents

Abstract

Plasmodium vivax is a major cause of febrile illness in endemic areas of Asia, Central and South America, and the horn of Africa. *P. vivax* infections are characterized by relapses of malaria arising from persistent liver stages of the parasite (hypnozoites), which can be prevented currently only by 8-aminoquinoline anti-malarials. Tropical *P. vivax* infections relapse at approximately 3-week intervals if rapidly eliminated anti-malarials are given for treatment, whereas in temperate regions and parts of the sub-tropics, *P. vivax* infections are characterized by either a long incubation or a long-latency period between illness and relapse – in both cases approximating 8–10 months. The epidemiology of the different relapse phenotypes has not been defined adequately despite obvious

Advances in Parasitology, Volume 80
ISSN 0065-308X, http://dx.doi.org/10.1016/B978-0-12-397900-1.00002-5

relevance to malaria therapeutic assessment, control, and elimination. The number of sporozoites inoculated by the anopheline mosquito is an important determinant of both the timing and the number of relapses. The intervals between *P. vivax* relapses display a remarkable periodicity which has not been explained. Evidence is presented that the proportion of patients who have successive relapses is relatively constant and that the factor which activates hypnozoites and leads to regular interval relapse in vivax malaria is the systemic febrile illness itself. It is proposed that in endemic areas, a large proportion of the population harbours latent hypnozoites which can be activated by a systemic illness such as vivax or falciparum malaria. This explains the high rates of vivax following falciparum malaria, the high proportion of heterologous genotypes in relapses, the higher rates of relapse in people living in endemic areas compared with artificial infection studies, and, by facilitating recombination between different genotypes, contributes to *P. vivax* genetic diversity particularly in low transmission settings. Long-latency *P. vivax* phenotypes may be more widespread and more prevalent than currently thought. These observations have important implications for the assessment of radical treatment efficacy and for malaria control and elimination.

1. INTRODUCTION

Plasmodium vivax malaria is characterised by relapses after resolution of the primary infection which derive from activation of dormant liver stage parasites 'hypnozoites'. It is this propensity to relapse, which makes *P. vivax* more difficult to control and eliminate than *Plasmodium falciparum*. In endemic areas, relapse of vivax malaria is a major cause of malaria in young children, and an important source of malaria transmission. Relapse also occurs in *Plasmodium ovale* infections and in several of the simian

Terminology

Recrudescence: The blood-stage infection declines initially following treatment but then increases again to produce a recurrent infection. Such recrudescent infections are genetically homologous (i.e. they are with one or more of the same genotypes which caused the original infection).

Relapse: A recurrent infection resulting from persistent liver stages or hypnozoites. In tropical areas, the interval from primary infection to relapse is short – typically 3 weeks following a rapidly eliminated drug treatment and 6 weeks following chloroquine or another slowly eliminated drug. These relapses can be either genetically homologous, or heterologous arising from activation of previously acquired hypnozoites (White, 2011).

In some sub-tropical areas, the interval from inoculation to first infection, or from primary illness to first relapse is approximately 9 months (long latency).

malarias, notably *Plasmodium cynomolgi*, which has often been used as an animal model of vivax malaria.

2. HISTORY

2.1. European and North American Vivax Malaria

The propensity of *P. vivax* malaria to relapse was recognised well over 100 years ago (White, 2011). Precise observations describing latencies of some 8–9 months between primary illness and relapse were complemented by detailed prospective epidemiological observations conducted in the village of Wormeveer in The Netherlands by Korteweg. His observations, and later entomological studies of Swellengrebel et al., indicated that the early summer peak of vivax malaria preceded abundance of vector mosquitoes, and therefore, had been acquired in the autumn of the previous year (Korteweg, 1902; Swellengrebel et al., 1936; Swellengrebel and De Buck, 1938; Winckel, 1955; Gill, 1938) (i.e. the incubation period was 9 months) (Fig. 2.1). A much clearer understanding of relapse in *P. vivax* malaria came from Julius Wagner-Jauregg's discovery that malaria could cure neurosyphilis. Between the 1920s and the 1950s, thousands of patients confined in mental hospitals with neurosyphilis were treated with malaria (malaria therapy). The majority of the malaria therapy experience was with a relatively small number of parasite 'strains' transmitted either by blood passage, or more usually by the bites of several infected anopheline mosquitoes. Recurrences of vivax malaria commonly occurred many months after apparently successful treatment of the primary infection. Furthermore, whereas recurrence in blood-transmitted vivax malaria could be prevented by curing the blood stage infection, recurrence in mosquito-transmitted vivax malaria could not (Hackett, 1937; Bignami, 1913; Yorke and MacFie, 1924; Yorke, 1925; James, 1931a; James et al., 1936; James and Shute, 1926). This pointed to an exoerythrocytic stage of malaria, but its location was unknown. The Dutch malariologists showed that their indigenous *P. vivax* could have a short incubation period if patients were bitten by a large number of infected mosquitoes, but by self-experimentation, they proved that bites by one or two infected mosquitoes were followed by vivax malaria 8–9 months later (Schuffner et al., 1929). Importantly, it was also noted in The Netherlands that relapse rates (proportion of incident infections which are followed by a relapse) in naturally acquired infections were higher than with mosquito-transmitted malaria therapy with local 'strains' despite their higher inoculum

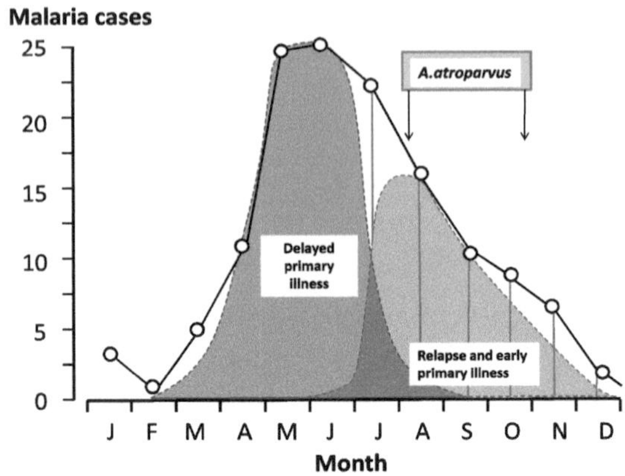

Figure 2.1 Long-latency *P. vivax* in the Netherlands a century ago. The mean monthly number of malaria cases in the village of Wormerveer, The Netherlands, recorded by Kortweg from 1902 to 1923 (black line) (White, 2011; Korteweg, 1902; Swellengrebel et al., 1936; Swellengrebel and De Buck, 1938; Winckel, 1955). Swellengrebel et al. showed anopheline vectors responsible for malaria transmission usually occurred between the first week of August and the end of October. From this, it can be deduced that the initial wave of malaria cases derived from inoculations the previous year (pink curve) and, by subtraction, that this was followed by relapses and primary cases with a short incubation period in the late summer and autumn (blue curve). For interpretation of the references to colour in this figure legend, the reader is referred to the online version of this book.

(Swellengrebel and De Buck, 1938). The pattern observed by James in England, Swellengrebel in The Netherlands, and Ciuca in Romania with the extensively studied Madagascar strain of *P. vivax* (Fig. 2.2) was very similar to the relapse patterns of the St. Elizabeth and McCoy strains, of indigenous origin, used for malaria therapy in the United States (Tiburskaja et al., 1968; Nikolaiev, 1949; Shute et al., 1978; Boyd, 1940a, 1947; Sinton, 1931; Boyd and Stratman-Thomas, 1933; Cooper et al., 1947, 1949). These strains were all chosen because of their suitability for malaria therapy, so they probably represented the upper end of the spectrum of abilities to produce early infection and relapse. Provided the initial illness was treated or attenuated by quinine, then relapse usually occurred approximately 9 months later. These single-isolate (presumably single or highly related genotype) infections eventually resulted in solid immunity such that after several episodes of protracted fever, reinfection with the same isolate was not possible (Boyd, 1947; Boyd et al., 1936). This acquired immunity suppressed later genetically homologous relapses. The malaria therapy patients often had no previous malaria, so any malaria recurrence must have arisen from

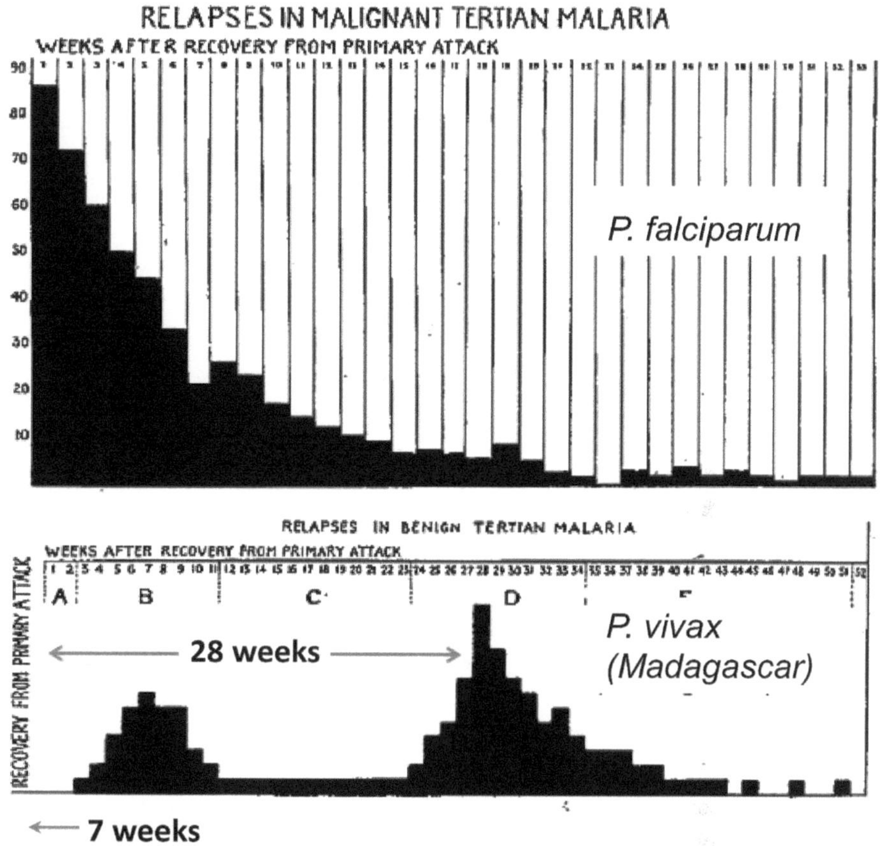

Figure 2.2 The temporal pattern of illness recurrence in patients with neurosyphilis artificially infected for malaria therapy with *Plasmodium falciparum* (87 patients) and the extensively studied 'Madagascar' strain of *P. vivax* (105 patients) studied by SP James et al. at the Horton Hospital, Epsom, England (James, 1931a; James et al., 1936; James and Shute, 1926) between 1925 and 1930. The vivax relapses had a bimodal pattern with the majority having a long-latent period (mode 28 weeks) before the relapse. For colour version of this figure, the reader is referred to the online version of this book.

the induced infection, and so would have been genetically homologous. In contrast, in endemic areas, multiple inoculations take place and relapses can arise from parasites, which are genetically different (heterologous), and were inoculated before the inoculation which caused the incident infection and were presumably activated by it (White, 2011). Boyd and Kitchen, who studied the McCoy strain in Florida, showed that (homologous) relapse did not follow infections which had terminated spontaneously (indicating an effective immune response) or infections, in which the primary illness lasted for ≥48 days (Boyd, 1947; Boyd et al., 1936). Importantly, asymptomatic parasitaemia (and gametocytaemia) tended to persist for weeks as disease controlling immunity was acquired.

During the malaria therapy experience, it became clear that both the incubation period and the number of relapses per infection were determined by the numbers of sporozoites inoculated (Boyd, 1940b). Low-sporozoite inocula often resulted in an extended incubation period of 7–10 months. The more sporozoites that were inoculated, the more likely was an early infection (incubation period – 2 weeks), and the more relapses that followed – provided that prompt anti-malarial treatment (quinine) was given each time. In order to ensure that there was a short incubation period preceding the malaria illness, malaria therapy infections were produced typically by the bites of 5–10 infected mosquitoes, and either no treatment or partial suppressive treatment was given. Theories that seasonal influences were important determinants of relapse were largely rejected as the intervals from primary illness to relapse in neurosyphilis patients were generally similar, whichever month the infection started.

Until the end of the Second World War, the long-latency infection was regarded as the 'usual' *P. vivax* phenotype. Throughout the endemic areas of Europe, vivax malaria peaked in the late spring and early summer (largely from inoculations the previous year) (Swellengrebel and De Buck, 1938; Winckel, 1955; Gill, 1938; Hackett, 1937). In southern Europe, there was often a bimodal pattern with a late summer peak of falciparum malaria (aestivo-autumnal malaria). During the Second World War observations on soldiers fighting in Europe and temperate areas of Asia, all pointed clearly to long-latency *P. vivax* with similar illness patterns to the Madagascar and St. Elizabeth strains (Findlay, 1951; Höring, 1947; Eyles and Young, 1948; Ebisawa, 1973; MRC, 1945; Mowrey, 1963; Wilson and Reid, 1949). In contrast, soldiers fighting in the Indo-Burman and South Pacific campaigns encountered vivax malaria with a very different relapse pattern. The relapse rate was very high – relapses were frequent and recurrent. Infections occurred at 3-week intervals if quinine was given, and 7-week intervals following treatment with mepacrine (quinacrine, atebrine) (Most, 1963; Noe et al., 1946; Bianco et al., 1947; Whorton et al., 1947; Craige et al., 1947; Jeffery, 1956). Initially, this disease pattern was thought simply to reflect the very high sporozoite inocula that the soldiers were exposed to but it soon became apparent that the phenotype was fundamentally different. Multiple relapses were very common and there was no evidence of long latency. The 'type strain' for this tropical frequent relapse *P. vivax* phenotype was the

'Chesson strain' isolated from a soldier who had fought in New Guinea (Whorton et al., 1947; Craige et al., 1947; Jeffery, 1956). In volunteers infected with the Chesson strain, 80% of relapses occurred within 30 days of initial treatment with quinine.

2.2. Discovery of the Liver Stages

In 1947, Garnham identified the pre-erythrocytic development of *Hepatocystis* (then *Plasmodium*) *kochi* in the hepatocytes of African monkeys (Garnham, 1947). Shortly afterwards, in England, definitive studies by Shortt and Garnham identified the site of pre-erythrocytic development in primate malarias as the liver, first in *P. cynomolgi*-infected Rhesus monkeys (Shortt and Garnham, 1948a, 1948b), and then in a very heavily infected volunteer who underwent open liver biopsy (Shortt et al., 1948a, 1948b; Garnham, 1967). This classic work still did not identify the persistent stage (Garnham, 1967; Shortt and Garnham, 1948c). Forty years later, Krotoski, working with Garnham et al. at Imperial College, finally identified the dormant stages or 'hypnozoites' of *P. cynomolgi* and *P. vivax* responsible for relapses in the liver (Krotoski et al., 1980; Krotoski et al., 1982; Krotoski, 1985; Garnham, 1988; Bray and Garnham, 1982). Although parasite bodies, which are probably hypnozoites, have since been demonstrated in liver cell cultures (Dembele et al., 2011), remarkably little is known of their biology. The term *relapse* is now used specifically to describe recurrences of malaria derived from persistent liver stages of the parasite (hypnozoites), whereas *recrudescence* refers to a recurrence of malaria derived from persistence of the blood-stage infection. The relapse arises after the 'awakening' of these hypnozoites and the subsequent intrahepatic schizogony followed by blood-stage multiplication.

3. PHENOTYPIC VARIATION IN *P. VIVAX*

Today, there is a tendency to regard all *P. vivax* together as a single homogenous species, indeed the temperate strains seem to have been all but forgotten, but the human malaria therapy and volunteer studies showed that there was substantial phenotypic variation between *P. vivax* 'strains'. Studies conducted over 50 years ago indicated that incubation periods, numbers of merozoites per blood schizont, antigenic relationships, intrinsic drug susceptibility, virulence, and relapse intervals all differed between 'strains'. At this time, the *P. vivax* with infection phenotypes similar to those of the 'Madagascar' and 'St Elizabeth' strains, which were prevalent in the United States and Southern Europe, were considered the 'typical' *P. vivax* infections. A primary illness followed approximately 2 weeks after mosquito inoculation, and, although relapse could follow some 3 weeks later, there was often a 7–10-month interval before a subsequent relapse. Sometimes, the latency could be as long as 1 year, and there were well-documented, but apparently unusual, cases reported of latencies greater than 2 years. Further north in the Netherlands, Northern Germany, Scandinavia, Finland and Central Russia, long-incubation phenotypes were prevalent (*P. vivax hibernans*) in which the primary infection occurred 8–10 months after inoculation (Swellengrebel and De Buck, 1938; Winckel, 1955; Gill, 1938; Hackett, 1937; Bignami, 1913; Yorke and MacFie, 1924; Yorke, 1925; James, 1931a; James et al., 1936; James and Shute, 1926; Schuffner et al., 1929; Tiburskaja et al., 1968; Nikolaiev, 1949; Shute et al., 1978; Boyd, 1940a; Coatney et al., 1950a; Tiburskaya, 1961; Moshkovsky, 1973; Huldén et al., 2005). It seemed that the proportion of infections, which had a short incubation period (*circa* 2 weeks), declined steadily with increasing latitude (and shorter summer mosquito breeding seasons).

There was one very important feature of the long-latency *P. vivax* infections, which was noted particularly in soldiers who had acquired vivax malaria in the Korean war; once the relapse had occurred (after a latency of 7–10 months), subsequent relapses would then usually occur with intervals of approximately 3–4 weeks following quinine or 6–8 weeks if chloroquine was given for treatment (Fig. 2.3), i.e. similar intervals in the tropical frequent relapse 'Chesson'-phenotype vivax malaria (Hankey et al., 1953; Coatney et al., 1950b; Hill and Amatuzio, 1949). It was also noted that in artificial infection studies, the duration of latency was independent of the season when the infection was induced (James, 1931a; James et al., 1936).

4. RELAPSE DETERMINANTS

The proportion of patients whose vivax malaria relapses depends on many factors including 'strain', sporozoite inoculum, immunity and drug treatment.

4.1. Effects of the Sporozoite Inoculum on Relapse Intervals

For long-latency vivax malaria, the sporozoite dose determined the clinical phenotype. In long-term observations of infections with the St Elizabeth strain and a North Korean strain used for malaria therapy in a Moscow

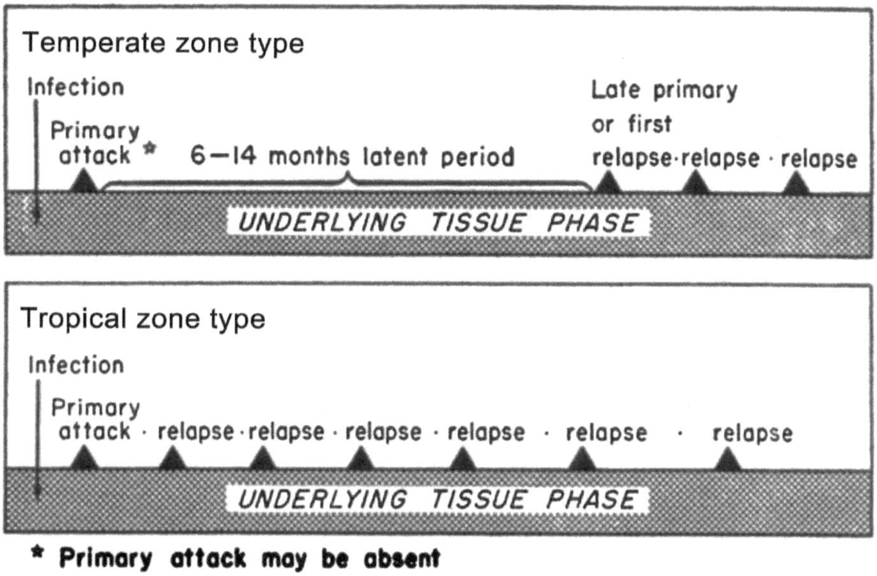

Figure 2.3 Schematic diagram by Hankey et al. (Sinha et al., 1953) of relapse patterns following Korean vivax malaria (upper panel) and tropical frequent-relapse *P. vivax* (lower panel). It is important to note the frequent-relapse pattern after a long interval with Korean vivax malaria.

hospital, there was a clear bimodal pattern in which a long pre-patent period (~300 days) only occurred following smaller sporozoite inocula (more reflective of natural infection) (Coatney et al., 1950a; Tiburskaya, 1961; Moshkovsky, 1973). For the Moscow strain, it required ≥1000 sporozoites to produce illness after a 'short' incubation period of 2 weeks. When increasing sporozoite doses of the tropical 'Chesson' strain were inoculated, the incubation period shortened slightly, but even with a single mosquito bite, there was no evidence for a long pre-patent period. These observations, and a series of experimental investigations in chimpanzees (Ungureanu et al., 1976), led Garnham to propose that the ratio of hypnozoites to immediately developing forms in the Korean *P. vivax* strain was 999:1 compared with the Chesson strain, where he estimated the ratio as 50:50 (Garnham, 1988; Bray and Garnham, 1982).

In contrast, the initial inter-relapse intervals of Chesson strain *P. vivax* in volunteers, and also *P. cynomolgi* in Rhesus monkeys, were remarkably regular although they gradually lengthened with each successive relapse (Krotoski et al., 1986; Schmidt, 1986) (Fig. 2.4). Two factors were probably responsible for gradually increasing inter-relapse intervals. The first is

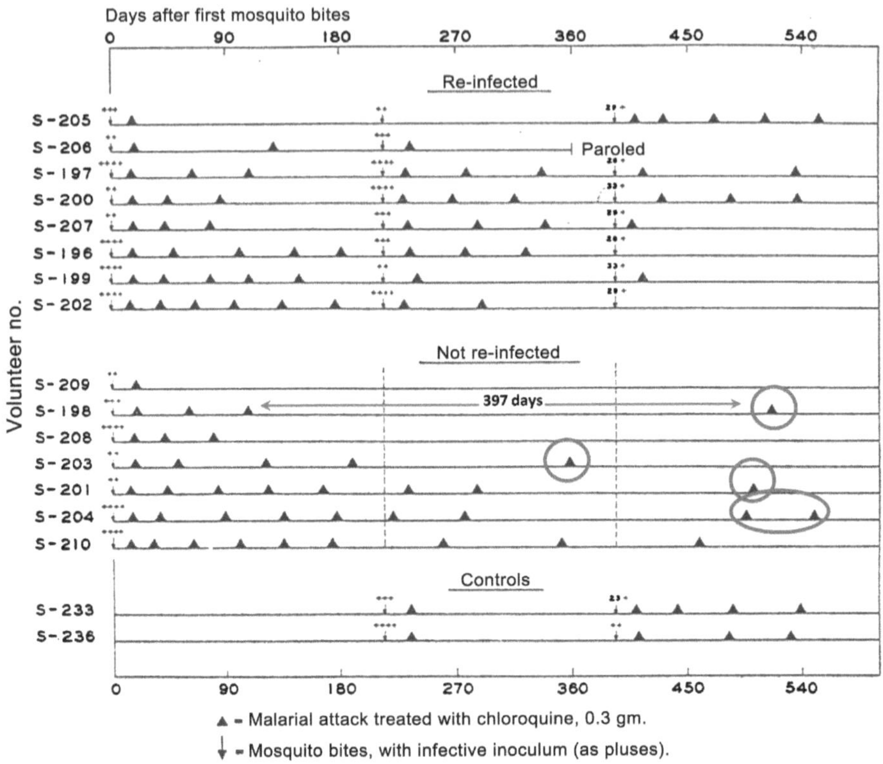

Figure 2.4 Relapse patterns in volunteers in the USA who were infected by a single mosquito bite with the Chesson strain of *P. vivax* studied by Coatney et al. (1950). Some were rechallenged as indicated. Late relapses are highlighted. For colour version of this figure, the reader is referred to the online version of this book.

simple numerical probability; simultaneous activation of several hypnozoites will shorten the relapse interval because the interval is measured until the progeny of the most rapidly growing parasites cause patent infection. There is natural phenotypic variation even amongst genetically identical organisms. It is the progeny of the earliest activated and most rapidly multiplying parasite that become patent first. In natural settings, multiple genotypically distinct (i.e. heterologous) hypnozoites may be activated but, on many occasions, only one or two genotypes will be detected subsequently at clinical relapse. The other hypnozoites' progeny may reach patency later, or in symptomatic relapses, asexual growth may be suppressed by fever, illness, immune responses, and treatment such that they never reach patency. The second factor lengthening inter-relapse intervals in artificial infections was the acquisition of blood-stage immunity against the single infecting genotype (homologous immunity). In the studies of the St. Elizabeth strain, there was clear evidence that late relapses were attenuated if there was an early infection, but this did not affect the interval to latency, whereas the greater the sporozoite inoculum, the shorter was the interval (Coatney et al., 1950a). These observations suggest that the size of the inoculum is more important than immunity in determining the duration of latency or the inter-relapse interval. Blood-stage immunity against homologous strains of *P. vivax*, which persists for many months, was a consistent observation in malaria therapy and challenge studies. Boyd noted that if the initial infection was allowed to run its natural course, then relapse did not occur and reinfection with the homologous strain was not possible (Boyd, 1947).

4.2. Artificial versus Natural Infections

Artificial infections were very informative but they differed from natural infections in several important respects (Swellengrebel and De Buck, 1938; Winckel, 1941). The artificial infections were only in non-immune adults, whereas the burden of vivax disease in endemic areas was in children. Adults in the malaria endemic areas have received multiple inoculations in their lifetime, and so have usually developed significant immunity to a broad range of local parasites, which controls symptoms and reduces parasite densities. The artificial infections in the majority of volunteer studies and in malaria therapy followed the bites of 5–10 infected anopheline mosquitoes selected for maximal infectivity based on salivary gland sporozoite loads. The optimisation of infectivity in malaria therapy and experimental studies contrasts with the natural setting where anopheline mosquitoes typically display a wide range of infectivities depending on sporozoite age and other factors. In artificial infection studies, mosquito mid guts often contained multiple oocyts, whereas in trapped wild anopheline vectors, oocyst numbers are typically low (median 2–3). Median sporozoite inocula in natural infections are estimated to be <10 sporozoites (Ponnudurai et al., 1991; Rosenburg and Wirtz, 1990; Beier et al., 1991). If Garnham's estimates (50:50 ratio of immediately developing parasites to hypnozoites) are correct, this corresponds to a median of five hypnozoites in tropical *P. vivax* infections. The inocula in artificial infections were therefore usually 'supranormal', which did not bring out the important stochastic component of *P. vivax* epidemiology resulting from low-sporozoite inocula in areas of low seasonal transmission. The 'strains' of *P. vivax* used in malaria therapy were likely to have been of a single genotype or very closely related interbreeding genotypes, which were passaged over many years. It is likely that with multiple passages in malaria therapy, the 'strains' became purified through successive interbreeding to a very closely related group of genotypes. In contrast, multiple unrelated genotype infections are common in natural infections.

4.3. Effects of Drugs on Relapse Intervals

Since the introduction of mepacrine (atebrine, quinacrine) in 1932, a drug which has a terminal elimination half-life of over 1 month, it was observed that early relapses were delayed by approximately 30 days compared with quinine (half-life in malaria 16 hours) treatment (i.e. from 3–7 weeks) (Most, 1963; Noe et al., 1946; Bianco et al., 1947; Whorton et al., 1947; Craige et al., 1947). Later, chloroquine, which is also very slowly eliminated, was found to delay early relapse appearance by 2–6 weeks (Jeffery, 1956). Larger doses of the anti-malarials resulted in longer intervals to relapse consistent with a concentration-dependent slowing of asexual growth rates. Slowly eliminated anti-malarials delayed the onset of *P. vivax* relapse, and consequently, reduced their frequency, but they did not appear to reduce the overall number of relapses experienced (Most, 1963). Only the 8-aminoquinolines reduced or prevented relapses. The relapse-preventing properties of these drugs, and their synergistic action with quinine, were demonstrated within years of their introduction in the mid-1920s (Sinton and Bird, 1928; Sinton et al., 1930; James, 1931b).

4.4. The Proportion of *P. vivax* Infections which Relapse

This is often considered an intrinsic property of the malaria parasites, which varies considerably by geographic region. Tropical 'strains' relapsed more than temperate 'strains'. But the relapse proportion is also clearly a function of the sporozoite inoculum and immunity (if any). In the Second World War, relapse rates were very high in soldiers who were generally non-immune and presumably experienced intense exposure. Overall, reported relapse rates have varied from 0 to 100%! (White, 2011). In many cases, relapse rates are likely to have been underestimated as follow-up periods, particularly in recent studies, were often 3 months or less, and in the majority of studies, radical treatments were being evaluated. In addition, studies conducted in endemic areas cannot reliably exclude reinfection. Without radical treatment, the proportion of patients who experience one relapse, the proportion of these patients who have a second relapse, and the proportion of these who have a third and so on, is constant (White, 2011; Höring, 1947; Eyles and Young, 1948; Most, 1963; Bianco et al., 1947; Hill and Amatuzio, 1949; Boyd and Kitchen, 1943; Mowrey, 1963). Thus if x is

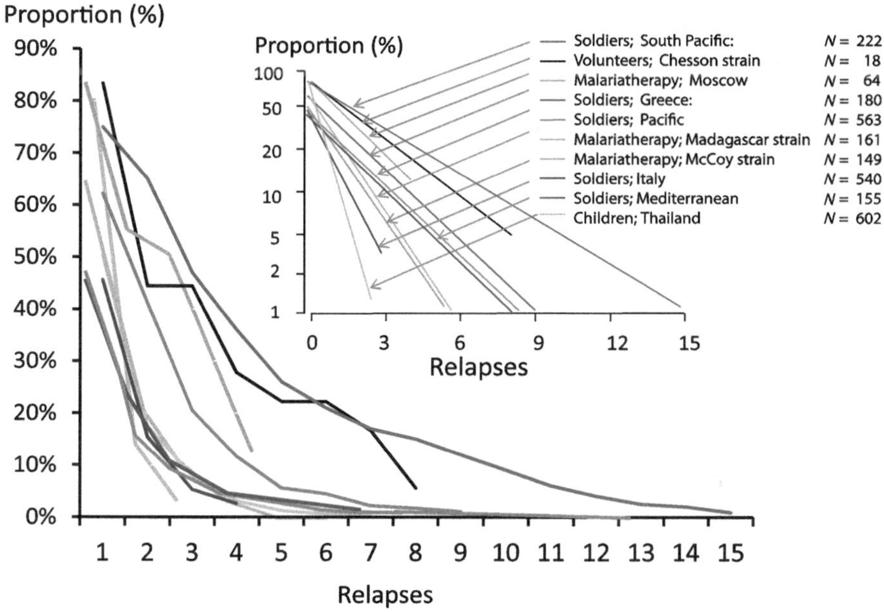

Figure 2.5 The proportions of patients experiencing successive *P. vivax* relapses taken from eight different clinical series, described in detail in White (2011). These were US soldiers with vivax malaria acquired in the South Pacific (2 series), German soldiers who acquired vivax malaria in Greece, US soldiers with vivax malaria acquired in the Mediterranean area (two series) and Italy, patients receiving malaria therapy with a local 'strain' in Moscow, British patients receiving malaria therapy with the Madagascar strain, patients receiving malaria therapy with the McCoy strain in the United States, volunteers infected with the Chesson strain in the United States and more recently children followed prospectively in an evaluation of an ineffective malaria vaccine (SPf66) in north-western Thailand. Inset shows proportions on a log scale and numbers of patients studied. For colour version of this figure, the reader is referred to the online version of this book.

the fraction of patients experiencing one or more relapses, then the fraction experiencing n or more relapses is approximately x^n (White, 2011) (Fig. 2.5). Thus, for areas with 50% relapse rates, approximately 12.5% of patients have to have three or more relapses per incident symptomatic infection. It follows that once symptomatic relapse rates exceed 50%, relapse becomes the predominant cause of vivax malaria illness.

5. GEOGRAPHIC DISTRIBUTION OF RELAPSE PHENOTYPES

Overall, there was good evidence for the presence of the long-latency 'Madagascar' relapse phenotypes in Europe, Southern Russia, North Africa, the horn of Africa, Madagascar, the Middle East, Central Asia, Afghanistan, Pakistan and Northern India, Central parts of China, Korea, North and Central America (Fig. 2.6). Long-incubation-period *P. vivax* (*hibernans*) was prevalent in Northern Europe and more Northern parts of Russia. Frequent relapse 'strains' were reported in parts of South America, India, South East Asia and Oceania. It is highly likely that where malaria persists, this geographical distribution of *P. vivax* phenotypes pertains today, although there is very little contemporary information apart from reports from South East Asia and the island of New Guinea (which has higher levels of malaria transmission than the majority of the other *P. vivax* endemic areas of the world). Both frequent relapse and long-latency phenotypes overlap in geographic distributions (Fig. 2.6). Where both are present together, it would be very easy for the long-latency infections to go unrecognised. The recent discovery of long-latency *P. vivax* in Kolkata is a case in point (Kim et al., 2012) (Fig. 2.7). The presence of a spring vivax malaria peak as was the case

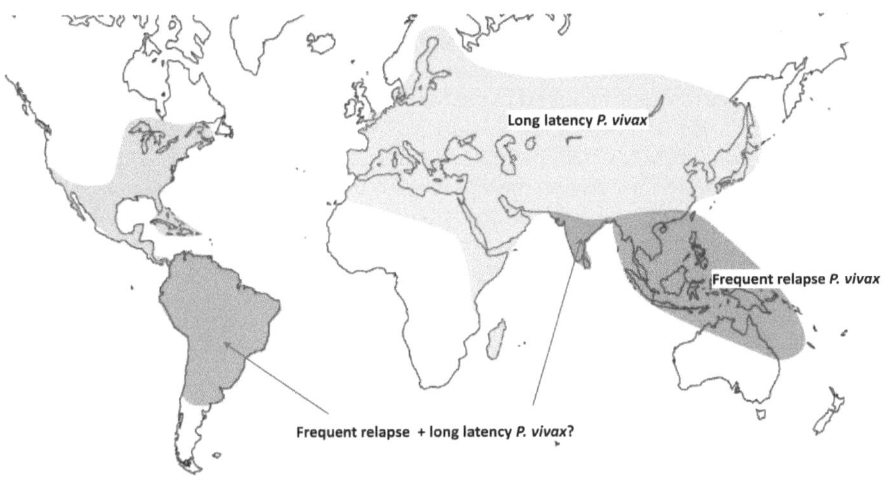

Figure 2.6 Approximate geographic distribution of *P. vivax*-latency phenotypes based upon historical data (White, 2011). Areas where tropical 'frequent-relapse phenotypes' are or were prevalent are shown in pink. Areas where both frequent-relapse and long-latency phenotypes have been reported are shown in purple, and areas where long-latency phenotypes were prevalent are shown in grey. Although both South America and India are generally considered to harbour frequent-relapse phenotypes predominantly, there is evidence that long-latency phenotypes are present in both areas (particularly across the North of India). Without genotyping, it may be difficult or impossible to distinguish the two phenotypes within an endemic area. For interpretation of the references to colour in this figure legend, the reader is referred to the online version of this book.

in Europe while vector densities are still low would be an epidemiological pointer. There has been remarkably little recent interest in this question despite its obvious importance. The greatest uncertainty is the epidemiology of relapse phenotypes in the Indian sub-continent, because the sub-continent harbours the majority of *P. vivax* in the world and clearly has both phenotypes.

Prospective studies from India over the past 25 years have recorded relapse rates following chloroquine treatment of between 8.6% (Orissa) and 8.9% (Madhya Pradesh) and 40.1% (Delhi) (Adak et al., 2001; Adak et al., 1998; Gogtay et al., 2000; Saifi et al., 2010). In Mumbai, 19 out of 150 patients with vivax malaria and treated with chloroquine only followed for 1 year had a relapse (17 within 6 months) (Gogtay et al., 2000). Reinfection was considered unlikely. Higher rates were reported from the Delhi area (Adak et al., 2001; Adak et al., 1998), where the authors concluded 'that the *P. vivax* population in northern India is polymorphic. Group I is the tropical or Chesson strain type of frequent relapsing *P. vivax* with a short period of latency between the primary attack and the first relapse, which is similar to other South East Asian strains such as those from Thailand and Vietnam.

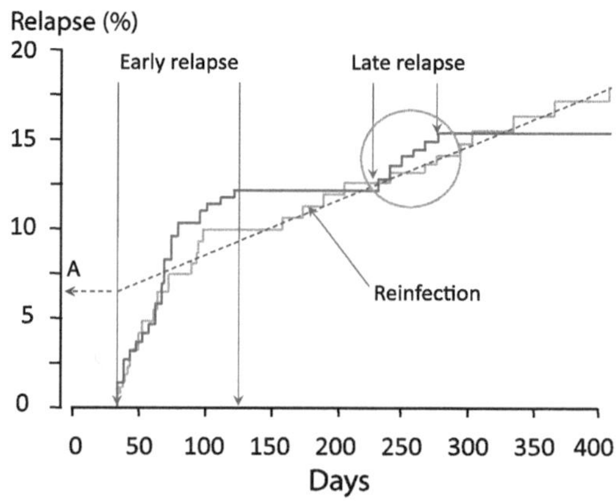

Figure 2.7 The temporal pattern of recurrences of *Plasmodium vivax* malaria in patients with acute vivax malaria treated in Kolkata (Kim et al., 2012). The recurrent infections are divided into genetically homologous relapses (shown in red) and genetically heterologous recurrences (shown in grey). The date of enrolment is shown as day 1. Heterologous recurrences occurring after 120 days were considered as reinfections allowing back-extrapolation of the proportion of all such recurrences, which were suspected reinfections (point A) and thus by subtraction, the proportion that were suspected relapses. The cluster of late genetically homologous relapses (green circle) presumably represent relapses of the long-latency *P. vivax* phenotype. For interpretation of the references to colour in this figure legend, the reader is referred to the online version of this book.

Group III is the temperate or St Elizabeth–type strain that has a long period of latency between the primary attack and the first relapse. Group II is intermediate between these two types'. This intermediate group has not been well characterised. The ratio of short- to long-latency relapse phenotypes in Aligarh, Uttar Pradesh was estimated as 4:1 (Saifi et al., 2010). A similar ratio was observed recently in Kolkata (Kim et al., 2012) (Fig. 2.7). Overall relapse rates from India have been relatively low by comparison with South East Asia (Gogtay et al., 1999; Singh et al., 1990; Prasad et al., 1991; Sinha et al., 1989; Srivastava et al., 1996; Luxemburger et al., 1999; Silachamroon et al., 2003). Data from travellers returning from the Indian sub-continent have suggested that long-incubation-period *P. vivax* may be common in the Punjab, which is consistent with evidence in the past from Northern India, Pakistan and Afghanistan (Warwick et al., 1980; Walker, 1983). There are few data on temporal patterns of relapse in travellers in recent years although studies in Eritrean immigrants to Israel, Turkish immigrants to Germany, and US soldiers returning from Somalia all suggest the presence of long-latency phenotypes in the countries of origin (Kopel et al., 2010; Eichenlaub et al., 1979; Smoak et al., 1997). Most travellers receive radical treatment for vivax malaria in temperate countries, which may be more effective against the long-incubation or long-latency infections than against the tropical frequent-relapse phenotypes (White, 2011).

In South East Asia, there is no evidence as yet for long-latency phenotypes. Luxemburger et al. studied 342 children with acute vivax malaria treated with chloroquine on the north-western border of Thailand (Luxemburger et al., 1999). The 28-day cure rate was 97% [95% confidence interval (CI) 95–99%] but by day 63, the relapse/re-infection rate was 63% (95% CI 57–69%). Most reappearances of parasitaemia (85%; 121/143) were symptomatic. Silachamroon et al. studied 316 adult patients in Thailand with acute vivax malaria who were treated with artesunate monotherapy (Silachamroon et al., 2003). Half the patients relapsed within 28 days. The timing of the relapses suggested that very few, if any, relapses emerged from the liver before the eighth day after starting anti-malarial treatment. In Papua, Indonesia, 38% of patients relapsed within 6 weeks following artemether–lumefantrine treatment (the total number may well have been higher because of later emerging relapses) (Ratcliff et al., 2007). In French Guiana, the relapse rate in children was 70% (nearly all relapses occurred within 3 months) but both recrudescence and reinfection could not be excluded (Hanf et al., 2009). Although South American *P. vivax* is generally regarded as 'tropical frequent relapsing' in phenotype, a recent report from

Rio de Janeiro that six of 80 travellers presenting with vivax malaria (who had returned from the Amazon region and not received chemoprophylaxis) had an incubation period of between 3–12 months, and another from Brasilia describing long latency in three patients suggest that long-latency forms may coexist with frequent-relapse phenotypes in Brazil (Brasil et al., 2011; Tauil et al., 2010). Long-latency *P. vivax* has been well documented in the Central American region.

6. THE EFFECTS OF AGE AND IMMUNITY ON RELAPSE

In malaria endemic areas, such as the north-western border of Thailand, the age profile of *P. vivax* malaria suggests much more rapid acquisition of immunity than for *P. falciparum* (Luxemburger et al., 1996). Entomological studies suggested similar transmission rates (at least in terms of measured entomological inoculation rates), so it is likely that relapse contributes to much of this age difference. It also suggests that relapses are probably partially suppressed in older patients. Thus, both the proportion of infections which relapse and the number of symptomatic relapses per mosquito sporozoite inoculation decline with age. Immunity, and therefore, age is likely to be a significant confounder in epidemiological assessments based on passive case reporting in many *P. vivax* endemic areas. In many areas, adults are more likely to present to malaria clinics than children. In India, the peak age of malaria presentation is often in young adults (and often with a predominance of young males). Yet in the indigenous population living in transmission areas, a significant degree of immunity should have been gained (by both sexes) by early adulthood, which reduces the number of relapses. Usually in endemic areas, it is children who bear the brunt of malaria. This applies to both falciparum and vivax malaria when malaria is uncontrolled, but with increasing control, falciparum declines more rapidly than vivax, and their epidemiology separates. The paucity of data from children may contribute to the low apparent relapse rates reported from the Indian subcontinent. Malaria clinic data may not be representative of disease epidemiology in some endemic areas. Studies of children living in the endemic areas of the Indian sub-continent might reveal higher relapse rates.

7. DRUG EFFECTS ON RELAPSE

Over 80 years ago, Sinton et al. in India provided the first evidence for the radical curative activity of plasmoquine (Sinton and Bird, 1928;

Sinton et al., 1930). Widely used in the 1920s and 1930s, plasmoquine was not well tolerated at the doses required for anti-malarial effects. During and after the Second World War, the enormous allied research effort to find new anti-malarial drugs gave new more effective and less toxic 8-aminoquinolines (Alving et al., 1948; Edgcomb et al., 1950; Cooper et al., 1953; Coatney et al., 1953). Pentaquine, isopentaquine, and then primaquine were developed and evaluated between 1944 and 1955 (Coatney et al., 1953; Robinson et al., 1953; Di Lorenzo et al., 1953). Primaquine took over as the standard radical treatment of vivax malaria (except in the USSR where the positional isomer quinocide was preferred initially). The 8-aminoquinolines also have significant blood-stage activity against *P. vivax* (and *Plasmodium knowlesi*) although resistance in the blood-stage infection can be induced experimentally (Arnold et al., 1961). The widely used chloroquine–primaquine regimen should, therefore, be considered a combination treatment for vivax malaria. The dose of primaquine recommended globally for radical cure (0.25 mg base/kg/day for 2 weeks) was chosen largely as a result of studies on the relatively drug-sensitive Korean *P. vivax* (Coatney et al., 1953; Robinson et al., 1953; Di Lorenzo et al., 1953). The Chesson strain had been shown to be more 'resistant' to 8-aminoquinolines (Cooper et al., 1953), but recommendations for a higher dose of primaquine (adult dose: 22.5 mg base/day) were applied initially only in Oceania. Sinton's work in India had suggested that quinine and plasmoquine were synergistic in the prevention of relapse in vivax malaria (Sinton and Bird, 1928; Sinton et al., 1930). Ruhe et al. showed that concurrent quinine and pamaquine (plasmoquine) were more effective in prevention of relapse with the St. Elizabeth strain than sequential administration (Ruhe et al., 1949). In the 1950s, Alving et al. conducted a formal interaction study, which provided evidence of marked synergy between both quinine and chloroquine and primaquine (Alving et al., 1955). The reason for this synergy, and whether it extends to other quinolines or related compounds, has not been explored further. Most countries adopted the 15 mg base/day primaquine radical curative regimen usually given for 14 days. Five-day regimens of primaquine (total adult dose 75 mg base) were recommended in the Indian sub-continent and some other areas; although there was no evidence, they were effective (Goller et al. 2007). It has been suggested that resistance to the radical curative activity of primaquine has emerged (Baird, 2009; Collins and Jeffery, 1996) – but it is not at all clear if there has been any significant change in susceptibility. As the tropical phenotype is usually associated with a greater number of relapses than the temperate phenotype, then it requires a greater proportional reduction in activatable hypnozoites to prevent all relapses.

8. VIVAX MALARIA FOLLOWING FALCIPARUM MALARIA

In East Asia, a remarkably high proportion (20–50%) of symptomatic infections with *P. falciparum* is followed by an infection with *P. vivax* (Looareesuwan et al., 1987; Mayxay et al., 2004; Douglas et al., 2011). The intervals between the onset of treatment for the acute *P. falciparum* infection and the subsequent *P. vivax* infection are very similar to the intervals between acute vivax malaria and the first relapse. As the treatments given for falciparum malaria are highly effective, and therefore, should be curative against the blood-stage infections of *P. vivax*, these recurrent malaria infections are highly likely to be relapses. Furthermore, as with relapses after vivax malaria, the probability of vivax recurrence after falciparum malaria is also age related (Douglas et al., 2011). Although some mixed infections may be acquired from a single doubly infected mosquito, studies of wild-anopheline vectors suggest that this is very unlikely, and the weight of evidence points to these vivax episodes being relapses of latent hypnozoites acquired *before* the *P. falciparum* inoculation.

9. THE PERIODICITY OF RELAPSE

Various theories to explain the remarkable periodicity of *P. vivax* infections have been proposed (Lysenko et al., 1977). These include reinfection of liver cells from released merozoites, intrinsic differences in latency periods of the inoculated sporozoites (tachyzoites, bradyzoites), and activation of dormant parasites by external stresses or seasonal stimuli (Shute, 1946; Cogswell, 1992). In bird malarias, there is reinfection of tissues from blood-stage parasites but there is no convincing evidence of this in the primate malarias.

The temperate-zone *P. vivax* usually had an incubation period of 8–9 months, so emergence of an infection acquired in late summer or autumn could coincide with vector emergence in the following late spring or summer. Further south in temperate climes, *P. vivax* infections had a primary illness 2–3 weeks after mosquito inoculation but the first relapse occurred 8–9 months later. Although the interval (latency) between the primary infection and first relapse for this 'Madagascar' phenotype was long

(8–10 months), the subsequent inter-relapse intervals were short (Fig. 2.3). In fact, they were similar to the intervals from primary infection to early relapse, which occurred in some Madagascar/St. Elizabeth phenotypes and all Chesson phenotypes. The generally accepted theory to explain the remarkable periodicity of vivax malaria relapses has been that 'sporozoite populations of all strains of *P. vivax, P. ovale*, and relapsing simian malarias such as *P. cynomolgi*, include a subpopulation that completes development promptly and is responsible for the early primary attack, and a group of subpopulations that undergo partial development to the resting hypnozoite stage. Subsets of these dormant pre-erythrocytic stages are preprogrammed to resume development at different times and, via this built-in time clock, evoke the sequence of relapses that characterize sporozoite-induced infections with the afore mentioned plasmodia' (Schmidt, 1986). However, it is difficult to understand how multiple relapses could occur at regular intervals with generally small inocula (median ~ 5 hypnozoites) and a simple clock mechanism. This remarkable efficiency suggests that some activation or feedback mechanism must operate in addition. A recent suggestion is that mosquito bites themselves might provide the trigger – perhaps by parasite sensing of a mosquito protein (Hulden and Hulden, 2011) cannot be reconciled with the similarity of latency periods in indigenous peoples living in malaria endemic areas and the latency periods observed in malaria therapy patients and returned soldiers (i.e. away from seasonal mosquitoes and independent of season). There are relatively few recent data on relapse patterns in frequent-relapse 'tropical' vivax malaria, but the available evidence confirms the remarkable periodicity documented in the volunteer studies with the 'Chesson strain' (Pukrittayakamee et al., 2000; Pukrittayakamee et al., 2010). Much of the early volunteer data with tropical 'strains' was confounded partly by slowly eliminated drug effects and inoculations of unnaturally large numbers of sporozoites. Robert Coatney realized this, and so in 204 sporozoite-induced infections with the Chesson strain (Coatney et al., 1950b), he made detailed longitudinal observations following a single infected mosquito bite (Fig. 2.3). The number of relapses varied considerably. In the seven volunteers, who were not reinfected, the median (range) number of relapses following a single bite was five (one to nine) and 11 of the 39 relapses in this group (28%) occurred more than 6 months after the initial infection. The interval from one relapse to the next was remarkably similar but overall, the inter-relapse intervals gradually lengthened. Importantly, there could then be very long intervals between relapses (4 relapses occurred with preceding latencies exceeding

6 months. The maximum documented interval was 397 days). This proves that long latency can and does occur with the tropical frequent-relapse phenotype.

These patterns of relapse are also illustrated well in the primate model; the Rhesus monkey infected with *P. cynomolgi* (the primate malaria 'equivalent' of *P. vivax*) (Schmidt, 1986). In Schmidt's detailed longitudinal monkey experiments, where different sporozoite inocula were evaluated, there was no clear difference in the number of relapses between monkeys inoculated with 5×10^2 sporozoites up to 5×10^6 sporozoites, although 'the intervals between relapses were related to size of inoculum, being distinctly shorter in monkeys inoculated with 5×10^6 sporozoites than in those challenged with 5×10^2 sporozoites, with recipients of 5×10^4 sporozoites occupying an intermediary position'. Taken together with the absence of any relapses following an inoculum of only five sporozoites in three monkeys, this argues for activation of a proportion of hypnozoites per relapse, and is consistent with the earlier observations in soldiers and malaria therapy patients of a fixed fraction of relapses following each illness episode in vivax malaria (Fig. 2.5).

9.1. Theories to Explain the Periodicity of Relapse

Any theory seeking to explain the remarkable biology of *P. vivax* relapse must accommodate the following (White, 2011):

1. Relapses show remarkable periodicity.
2. Early relapses reach patency around 3 weeks after starting treatment, which suggests emergence from the liver at least 1 week earlier.
3. Not all *P. vivax* primary infections are followed by a relapse. In Thailand, approximately 50% of infections are followed by a subsequent relapse within 28 days if a rapidly eliminated anti-malarial drug (artesunate) is given for treatment of the primary infection and primaquine is not given (Silachamroon et al., 2003). Elsewhere, the probability of relapse generally varies between 20% and 80%. Animal experiments, the malaria therapy experience, and volunteer studies all suggest this proportion is a function of sporozoite inoculum.
4. Multiple relapses are common, particularly in young children, even though sporozoite inocula are thought to be relatively small (median 6–10 sporozoites). It is not uncommon in tropical areas for children to have 4–6 relapses at 4–6 week intervals and sometimes more following an incident infection. Even larger numbers of relapses were observed in soldiers following intense exposure and in Rhesus monkeys receiving

very large sporozoite inocula (Schmidt, 1986). Importantly, the fraction of people experiencing a relapse after each illness episode in a particular location appears constant (Fig. 2.5).

5. In long-latency phenotypes, there is commonly a period of 8–9 months either before the first symptomatic infection or between the first symptomatic infection and the first relapse. This long-latency interval appears to be normally distributed (mode 28 weeks for the Madagascar strain) (Fig. 2.2). Sometimes, there are several short interval relapses followed by a long interval. Conversely, long latencies may also occur after multiple relapses in the tropical frequent-relapse phenotype (Fig. 2.3).
6. If there are further relapses after the long-latent period, then they occur frequently with short intervals, which are very similar to those observed in the tropical 'strains' (Fig. 2.3).
7. The relapses in clinical studies conducted in endemic areas are commonly with a genotype *which is different* to that identified in the primary infection (48% in Columbian isolates, 55% in Indian isolates, 61% in Thai and Burmese isolates, and 71% in East Timor isolates) (Imwong et al., 2007; Chen et al., 2007; Restrepo et al., 2011).
8. A remarkably high proportion of acute infections with *P. falciparum* are followed by an episode of *P. vivax* infection. The proportion is currently 30% in Thailand (Looareesuwan et al., 1987; Mayxay et al., 2004; Douglas et al., 2011) and 50% in Myanmar (Smithuis et al., 2010). The intervals between the acute *P. falciparum* malaria illness and the subsequent *P. vivax* malaria are similar to those between acute *P. vivax* malaria and the subsequent *P. vivax* relapse (White, 2011). The epidemiological characteristics suggest that these are all relapses.

It is also interesting and probably relevant that in endemic areas, despite very low seasonal transmission, *P. vivax* maintains a high degree of genetic diversity.

9.2. Relapses of Vivax Malaria Arise from Activation of Latent Hypnozoites (ALH)

The combined evidence suggests that *in endemic areas, a high proportion of the population have latent* P. vivax *hypnozoites, which can be activated by a sufficient stimulus (White, 2011).*

Four lines of evidence support this ALH hypothesis.

1. Mixed species infections

Mixed blood-stage infections with *P. falciparum* and *P. vivax* are underestimated by microscopy, but, even with sensitive PCR techniques, this proportion does not explain the 30–50% of patients in South East Asia who experience vivax malaria following falciparum malaria (Douglas et al., 2011). The interval between the primary *P. falciparum* infection and the subsequent *P. vivax* malaria strongly suggests that this is a relapse (White, 2011). Supporting evidence that these are *P. vivax* relapses and not simultaneously acquired infections comes from entomology studies, in which single anopheline mosquitoes have been examined for both species (Imwong et al., 2011). If these mixed species infections resulted from simultaneous inoculation, then 30–50% of anopheline vectors carrying one species should also carry the other. In fact, finding vectors with both *P. falciparum* and *P. vivax* sporozoites is relatively uncommon (overall in Asia, 6.6% of *P. falciparum*-sporozoite-positive mosquitoes (17 of 258) also contained *P. vivax*). In the published literature, not one of the 45 individually examined malaria-positive wild-anopheline vectors trapped in Thailand contained both malaria species (Imwong et al., 2011). This is also supported by the rarity of finding patients or healthy subjects with gametocytes of both species in the blood at the same time. Development rates in the mosquito are also slightly different (*P. vivax* being more rapid). Although space-time clustering of infections may occur in low-transmission areas, it is implausible that over 20% of *P. falciparum* inoculations would be associated with a separate *P. vivax* inoculation within 1 or 2 days, particularly when individuals living in these areas receive on average less than one infectious bite per year. There is other supporting anecdotal evidence from travellers and from soldiers with brief periods of exposure in South East Asia. These groups have much lower rates of *P. vivax* following *P. falciparum* presumably because they do not have a 'bank' of activatable hypnozoites acquired from earlier inoculations. The most plausible explanation for these findings is that the majority of these *P. vivax* episodes arise from hypnozoites, which were latent in the liver of the patient at the time of acquiring *P. falciparum* (ALH). The remarkable similarity of both the timing of the *P. vivax* recurrences and their variance strongly suggest that latent *P. vivax* hypnozoites are activated by acute falciparum malaria.

2. Heterologous genotypes

If *P. falciparum* malaria activates latent *P. vivax* hypnozoites, then *P. vivax* malaria should do the same. This explains the finding of heterologous genotypes in one-half to two-thirds of *P. vivax* relapses (Imwong et al., 2007; Chen et al., 2007; Restrepo et al., 2011). In these cases, where the original genotype was not detected in the malaria recurrence, then either the inoculated infection did not relapse or its hypnozoite(s) were activated but their progeny was outcompeted by the earlier activation or more rapid growth of the progeny of the activated latent hypnozoite(s). In one-third of the patients in whom the relapse is homologous or highly related with the primary infection, either there were no latent hypnozoites (as in volunteer studies) or the homologous infection's hypnozoites' progeny won the race to patency against the heterologous hypnozoites' progeny. It is evident then that close 'races' between different genotypes to reach patency commonly result in gametocyte genotype mixtures in relapses (which may then recombine in the mosquito, providing genetic diversity despite very low levels of transmission). Strong support for the ALH hypothesis comes from observations in mothers and their infants living in a malaria endemic area on the north-west border of Thailand (Fig. 2.9). The relapses of vivax malaria in the babies' mothers were usually genetically different to those which caused the primary infection, as in other patients studied in this area, whereas the relapses which followed the first *P. vivax* infection of life in their babies were usually of the *same* genotypes as those which caused the initial infection (Imwong et al., 2012). Obviously, the infants could not have any previously acquired latent hypnozoites in their livers to be activated by the illness.

3. Natural versus artificial infection relapse rates

Higher rates of vivax relapse in indigenous compared with artificial infection (Swellengrebel & De Buck 1938) would also be explained by the ALH hypothesis. The incidence and number of relapses depends on the number of sporozoites inoculated. If all relapses derived from the inoculated infection, then artificial infections which follow inoculations with 5–10 times more sporozoites should have much higher, not lower, rates of relapse. Relapse rates were particularly high in soldiers who were immunologically naïve and underwent intense exposure. If all relapses derived from the most recent inoculum, then there should be no relationship between intensity of exposure and number of relapses.

4. Long-latency also occurs in the tropical frequent-relapse 'Chesson' *P. vivax* phenotypes

Four of seven volunteers receiving a single infected mosquito bite in Coatney's series had relapses of the Chesson strain of *P. vivax* with variable but long-latency periods – all exceeding 6 months after their preceding relapse (Coatney et al., 1950b). It is not uncommon to see patients returning from malaria endemic areas, where tropical phenotypes are prevalent with relapses more than 3 months after either a primary infection or return to the non-endemic area (of course, these might also be long-latency phenotypes, particularly if the interval is ~8–10 months). This supports the

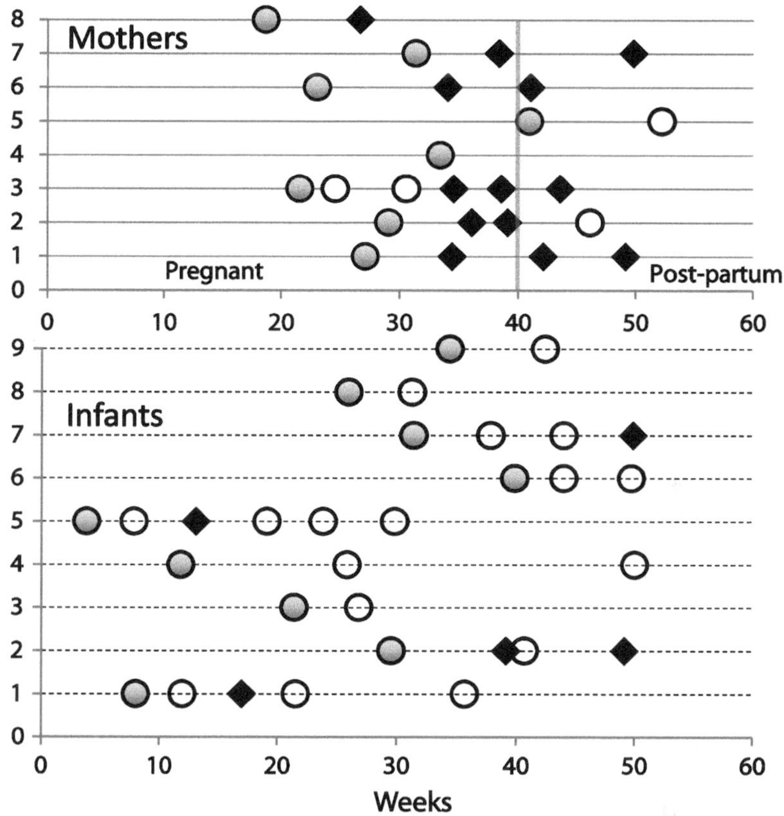

Figure 2.8 Temporal pattern and genotyping of recurrences of vivax malaria in eight mothers (upper panel) and nine infants (lower panel) studied on the NE border of Thailand (Imwong et al., 2012). Each number on the vertical axis represents a patient. Filled circles denote the initial infection. An open circle represents a genetically homologous recurrence, and a black diamond, a genetically heterologous recurrence. Recurrences where genotyping was indeterminate (two in infants, three in mothers) are not shown. One mother's sole relapse was indeterminate. For colour version of this figure, the reader is referred to the online version of this book.

earlier experimental observations that some hypnozoites of frequent-relapse phenotypes can remain dormant for long periods (Coatney et al., 1950b).

9.3. The Stimulus for Hypnozoite Activation

If acute malaria activates latent hypnozoites and thereby causes vivax malaria relapses, then it seems that a significant systemic illness is necessary for this activation. Formal studies to provoke relapses of vivax malaria, which included forced route marches, simulated altitude, and induced hypoxia, were unsuccessful (Ross, 1918; Howe and Duff, 1946). Nevertheless, soldiers in field hospitals in the Mediterranean region between 1940 and 1945 (who had acquired *P. vivax* in North Africa or Italy) were apparently more likely to experience vivax malaria (presumably a relapse) if they had pneumonia or hepatitis compared with trench foot – a local rather than systemic illness (Levine, 1963; McLester, 1945). In a series of children seen daily for over 18 months on the Thai–Burmese border during a study of the SPf66 malaria vaccine (Nosten et al., 1996), *P. vivax* episodes were associated with malaria, but not with minor illnesses (S Lee; personal communication), suggesting that a substantial systemic stimulus is required for activation. In endemic areas of South East Asia, immunity to *P. vivax* is acquired more rapidly than to *P. falciparum* partly because of relapses, so in adults, the systemic illness associated with acute falciparum malaria may be a greater stimulus to *P. vivax* hypnozoite activation than acute vivax malaria.

How can periodicity in vivax relapse be explained? Periodicity can be generated in biological systems in several ways. Illness, generally (and the associated cytokine responses), or malaria, specifically, could provide the activating or inhibitory necessary stimuli to generate periodic relapses in vivax malaria. It is suggested (White, 2011) that the illness associated with the infection itself is the hypnozoite activator that results in *P. vivax* relapses (Fig. 2.8). Each symptomatic episode provides an activation stimulus, which may give rise to the next relapse. Iterations continue until the stimulus fails or there are no more hypnozoites. A simple clock mechanism with a single unimodal distribution of latencies does not explain the observed phenomena. Equally, a multiplicity of different latencies with a single harmonic seems more complex and does not explain the efficient use of relatively small sporozoite inocula (median ~10 sporozoites). For efficient yet periodic activation of small numbers of hypnozoites (i.e. so that they do not all activate together), it is necessary to hypothesise either a negative feedback mechanism, with illness perhaps being the negative feedback, or a relatively broad temporal distribution of latencies, and a *low individual susceptibility* to activation. Either results in several relapses at regular intervals, each activated by the preceding illness. The proportion of susceptible hypnozoites is fractional, at least initially, which is consistent with the observed fixed proportions of patients who experience multiple relapses (Fig. 2.5). It may leave behind unactivated (but activatable) hypnozoites, which remain dormant until either they die (when the host hepatocyte dies), or they are activated by a systemic illness such as malaria. The size of this 'bank' depends on the intensity of exposure. This would

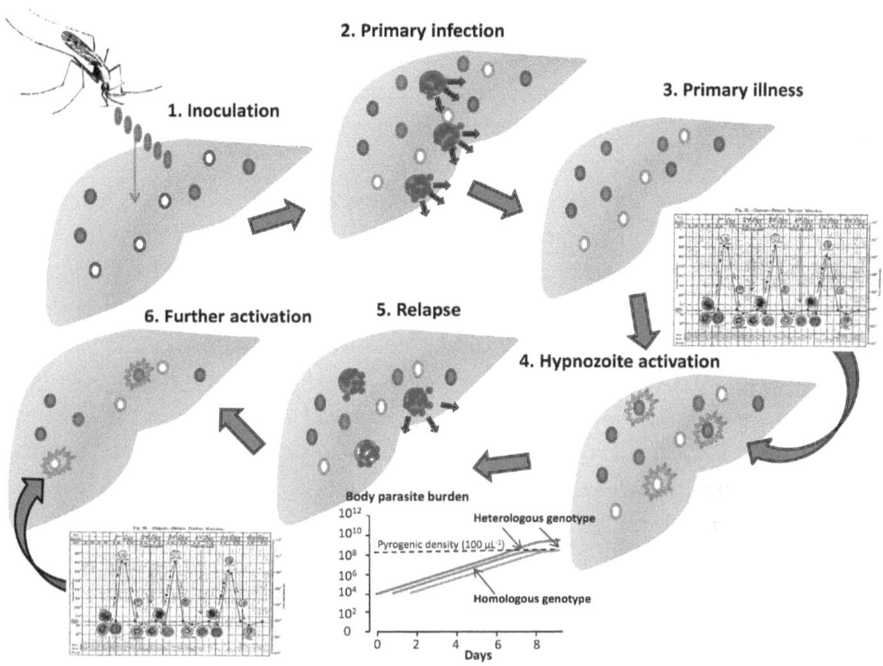

Figure 2.9 Proposed mechanism of *Plasmodium vivax* relapse activation in a malaria endemic area (White, 2011). In this example, at the time of infection (sporozoite inoculation) the individual already has hypnozoites of two different genotypes acquired from two previous inoculations, which are latent in the liver. The different genotypes are denoted by different colours (red and white). Half the newly acquired infection sporozoites (blue) develop into pre-erythrocytic schizonts and half become dormant as hypnozoites (the estimated proportions for tropical 'strains'). Illness associated with the blood-stage infection activates a small fraction of the hypnozoites (inset shows a 'classic' *P. vivax* fever pattern in relation to asexual parasite development). In this example, one hypnozoite of each genotype is activated by the illness and develops into pre-erythrocytic schizonts. By chance, the progeny of one of the pre-existing latent hypnozoites reach pyrogenic densities before the progeny of the recently inoculated hypnozoite (inset shows the logarithmic growth of the three genotypically different infections; vertical axis shows the number of parasites in the body). The consequent febrile illness then suppresses further multiplication of the blood-stage infection, so that the progeny of the other two pre-erythrocytic schizonts may not reach transmissible densities. The ensuing illness activates some of the remaining hypnozoites (the same fraction as were activated initially) and relapses continue until either the number of hypnozoites is exhausted or some fail to be activated. If there are some hypnozoites which fail to be activated, these may be activated at a later date by a subsequent malaria infection. For interpretation of the references to colour in this figure legend, the reader is referred to the online version of this book.

explain satisfactorily the conundrum of the high rate of vivax malaria (presumably relapses), which follows falciparum malaria in endemic areas of East Asia. It follows from this that the number of early relapses per mosquito inoculation will decrease with increasing age in endemic areas as the stimulus to hypnozoite activation declines with increasing disease controlling immunity, and immunity increases the probability that the relapse is asymptomatic. In Thailand, the incidence of symptomatic *P. vivax* malaria peaks in childhood, but *P. vivax* following *P. falciparum* malaria shows a weaker age relationship. Thus, *P. falciparum* may contribute significantly to *P. vivax* transmission, particularly in young adults, through this mechanism. This emphasises the importance of considering the two species together and not in isolation. With small-sporozoite inocula, it is quite possible for all sporozoites to develop immediately (early infection, no relapses) or for all to result in hypnozoites, and all the hypnozoites to become latent (no early infection as no activation stimulus, the first infection follows a later activation stimulus such as a malaria illness).

For temperate strains of *P. vivax*, there are clearly at least two populations of hypnozoites, one becoming activatable early (as for tropical *P. vivax*) and another remaining latent and not immediately activatable. To study activation during the latent period, Cooper et al. gave homologous (St. Elizabeth strain) infections by blood 120 days after subjects had been infected with the same strain by mosquito bites (Cooper et al., 1947). The blood inoculations reliably gave symptomatic malaria but critically, they *did not* affect the timing of the subsequent relapse. This shows that the hypnozoites half way through their sleep were absolutely refractory. This supports the concept of a biological clock for the long-latency *P. vivax*. In order to account for the distribution of long latencies, it is likely that once hypnozoites do become activatable, that there is a background relatively low rate of spontaneous activation. Thus, for the long-latency *P. vivax*, the first relapse after the long-latent period is usually spontaneous (i.e. there is no external activation stimulus), but the illness then activates further hypnozoites, accounting for the subsequent short inter-relapse intervals. If this is correct, it follows that *first* relapses, which occur with a long-latency interval (i.e. 8–10 months after the primary infection), should usually be of a *similar* genotype to the initial infection, whereas relapses at other times could be heterologous. This has been observed recently in Kolkata (Kim et al., 2012) (Fig. 2.7). If there are subsequent relapses, i.e. which follow the first relapse after the latent period (i.e. with a short periodicity), then these could be genetically heterologous.

P. vivax in temperate zones has evolved to adapt to the long winters across the Northern Hemisphere. The large proportion of sporozoites dedicated to latency, and the relatively low number of relapses compared with the tropical strains may reflect the high natural wastage of hypnozoites in hepatocytes, which die before the hypnozoites become activatable. In experimental *P. cynomolgi bastianelli* infection in the Rhesus monkey, Garnham observed a 10-fold reduction in the number of hypnozoites in serial liver biopsies over a 9-month period, but it is unclear how much of this reduction resulted from activation and how much was from cell death (James, 1931b).

The hypnozoite can be considered as an unactivated sporozoite. If the duration of pre-erythrocytic development of the liver stage is similar for sporozoites and hypnozoites, as seems likely, then activation occurs around the time of presentation with acute malaria illness (of any species). If the ALH hypothesis is correct, then the key biological difference between frequent-relapse and long-latency *P. vivax* is in the temporal distribution of susceptibility to activation among the sporozoites.

Thus, the ALH hypothesis proposes that there is a biological clock, that is most evident in temperate strains, which determines latency in *P. vivax*. This clock, which could be an intrinsic parasite clock or could reflect a host–parasite interaction, determines the duration of the interval before the hypnozoite becomes susceptible to activation. Once the hypnozoite becomes susceptible, there is then a separate sensing mechanism which determines whether or not activation does occur. The trigger could be either a positive-activation stimulus or removal of inhibition. This mechanism may activate spontaneously once the hypnozoite has become susceptible, and spontaneous activation presumably usually explains the first relapse after a long-latent period, but, once susceptible, activation is much more likely with an external systemic trigger such as malaria illness. Activation must involve signalling via the infected hepatocyte (which is very sensitive to systemic inflammatory responses). Importantly, the individual probability of activation for each hypnozoite is low, allowing accumulation of latent but 'activatable' hypnozoites after each sporozoite inoculum. This implies that people living in vivax endemic areas commonly harbour latent but 'activatable' hypnozoites. In endemic areas of South East Asia, if patients who acquire falciparum malaria are representative of the population, this is approximately one quarter to one-third. The periodicity of *P. vivax* relapses is derived from the sequential iterative activation of hypnozoites by illness.

10. IMPLICATIONS FOR EPIDEMIOLOGICAL ASSESSMENT

If this ALH theory is correct, it explains why relapse phenotypes may be difficult to characterize in malaria endemic areas, and why in areas with long-latency *P. vivax* frequent-relapse patterns may still be observed. This is because a primary illness with long-latency *P. vivax* of the 'Madagascar' or 'St. Elizabeth' phenotype may activate previously acquired latent hypnozoites (of similar phenotype) – giving an early relapse. The illness caused by the relapse may then activate further latent homologous or heterologous hypnozoites. Thus, several relapses may follow at short intervals even though all the parasites are of the long-latency type. The net result would be indistinguishable from the epidemiological pattern of frequent-relapse vivax malaria [for further explanation see White, 2011]. The converse may also occur; for example, long latency with the tropical frequent-relapse phenotypes was demonstrated clearly in the studies of the Chesson strain following single mosquito bites (Coatney et al., 1950b) (Fig. 2.3). If both short- and long-latency types are prevalent in the same area, then identifying phenotypes from illness patterns become even more difficult, and it may be impossible to discern the long-latency phenotypes amongst the frequent relapses. The presence of long-latency phenotypes may only be evident in travellers and soldiers who spend a brief period in the endemic area, do not receive primaquine, and then return to a non-endemic area and relapse many months later without re-exposure (Warwick et al., 1980; Walker, 1983; Kopel et al., 2010; Eichenlaub et al., 1979; Smoak et al., 1997; Jiang et al., 1982). There is uncertainty over the true epidemiology of relapse patterns over a large proportion of the *P. vivax* endemic world. This large knowledge gap seems to have gone unnoticed. Long-latency phenotypes may be present in many tropical areas (Fig. 2.6). Identifying the relapse phenotypes is an essential prerequisite for therapeutic assessments, control and elimination planning, and the evaluation of novel interventions such as vaccines.

There is an additional point in biological interest. Activation of hypnozoites from different preceding inoculations will commonly result in two or more genotypes reaching patent parasitaemia at similar times. As gametocytogenesis in *P. vivax* occurs simultaneously with asexual stage development, this provides an effective method of ensuring that a mosquito will ingest gametocytes of different genotypes, thereby facilitating meiotic recombination between genetically unrelated *P. vivax* parasites.

This must be an important contributor to the relatively high degree of genetic diversity in *P. vivax* often found in areas with very low seasonal transmission.

10.1. Practical Implications

If long-latency *P. vivax* still contributes a significant proportion of vivax malaria in the Indian sub-continent and further west, then current methods of assessing drugs and vaccines may need reconsideration.

In therapeutic assessments, a follow-up period of 6 months or less may miss a significant proportion of the relapses. If the majority of relapses in endemic areas derive from heterologous latent hypnozoites, then malaria control interventions, which are effective, will not prevent relapses emerging for months or years, although the number of relapses will decline as the reduced transmission will result in less illness and, therefore, less-hypnozoite activation. Mass drug administration with radical curative regimens (currently primaquine is the only option) would be the only way to eliminate this reservoir of infection quickly. Epidemiological assessments in older children and adults in endemic areas may underestimate the burden of vivax malaria as partial immunity (and premunition) will ameliorate disease severity and may lead to reduced activation of relapses. This would result in relatively low relapse rates. The proportion of acute falciparum malaria infections, which are followed by *P. vivax*, may be a better indicator of the prevalence of latent hypnozoite carriage.

It is possible that much of the variance in responses to radical curative primaquine regimens is explained by differences in rates and burdens of latent hypnozoite carriage, and degree of immunity, and not to variation in intrinsic drug susceptibility. The same factors, which affect therapeutic responses in the blood-stage infection, affect the responses to hypnozoitocidal treatment, i.e. organism load and immunity. We should take a quantitative approach to assessing 8-aminoquinoline radical curative activity based on hypnozoite burdens. Patients with very large liver burdens of hypnozoites from either a very heavy inoculation or multiple inoculations and little or no immunity (such as soldiers) would be expected to have a larger number of relapses than travellers who have a brief period of exposure. The apparent radical curative activity of primaquine would be expected to improve as malaria transmission falls. Long-term follow-up (minimum 1 year) of well-characterised patients with parasite genotyping in low-transmission settings should help to dissect the contributions of pre-existing versus recently inoculated hypnozoites to relapse.

This review suggests that there are major questions about the basic biology and the epidemiology of *P. vivax* relapse, which need answering, if we are to address seriously controlling and eliminating this important malaria parasite (Garnham, 1966; Mendis et al., 2001).

ACKNOWLEDGEMENTS

NW is a Wellcome Trust Principal Fellow and MI is a Wellcome Trust Intermediate Fellow.

We are very grateful to our colleagues past and present in the Wellcome-Trust Mahidol University Oxford Tropical Medicine Research Programme who have contributed to this research and the development of these concepts.

REFERENCES

Adak, T., Sharma, V.P., Orlov, V.S., 1998. Studies on the *Plasmodium vivax* relapse pattern in Delhi, India. Am. J. Trop. Med. Hyg. 59, 175–179.

Adak, T., Valecha, N., Sharma, V.P., 2001. *Plasmodium vivax* polymorphism in a clinical drug trial. Clin. Diagn. Lab. Immunol. 8, 891–894.

Alving, A.S., Craige, B., Jones, R., Whorton, C.M., Pullman, T.N., Eichelberger, L., 1948. Pentaquine (SN-13,276) a therapeutic agent effective in reducing the relapse rate in vivax malaria. J. Clin. Invest. 27, 25–33.

Alving, A.S., Arnold, J., Hockwald, R.S., Clayman, C.B., Dern, R.J., Beutler, E., Flanagan, C.L., 1955. Potentiation of the curative action of primaquine in vivax malaria by quinine and chloroquine. J. Lab. Clin. Med. 46, 301–306.

Arnold, J., Alving, A.S., Clayman, C.B., 1961. Induced primaquine resistance in vivax malaria. Trans. R. Soc. Trop. Med. Hyg. 55, 345–350.

Baird, J.K., 2009. Resistance to therapies for infection by *Plasmodium vivax*. Clin. Microbiol. Rev. 22, 508–534.

Beier, J.C., Davis, J.R., Vaughan, J.A., Noden, B.H., Beier, M.S., 1991. Quantitation of *Plasmodium falciparum* sporozoites transmitted in vitro by experimentally infected *Anopheles gambiae* and *Anopheles stephensi*. Am. J. Trop. Med. Hyg. 44, 564–570.

Bianco, A.A., Saunders, G.M., Levine, A.S., Cohn, R.V., 1947. Long term observation of *Plasmodium vivax* malaria. U. S. Naval Med. Bull. 47, 550.

Bignami, A., 1913. Concerning the pathogenesis of relapses in malarial fevers. South. Med. J. 6, 79–89.

Boyd, M.F., 1940a. On strains or races of the malaria parasites. Am. J. Trop. Med. 20, 69–80.

Boyd, M.F., 1940b. The influence of sporozoite dosage in vivax malaria. Am. J. Trop. Med. 20, 279–286.

Boyd, M.F., 1947. A review of studies on immunity to vivax malaria. J. Natl. Malar. Soc. 6, 12–31.

Boyd, M.F., Kitchen, S.F., 1943. Renewed clinical activity in vivax malaria. Am. J. Trop. Med. Hyg. 23, 221–234.

Boyd, M.F., Stratman-Thomas, W.K., 1933. A controlled technique for the employment of naturally induced malaria in the therapy of paresis. Am. J. Hyg. 17, 37–43.

Boyd, M.F., Stratman-Thomas, W.K., Kitchen, S.F., 1936. On the duration of homologous immunity to *Plasmodium vivax*. Am. J. Trop. Med. 21, s1–s16 311–315.

Brasil, P., de Pina Costa, A., Saraiva Pedro, R., da Silveira Bressan, C., da Silva, S., Tauil, P.L., Daniel-Ribeiro, C.T., 2011. Unexpectedly long incubation period of *Plasmodium vivax* malaria, in the absence of chemoprophylaxis, in patients diagnosed outside the transmission area in Brazil. Malar. J. 10, 122.

Bray, R.S., Garnham, P.C., 1982. The life-cycle of primate malaria parasites. Br. Med. Bull. 38, 117–122.
Chen, N., Auliff, A., Rieckmann, K., Gatton, M., Cheng, Q., 2007. Relapses of *Plasmodium vivax* infection result from clonal hypnozoites activated at predetermined intervals. J. Infect. Dis. 195, 934–941.
Coatney, G.R., Cooper, W.C., Ruhe, D.S., Young, M.D., Burgess, R.W., 1950a. Studies in human malaria. XVIII. The life pattern of sporozoite-induced St. Elizabeth strain vivax malaria. Am. J. Hyg. 51, 200–215.
Coatney, G.R., Cooper, W.C., Young, M.D., 1950b. Studies in human malaria. XXX. A summary of 204 sporozoite-induced infections with the Chesson strain of Plasmodium vivax. J. Natl. Malar. Soc. 9, 381–396.
Coatney, G.R., Alving, A.S., Jones Jr., R., Hankey, D.D., Robinson, D.H., Garrison, P.L., Coker, W.G., Donovan, W.N., Di Lorenzo, A., Marx, R.L., Simmons, I.H., 1953. Korean vivax malaria. V. Cure of the infection by primaquine administered during long-term latency. Am. J. Trop. Med. Hyg. 2, 985–988.
Cogswell, F.B., 1992. The hypnozoite and relapse in primate malaria. Clin. Microbiol. Rev. 5, 26–35.
Collins, W.E., Jeffery, G.M., 1996. Primaquine resistance in *Plasmodium vivax*. Am. J. Trop. Med. Hyg. 55, 243–249.
Cooper, W.C., Coatney, G.R., Ruhe, D.S., 1947. Studies in human malaria. V. Homologous strain superinfection during latency in subjects with sporozoite-induced vivax malaria (St. Elizabeth strain). Am. J. Hyg. 46, 141–148.
Cooper, W.C., Ruhe, D.S., Coatney, G.R., 1949. Studies in human malaria. XVI. Results of massive sub-inoculation during latency from patients infected with St. Elizabeth strain vivax malaria. Am. J. Trop. Med. Hyg. 50, 189–193.
Cooper, W.C., Myatt, A.V., Hernandez, T., Jeffery, G.M., Coatney, G.R., 1953. Studies in human malaria. XXXI. Comparison of primaquine, isopentaquine, SN-3883, and pamaquine as curative agents against Chesson strain vivax malaria. Am. J. Trop. Med. Hyg. 2, 949–957.
Craige Jr., B., Alving, A.S., Jones Jr., R., Whorton, C.M., Pullman, T.N., Eichelberger, L., 1947. The Chesson strain of *Plasmodium vivax* malaria. I. Relationship between prepatent period, latent period and relapse rate. J. Infect. Dis. 80, 228–236.
Dembele, L., Gego, A., Zeeman, A.M., Franetich, J.F., Silvie, O., Rametti, A., Le Grand, R., Dereuddre-Bosquet, N., Sauerwein, R., van Gemert, G.J., Vaillant, J.C., Thomas, A.W., Snounou, G., Kocken, C.H., Mazier, D., 2011. Towards an in vitro model of *Plasmodium* hypnozoites suitable for drug discovery. PLoS One 6, e18162.
Di Lorenzo, A., Marx, R.L., Alving, A.S., Jones, R., 1953. Korean vivax malaria. IV. Curative effect of 15 milligrams of primaquine daily for 7 days. Am. J. Trop. Med. Hyg. 2, 983–984.
Douglas, N.M., Nosten, F., Ashley, E.A., Phaiphun, L., van Vugt, M., Singhasivanon, P., White, N.J., Price, R.N., 2011. *Plasmodium vivax* recurrence following falciparum and mixed species malaria: risk factors and effect of anti-malarial kinetics. Clin. Infect. Dis. 52, 612–620.
Ebisawa, I., 1973. Malaria in China and the Equatorial Pacific Area from 1937 to 1943. Yale J. Biol. Med. 46, 94–101.
Edgcomb, J.H., Arnold, J., Yount, E.H., Alving, A.S., Eichelberger, L., Jeffery, G., Eyles, D., Young, M.D., 1950. Primaquine, SN 13272, a new curative agent in vivax malaria; a preliminary report. J. Natl. Malar. Soc. 9, 285–292.
Eichenlaub, D., Reimann, E., Bunjes, R., 1979. Malaria tertiana bei Kindern und Erwachsenen aus dem Epidemiegebiet der südlichen Türkei. Dtsch. Med. Wschr. 104, 288–292.
Eyles, D.E., Young, M.D., 1948. Studies on imported malarias the parasitological pattern of relapsing *Plasmodium vivax* in military patients. J. Natl. Mal Soc. 7, 23–37.

Findlay, G.M., 1951. Recent Advances in Chemotherapy, third ed.. Malaria. vol. 2. The Blakiston Company, Philadelphia.
Garnham, P.C.C., 1947. Exoerythrocytic schizogony in *Plasmodium kochi* Laveran. A preliminary note. Trans. R. Soc. Trop. Med. Hyg. 40, 719–722.
Garnham, P.C.C., 1966. Malaria Parasites and Other Haemosporidiae. Blackwell, Oxford.
Garnham, P.C.C., 1967. Relapses and latency in malaria. Protozoology 2, 55–64.
Garnham, P.C., 1988. Swellengrebel lecture. Hypnozoites and 'relapses' in *Plasmodium vivax* and in vivax-like malaria. Trop. Geogr. Med. 40, 187–195.
Gill, C.A., 1938. The Seasonal Periodicity of Malaria and the Mechanism of the Epidemic Wave. JA Churchill Ltd, London.
Gogtay, N.J., Desai, S., Kamtekar, K.D., Kadam, V.S., Dalvi, S.S., Kshirsagar, N.A., 1999. Efficacies of 5- and 14-day primaquine regimens in the prevention of relapses in *Plasmodium vivax* infections. Ann. Trop. Med. Parasitol. 93, 809–812.
Gogtay, N.J., Desai, S., Kadam, V.S., Kamtekar, K.D., Dalvi, S.S., Kshirsagar, N.A., 2000. Relapse pattern of *Plasmodium vivax* in Mumbai: a study of 283 cases of vivax malaria. J. Assoc. Phys. India 48, 1085–1086.
Goller, J.L., Jolley, D., Ringwald, P., Biggs, B.A., 2007. Regional differences in the response of *Plasmodium vivax* malaria to primaquine as anti-relapse therapy. Am. J. Trop. Med. Hyg. 76, 203–207.
Hackett, L.W., 1937. Malaria in Europe. An Ecological Study. Oxford University Press, London.
Hanf, M., Stéphani, A., Basurko, C., Nacher, M., Carme, B., 2009. Determination of the *Plasmodium vivax* relapse pattern in Camopi, French Guiana. Malar. J. 8, 278.
Hankey, D.D., Jones Jr., R., Coatney, O.R., Alving, A.S., Coker, W.O., Garrison, P.L., Donovan, W.N., 1953. Korean vivax malaria. I. Natural history and response to chloroquine. Am. J. Trop. Med. Hyg. 2, 958–969.
Hill, E., Amatuzio, D.S., 1949. Southwest Pacific vivax malaria; clinical features and observations concerning duration of clinical activity. Am. J. Trop. Med. Hyg. 29, 203–214.
Höring, R.O., 1947. Induced and war malaria. J. Trop. Med. Hyg. 50, 150–159.
Howe, C.D., Duff, F.L., 1946. Effect of altitude anoxia in provoking relapse in malaria. Science 103, 223.
Hulden, L., Hulden, L., 2011. Activation of the hypnozoite: a part of *Plasmodium vivax* life cycle and survival. Malar. J. 10, 90.
Huldén, L., Huldén, L., Heliövaara, K., 2005. Endemic malaria: an 'indoor' disease in northern Europe. Historical data analysed. Malar. J. 4, 19.
Imwong, M., Snounou, G., Pukrittayakamee, S., Tanomsing, N., Kim, J.R., Nandy, A., Guthmann, J.P., Nosten, F., Carlton, J., Looareesuwan, S., Nair, S., Sudimack, D., Day, N.P., Anderson, T.J., White, N.J., 2007. Relapses of *Plasmodium vivax* infection usually result from activation of heterologous hypnozoites. J. Infect. Dis. 195, 927–933.
Imwong, M., Nakeesathit, S., Day, N.P.J., White, N.J., 2011. A review of mixed malaria species infections in anopheline mosquitoes. Malar. J. 10, 253.
Imwong, M., Boel, M.E., Pagornrat, W., Pimanpanarak, M., McGready, R., Day, N.P., Nosten, F., White, N.J., 2012. The first *Plasmodium vivax* relapses of life are usually genetically homologous. J. Infect. Dis. 205, 680–683.
James, S.P., 1931a. Some general results of a study of induced malaria in England. Trans. R Soc. Trop. Med. Hyg. 24, 477–538.
James, S.P., 1931b. The use of plasmoquine in the prevention of malarial infections. Proc. R. Acad. Sci. Amsterdam 34, 1424–1425.
James, S.P., Shute, P.G., 1926. Report on the First Results of Laboratory Work on Malaria in England. C. H. Malaria. vol. 57. League of Nations. Health Organisation, Geneva, p. 30.
James, S.P., Nicol, W.D., Shute, P.G., 1936. Clinical and parasitological observations on induced malaria. Proc. R. Soc. Med. 29, 27–42.

Jeffery, G.M., 1956. Relapses with Chesson strain *Plasmodium vivax* following treatment with chloroquine. Am. J. Trop. Med. Hyg. 5, 1–13.

Jiang, J.B., Huang, J.C., Liang, D.S., Liu, J.X., Zhang, S.W., Cheng, F.C., 1982. Long incubation of *Plasmodium vivax multinucleatum* as demonstrated in three experimental human cases. Trans. R. Soc. Trop. Med. Hyg. 76, 845–847.

Kim, J.R., Nandy, A., Maji, A.K., Addy, M., Dondorp, A.M., Day, N.P., Pukrittayakamee, S., White, N.J., Imwong, M., 2012. Genotyping of *Plasmodium vivax* reveals both short and long latency relapse patterns in Kolkata. PLoS One 7, e39645.

Kopel, E., Schwartz, E., Amitai, Z., Volovik, I., 2010. Relapsing vivax malaria cluster in Eritrean refugees, Israel. Euro Surveill. 15, 19601.

Korteweg, P.C., 1902. Hereinneringsbundel voor Prof SS Rosenstein, Leiden.

Krotoski, W.A., 1985. Discovery of the hypnozoite and a new theory of malarial relapse. Trans. R. Soc. Trop. Med. Hyg. 79, 1–11.

Krotoski, W.A., Krotoski, D.M., Garnham, P.C.C., Bray, R.S., Killick-Kendrick, R., Draper, C.C., Targett, O.A.T., Guy, M.W., 1980. Relapses in primate malaria: discovery of two populations of exoerythrocytic stages. Preliminary note. Brit. Med. J. 1, 153–154.

Krotoski, W.A., Collins, W.E., Bray, R.S., Garnham, P.C.C., Cogswell, F.B., Gwadz, R.W., Killick-Kendrick, R., Wolf, R., Sinden, R., Koontz, L.C., Stanfill, P.S., 1982. Demonstration of hypnozoites in sporozoite-transmitted *Plasmodium vivax* infection. Am. J. Trop. Med. Hyg. 31, 1291–1293.

Krotoski, W.A., Garnham, P.C., Cogswell, F.B., Collins, W.E., Bray, R.S., Gwasz, R.W., Killick-Kendrick, R., Wolf, R.H., Sinden, R., Hollingdale, M., Lowrie, R.C., Koontz, L.C., Stanfill, P.S., 1986. Observations on early and late post-sporozoite tissue stages in primate malaria. IV. Pre-erythrocytic schizonts and/or hypnozoites of Chesson and North Korean strains of *Plasmodium vivax* in the chimpanzee. Am. J. Trop. Med. Hyg. 35, 263–274.

Levine, H.D., 1963. Clinical aspects of malaria. Internal Medicine in World War II, Infectious Diseases, vol. 2. Office of the Surgeon General, Department of the Army, pp. 479–492.

Looareesuwan, S., White, N.J., Chittamas, S., Bunnag, D., Harinasuta, T., 1987. High rate of *Plasmodium vivax* relapse following treatment of falciparum malaria in Thailand. Lancet 2, 1052–1055.

Luxemburger, C., Thwai, K.L., White, N.J., Webster, H.K., Kyle, D.E., Maelankirri, L., Chongsuphajaisiddhi, T., Nosten, F., 1996. The epidemiology of malaria in a Karen population on the western border of Thailand. Trans. R. Soc. Trop. Med. Hyg. 90, 105–111.

Luxemburger, C., van Vugt, M., Jonathan, S., McGready, R., Looareesuwan, S., White, N.J., Nosten, F., 1999. Treatment of vivax malaria in an endemic area on the western border of Thailand. Trans. R. Soc. Trop. Med. Hyg. 93, 433–438.

Lysenko, A.J., Beljaev, A.E., Rybalka, V.M., 1977. Population studies of *Plasmodium vivax*. 1. The theory of polymorphism of sporozoites and epidemiological phenomena of tertian malaria. Bull. World Health Organ. 55, 541–549.

Mayxay, M., Pukrittayakamee, S., Newton, P.N., White, N.J., 2004. Mixed-species malaria infections in humans. Trends Parasitol. 20, 233–240.

McLester, J.B., 1945. Relapsing malaria. M Bull. Mediterranean Theat. 3, 111–113.

Mendis, K., Sina, B.J., Marchesini, P., Carter, R., 2001. The neglected burden of *Plasmodium vivax* malaria. Am. J. Trop. Med. Hyg. 64 (1–2 Suppl), 97–106.

Moshkovsky, S.D., 1973. Explanation for the difference of incubation type and features of alteration of acute periods of tertian malaria associated with different strains of *Plasmodium vivax*. Med. Parazitol. 42, 393–400 (in Russian).

Most, H., 1963. Clinical trials of anti-malarial drugs. Internal Medicine in World War II, Infectious Diseases, vol. 2Office of the Surgeon General, Department of the Army, pp. 525–598.

Mowrey, F.H., 1963. Statistics of malaria. Internal Medicine in World War II, Infectious Diseases, vol. 2Office of the Surgeon General, Department of the Army, pp. 449–463.
MRC Malaria sub-committee, 1945. Report on relapsing vivax malaria. MLE 30.
Nikolaiev, B.P., 1949. Subspecies of the parasite of tertian malaria (*Plasmodium vivax*). Dokl. Akad. Navk. USSR 67, 201–210.
Noe, W.L., Green, C.C., Cheney, G., 1946. Natural course of Southwest Pacific malaria. Am. J. Med. Sci. 211, 215–228.
Nosten, F., Luxemburger, C., Kyle, D.E., Ballou, W.R., Wittes, J., Heh Wah, Chongsuphajaisiddhi, T., Gordon, D.M., White, N.J., Sadoff, J.C., Heppner, D.G., 1996. Randomised double-blind placebo-controlled efficacy trial of the SPf66 vaccine against falciparum malaria in children in north-western Thailand. Lancet 348, 701–707.
Ponnudurai, T., Lensen, A.H.W., van-Gemart, G.J.A., Bolmer, M.G., Meuwissen, J.H., 1991. Feeding behaviour and sporozoite ejection by infected *Anopheles stephensi*. Trans. R. Soc. Trop. Med. Hyg. 85, 175–180.
Prasad, R.N., Virk, K.J., Sharma, V.P., 1991. Relapse/reinfection patterns of *Plasmodium vivax* infection: a four year study. Southeast. Asian J. Trop. Med. Public Health 22, 499–503.
Pukrittayakamee, S., Chantra, A., Simpson, J.A., Vanijanonta, S., Clemens, R., Looareesuwan, S., White, N.J., 2000. Therapeutic responses to different anti-malarial drugs in vivax malaria. Antimicrob. Agents Chemother. 44, 1680–1685.
Pukrittayakamee, S., Imwong, M., Chotivanich, K., Singhasivanon, P., Day, N.P., White, N.J., 2010. A comparison of two short-course primaquine regimens for the treatment and radical cure of *Plasmodium vivax* malaria in Thailand. Am. J. Trop. Med. Hyg. 82, 542–547.
Ratcliff, A., Siswantoro, H., Kenangalem, E., Maristela, R., Wuwung, R.M., Laihad, F., Ebsworth, E.P., Anstey, N.M., Tjitra, E., Price, R.N., 2007. Two fixed-dose artemisinin combinations for drug-resistant falciparum and vivax malaria in Papua, Indonesia: an open-label randomised comparison. Lancet 369, 757–765.
Restrepo, E., Imwong, M., Rojas, W., Carmona-Fonseca, J., Maestre, A., 2011. High genetic polymorphism of relapsing *P. vivax* isolates in northwest Colombia. Acta Trop. 119, 23–29.
Robinson, D.H., Di Lorenzo, A., Marx, R.L., Simmons, I.H., Alving, A.S., Hankey, D.D., Coatney, G.R., Jones, R., Coker, W.G., Garrison, P.L., Donovan, W.N., 1953. Korean vivax malaria. II. Curative treatment with pamaquine and primaquine. Am. J. Trop. Med. Hyg. 2, 970–976.
Rosenburg, R., Wirtz, R.A., 1990. An estimation of the number of sporozoites ejected by a feeding mosquito. Trans. R. Soc. Trop. Med. Hyg. 84, 209–212.
Ross, R., 1918. An interim report on the treatment of malaria. Trans. R. Soc. Trop. Med. Hyg. 11, 179–204.
Ruhe, D.S., Cooper, W.C., Coatney, G.R., Josephson, E.S., 1949. Studies in human malaria. XV. The therapeutic action of pamaquine against St. Elizabeth strain vivax malaria. Am. J. Hyg. 49, 367–373.
Saifi, M.A., Wajihullah, Siddiqui, M.I., Al-Khalifa, M.S., 2010. Malaria: patterns of relapse and resistance. J. King Saud Univ. (Science) 22, 31–36.
Schmidt, L.H., 1986. Compatibility of relapse patterns of *Plasmodium cynomolgi* infections in Rhesus monkeys with continuous cyclical development and hypnozoite concepts of relapse. Am. J. Trop. Med. Hyg. 35, 1077–1099.
Schuffner, W.A.C., Kortweg, P.C., Swellengrebel, N.H., 1929. Pathology – experimental malaria with protracted incubation. Proc. R. Acad. Sci. Amsterdam 32, 903–911.
Shortt, H.E., Garnham, P.C.C., 1948a. Pre-erythrocytic stage in mammalian malaria parasites. Nature 161, 126.
Shortt, H.E., Garnham, P.C.C., 1948b. Demonstration of a persisting exo-erythrocytic cycle in *Plasmodium cynomolgi* and its bearing on the production of relapses. Brit. Med. J. 1, 1225–1228.

Shortt, H.E., Garnham, P.C.C., 1948c. The pre-erythrocytic development of *Plasmodium cynomolgi* and *Plasmodium vivax*. Trans. R Soc. Trop. Med. Hyg. 41, 785–795.
Shortt, H.E., Garnham, P.C.C., Covell, G., Shute, P.G., 1948a. The pre-erythrocytic stage of human malaria, *Plasmodium vivax*. Brit. Med. J. 1, 547.
Shortt, H.E., Garnham, P.C.C., Malamos, B., 1948b. The pre-erythrocytic stage of mammalian malaria. Brit. Med. J. 1, 192–194.
Shute, P.O., 1946. Latency and long-term relapses in benign tertian malaria. Trans. R Soc. Trop. Med. Hyg. 40, 189–200.
Shute, P.G., Garnham, P.C.C., Maryon, M., 1978. The Madagascar strain of *Plasmodium vivax*. Arch. Inst. Pasteur Madagascar 47, 173–183.
Silachamroon, U., Krudsood, S., Treeprasertsuk, S., Wilairatana, P., Chalearmrult, K., Mint, H.Y., Maneekan, P., White, N.J., Gourdeuk, V.R., Brittenham, G.M., Looareesuwan, S., 2003. Clinical trial of oral artesunate with or without high-dose primaquine for the treatment of vivax malaria in Thailand. Am. J. Trop. Med. Hyg. 69, 14–18.
Singh, N., Mishra, A.K., Sharma, V.P., 1990. Radical treatment of vivax malaria in Madhya Pradesh, India. Indian J. Malariol. 27, 55–56.
Sinha, S., Dua, V.K., Sharma, V.P., 1989. Malaria relapses and chloroquine resistance at the BHEL industrial complex, Hardwar, India. Trans. R. Soc. Trop. Med. Hyg. 83, 606–610.
Sinton, J.A., 1931. Studies on malaria with special reference to treatment. XV. Does the strain of parasite influence cure?. Indian J. Med. Res. 18, 845–853.
Sinton, J.A., Bird, W., 1928. Studies in malaria with special reference to treatment; plasmoquine in treatment of malaria. Indian J. Med. Res. 16, 159.
Sinton, J.A., Smith, S., Potinger, D., 1930. Studies in malaria, with special reference to treatment. XII. Further researches into treatment of chronic benign tertian malaria with plasmoquine and quinine. Indian J. Med. Res. 17, 793.
Smithuis, F., Kyaw, M.K., Phe, O., Win, T., Aung, P.P., Oo, A.P., Naing, A.L., Nyo, M.Y., Myint, N.Z., Imwong, M., Ashley, E., Lee, S.J., White, N.J., 2010. Effectiveness of five artemisinin combination regimens with or without primaquine in uncomplicated falciparum malaria: an open-label randomised trial. Lancet Infect. Dis. 10, 673–681.
Smoak, B.L., DeFraites, R.F., Magill, A.J., Kain, K.C., Wellde, B.T., 1997. *Plasmodium vivax* infections in U.S. Army troops: failure of primaquine to prevent relapse in studies from Somalia. Am. J. Trop. Med. Hyg. 56, 231–234.
Srivastava, H.C., Sharma, S.K., Bhatt, R.M., Sharma, V.P., 1996. Studies on *Plasmodium vivax* relapse pattern in Kheda district, Gujarat. Indian J. Malariol. 33, 173–179.
Swellengrebel, N.H., De Buck, A., 1938. Malaria in the Netherlands. Scheltema & Holkema, Amsterdam.
Swellengrebel, N.H., De Buck, A., Schoute, E., Kraan, M.H., 1936. Investigations on the transmission of malaria in some villages north of Amsterdam. Quart. Bull. Health Org. League Nations 3, 295–352.
Tauil, P.L., Luz, F.C., Oliveira, A.P., Deckers, F.A., Santos, J.B., 2010. Vivax malaria with long incubation period, detected in the Federal District: three case reports. Rev. Soc. Bras. Med. Trop 43, 213–214.
Tiburskaja, N.A., Sergiev, P.O., Vrublevskaja, O.S., 1968. Dates of onset of relapses and the duration of infection in induced tertian malaria with short and long incubation periods. Bull. World Health Organ. 38, 447–457.
Tiburskaja, N.A., 1961. Features specific to the Moscow strain of *P. vivax*. Medskayta Parazit-Moscow 30, 446–454.
Ungureanu, E., Killick-Kendrick, R., Garnham, P.C.C., Branzei, P., Shute, P.G., 1976. Prepatent periods of a tropical strain of *Plasmodium vivax* after inoculations of tenfold dilutions of sporozoites. Trans. R. Soc. Trop. Med. Hyg. 70, 482–483.
Walker, E., 1983. The seasonal pattern of *Plasmodium vivax* malaria in Glasgow. J. Infect. 7, 227–230.

Warwick, R., Swimer, G.J., Britt, R.P., 1980. Prolonged incubation period of imported *P. vivax* malaria in London. J. R. Soc. Med. 73, 333–336.

White, N.J., 2011. Determinants of relapse periodicity in *Plasmodium vivax* malaria. Malar. J. 10 e 297.

Whorton, C.M., Yount, E., Jones, R., Alving, A.S., Pullman, T.N., Craige, B., Eichelberge, L., 1947. The Chesson strain of *Plasmodium vivax* malaria; clinical aspects. J. Infect. Dis. 80, 237–249.

Wilson, T., Reid, J.A., 1949. Malaria among prisoners of war in Siam (F Force). Trans. R. Soc. Trop. Med. Hyg. 43, 257–272.

Winckel, C.W.F., 1941. Are the experimental data of therapeutic malaria applicable to conditions obtaining in nature. Am. J. Trop. Med. Hyg., 789–794.

Winckel, C.W.F., 1955. Long-latency in *Plasmodium vivax* infections in a temperate zone. Doc. Med. Geogr. Trop. 7, 292–298.

Yorke, W., 1925. Further observations on malaria made during the treatment of general paralysis. Trans. R. Soc. Trop. Med. Hyg. 19, 108–122.

Yorke, W., MacFie, J.W.S., 1924. Observations on malaria made during the treatment of general paralysis. Trans. R. Soc. Trop. Med. Hyg. 18, 13–37.

CHAPTER THREE

Plasmodium vivax: Clinical Spectrum, Risk Factors and Pathogenesis

Nicholas M. Anstey[*,†,1], Nicholas M. Douglas[*,†], Jeanne R. Poespoprodjo[‡,§], Ric N. Price[*,†,¶]

[*]Global Health Division, Menzies School of Health Research and Charles Darwin University, Darwin, Australia
[†]Department of Infectious Diseases, Royal Darwin Hospital, Darwin, Australia
[‡]Papuan Community Health and Development Foundation, Timika, Papua, Indonesia
[§]Department of Pediatrics, Timika General Hospital, Timika, Papua, Indonesia
[¶]Centre for Tropical Medicine, Nuffield Department of Clinical Medicine, University of Oxford, Oxford, UK
[1]Corresponding author: Nicholas M. Anstey, Global Health Division, Menzies School of Health Research, PO Box 41096, Casuarina, Darwin, NT 08111, Australia. E-mail: nicholas.anstey@menzies.edu.au

Contents

Advances in Parasitology, Volume 80
ISSN 0065-308X, http://dx.doi.org/10.1016/B978-0-12-397900-1.00003-7

Abstract

Vivax malaria was historically described as 'benign tertian malaria' because individual clinical episodes were less likely to cause severe illness than *Plasmodium falciparum*. Despite this, *Plasmodium vivax* was, and remains, responsible for major morbidity and significant mortality in vivax-endemic areas. Single infections causing febrile illness in otherwise healthy individuals rarely progress to severe disease. Nevertheless, in the presence of co-morbidities, *P. vivax* can cause severe illness and fatal outcomes. Recurrent or chronic infections in endemic areas can cause severe anaemia and malnutrition, particularly in early childhood. Other severe manifestations include acute lung injury, acute kidney injury and uncommonly, coma. Multiorgan failure and shock are described but further studies are needed to investigate the role of bacterial and other co-infections in these syndromes. In pregnancy, *P. vivax* infection can cause maternal anaemia, miscarriage, low birth weight and congenital malaria. Compared to *P. falciparum*, *P. vivax* has a greater capacity to elicit an inflammatory response, resulting in a lower pyrogenic threshold. Conversely, cytoadherence of *P. vivax* to endothelial cells is less frequent and parasite sequestration is not thought to be a significant cause of severe illness in vivax malaria. With a predilection for young red cells, *P. vivax* does not result in the high parasite biomass associated with severe disease in *P. falciparum*, but a four to fivefold greater removal of uninfected red cells from the circulation relative to *P. falciparum* is associated with a similar risk of severe anaemia. Mechanisms underlying the pathogenesis of severe vivax syndromes remain incompletely understood.

1. INTRODUCTION

Globally, *Plasmodium falciparum* is responsible for the majority of uncomplicated febrile illness, as well as severe and fatal malaria, and has therefore overshadowed the clinical and public health importance of vivax malaria (Baird, 2007; Price et al, 2007b). Despite this, *Plasmodium vivax* is a major cause of morbidity, accounting for almost half of all malaria cases outside of Africa. Each year, there are between 70 and 390 million clinical vivax infections (Price et al., 2007b) (reviewed in Chapter 1).

Because individual clinical episodes of malaria due to *P. vivax* infection were less likely to cause severe illness than *P. falciparum* (Marchiafava and Bignami, 1894), vivax malaria was historically described as 'benign tertian malaria' (Manson, 1900). However this relative term is misleading since *P. vivax* was (Dobson, 1997), and remains (Price et al., 2009), responsible for major morbidity and attributable mortality in vivax-endemic areas. The spectrum of disease associated with *P. vivax* infection ranges from asymptomatic parasitaemia (Luxemburger et al., 1996; Gamage-Mendis et al., 1991; Karyana et al., 2008; Harris et al.,

2010) and uncomplicated febrile illness (Ross and Thomson, 1911; Kitchen, 1949b) through to severe and fatal malaria (Barcus et al., 2007; Rodriguez-Morales et al., 2008; Tjitra et al., 2008; Kochar et al., 2009, 2010; Andrade et al., 2010; Alexandre et al., 2010; Yadav et al., 2012; Lacerda et al., 2012).

2. PRE-PATENT AND INCUBATION PERIODS

Kitchen in 1949 reviewed the pre-patent period (time from infection to first appearance of parasites in peripheral blood) and incubation period (time from infection to first fever >100 °F [37.8 °C]) in 428 naturally induced (mosquito-borne) infections in adults with nine strains of *P. vivax*. The mean pre-patent and incubation periods were 12.2 (range 8–23) and 13.4 (range 8–31) days, respectively (Kitchen, 1949a). The incubation periods were similar to those (13.6 days for each strain) seen with the Madagascar (mean 13.6 days) (James, 1931) and McCoy (mean 13.6 days) (Boyd, 1941) strains. In contrast to *P. vivax* strains of tropical origin, prolonged incubation periods are often seen with temperate-zone strains (Hankey et al., 1953; Coatney et al., 1971), including a mean of 282 days for a Rumanian strain (Shute, 1939); 22% of patients infected with Russian strains had prolonged incubation periods ranging from 254 to 360 days (Nicolajev, 1939).

In a primary *P. vivax* infection of a non-immune host, the first fever occurs at very low parasite densities and can occur up to two (Boyd, 1938) to three days (Kitchen, 1949a) before parasites are detectable in the peripheral blood. In a low-endemicity area of Thailand, 4/151 (2.6%) of *P. vivax*-infected children presenting to a primary care centre with fever, did not have parasites detected on their first film and were only detected on a subsequent film within the following week (Luxemburger et al., 1998). In both adults and children, *P. vivax* has a lower pyrogenic threshold (the parasite density required to evoke a fever) compared to *P. falciparum*, with median parasitaemia in uncomplicated vivax malaria lower than that seen in uncomplicated falciparum malaria (Ross and Thomson, 1911; Kitchen, 1949a). There have been similar findings in endemic areas, with the pyrogenic threshold for *P. vivax* (181/μl) in Thailand being eightfold lower than that for *P. falciparum* (1460/μl) (Luxemburger et al., 1996). Fever thresholds rise with time: in untreated pauci-immune adults, parasite densities by the time of natural termination of fever are rarely <500/μl (Boyd, 1938).

3. SYMPTOMS AND SIGNS OF VIVAX MALARIA

3.1. Fever

Infection of a non-immune host results in a prodromal period followed by an acute fever. Prodromal symptoms can include increasing headache, anorexia, malaise, myalgias and/or gastrointestinal symptoms for one or more days, sometimes with periodicity (Kitchen, 1949b). A low-grade prodromal fever without paroxysmal symptoms (below) may also occur. Kitchen describes the prodromal phase being followed by a remittent fever (continuous fever with daily exacerbations) and then an intermittent daily fever (Kitchen, 1949b). Observations in non-immune hosts indicate that 'tertian fever' is frequently a misnomer, with Kitchen reporting that 'it should be emphasised…that the intermittency so developed is invariably quotidian' (once daily) rather than tertian (every second day) because 'the great majority of the vivax plasmodia are segregated into two pyrogenic broods' with shizogony occurring approximately 24 h apart (Kitchen, 1949b). With the Madagascar strain, intermittent fevers were also more frequently quotidian than tertian (James, 1926), but each pattern was seen equally frequently with the Chesson strain (Coatney et al., 1971).

Tertian fevers have been reported in vivax-endemic areas since ancient times, with first descriptions attributed to Hippocrates (Adams, 1849); these are more likely in those with some pre-existing exposure and immunity (Kitchen, 1949b), but even with past infection with heterologous strains, quotidian fever can occur more frequently than a tertian pattern (Kitchen, 1949b).

The 'paroxysm' has long been recognised as the periodic febrile response to *Plasmodium* infection and was shown by Golgi in the 1880s to occur following rupture of schizont-infected red cells (Golgi, 1886, 1889). The paroxysm is classically, but not invariably, preceded by a 'cold stage' of chills and a rigor (Kitchen, 1949a). A chilly sensation proceeds to shivering and then intense muscle tremors, chattering of teeth and sometimes violent shaking. When present, Kitchen observed the duration of the rigor to last from 5 to over 90 min; the mean duration was 50–55 min (Kitchen, 1949a). The temperature rises before the rigor ceases, usually reaching a peak 1–3 h after the rigor finishes, commonly 39.5–40.5 °C but occasionally over 41.5 °C (Kitchen, 1949b). Defervescence is accompanied by sweating and fatigue. Primary infection in adults results in fevers continuing for a minimum of 3–4 weeks (reviewed by Kitchen (1949a) and in Chapter 5).

In adults with previous exposure to heterologous strains and partial immunity, prodromal symptoms are often absent, with an abrupt onset of fever with intermittent pattern, and either quotidian or tertian frequency (Kitchen, 1949b). Kitchen also observed that the pyrogenic threshold with heterologous reinfection is often higher, depending on the degree of tolerance achieved from the preceding infection; reinfection with a homologous strain usually evokes no febrile response (Kitchen, 1949b). Clinical features of temperate strains with long incubation periods have been reported to be no different to short-incubation tropical strains (Hankey et al., 1953).

In vivax-endemic areas, parasitaemia without fever is common (Luxemburger et al., 1996; Gamage-Mendis et al., 1991; Karyana et al., 2008; Harris et al., 2010), even in areas of low endemicity (Luxemburger et al., 1996). In co-endemic areas, the pattern of fever (or indeed any other clinical feature) cannot be used to reliably differentiate *Plasmodium* species in children (Luxemburger et al., 1998). In Thailand, median duration of fever and proportion with fever >38 °C were similar between vivax malaria (2.1 days; 48%) and falciparum malaria (2.4 days; 60%) (Luxemburger et al., 1998).

3.2. Other Symptoms and Signs

Headache, anorexia, vomiting, myalgias, abdominal pain and diarrhoea commonly occur during the paroxysm of vivax malaria (Kitchen, 1949a, 1949b). Restlessness and delirium can occur at the time of maximum temperature (Kitchen, 1949b). In primary attack in adults, the spleen is not usually palpable until the end of the first week of symptoms and rarely reaches the umbilicus in untreated primary infection (Kitchen, 1949a). Similar symptoms and signs are seen in children. In a series of 151 Karen children with vivax malaria (mean age of 8 years), prominent symptoms during acute vivax malaria were headache (87%), cough (42%), muscle pain (35%), nausea (29%), vomiting (24.5%), abdominal pain (25%) and diarrhoea (4%) (Luxemburger et al., 1998). In this area of low endemicity, clinical anaemia (5%), a palpable spleen (8%) and palpable liver (3%) were less frequently seen (Luxemburger et al., 1998). Anaemia is more frequently seen in infancy, and in prolonged, untreated or recurrent infections (Section 8.1.1). No clinical signs or symptoms reliably differentiate *P. vivax* from *P. falciparum* or either species from non-malarial causes of fever (Luxemburger et al., 1998).

Respiratory symptoms have long been recognized as common in vivax malaria. World War I and II series describe clinical presentations with cough, dyspnoea and 'bronchitis' (Falconer, 1919; Stirk, 1943; Hyman, 1945). Cough, usually non-productive, occurs in approximately half of the adults with vivax malaria, in residents of both vivax-endemic and -non-endemic areas (Anstey et al., 2002; Anstey et al., 2007). Tachypnoea may occur because of high fever, anaemia and/or pulmonary pathology (Kitchen, 1949a; Taylor et al., 2012). As with falciparum malaria (O'Dempsey et al., 1993), there is a clinical overlap with symptoms suggestive of lower respiratory tract infection (Hyman, 1945; Anstey et al., 2007).

3.3. Malnutrition and Impaired Growth

In the pre-antibiotic era, untreated vivax malaria was associated with chronic relapsing-remitting debilitating fevers, weight loss, cachexia, hypoproteinemia and oedema in adults and children (reviewed in Dobson, 1997; Kitchen, 1949a). In the antimalarial era, vivax malaria has also been clearly associated with malnutrition in early childhood (Williams et al., 1997). A longitudinal Peruvian study has recently shown that a single episode of vivax malaria results in impaired weight gain and growth persisting for at least 6 months, with greater deficits per incident episode than either diarrhoeal disease or non-specific febrile illness (Lee et al., 2012). The interaction between *P. vivax* infection and malnutrition appears to be bidirectional. As well as being caused by vivax malaria (Lee et al., 2012), malnutrition is associated with increased risk of severe and fatal outcome in those infected with *P. vivax* (Section 7).

3.4. Impaired School Performance

Findings from Sri Lanka showed that acute malaria from either *P. vivax* or *P. falciparum* resulted in a median of 5.4 days school absenteeism per episode of malaria (Fernando et al., 2003b). Furthermore, acute malaria was associated with significant impairment of cognitive performance, persisting for at least 2 weeks after the febrile episode. This effect was greater than that due to non-malarial fever and independent of confounding socioeconomic factors. The effects were cumulative, with repeated attacks of malaria having a greater adverse impact on school performance (Fernando et al., 2003a). Similar findings have been reported from Brazil (Vitor-Silva et al., 2009). In view of the relapsing nature of *P. vivax* and its propensity to occur in children, the long-term social implications on education and development could be substantial.

4. CLINICAL ILLNESS REQUIRING HOSPITALISATION

Whilst most infections are uncomplicated, this does not imply that *P. vivax* is benign (see Price et al., 2007b, 2009; Baird, 2009; Bassat and Alonso, 2011 for reviews). In vivax-endemic countries, infection with *P. vivax* can account for up to 40% of patients hospitalized with malaria, including 17% (34/204) of all hospital malaria admissions in Vanuatu (Maitland et al., 1997), 38% (382/1000) in Manila (Buck et al., 1983), 32% (237/740) in northeast India (Gopinathan and Subramanian, 1986) and 24% (1002/4140)–31% (799/2537) in western Thailand (Carrara et al., 2006). In Papua, Indonesia, *P. vivax* accounted for 24% of malaria admissions in all age groups, but 47% (415/887) are children under 1 year of age (Tjitra et al., 2008). In Papua, the risk of illness requiring hospitalisation has been estimated at 1 in 84 infections with *P. vivax* (Douglas et al., unpublished). The need for hospitalization indicates at least moderately severe disease and significant morbidity. This includes patients with vomiting, who cannot tolerate oral therapy, through to prostration and those with disease manifestations fulfilling the severity criteria used to define severe falciparum malaria (World Health Organization, 2010) as outlined below.

5. SEVERE MALARIA

Plasmodium vivax was recognized in the malariotherapy era as being associated with significant risk of death, with 5–10% of patients with neurosyphilis dying during *P. vivax* therapy (Swellengrebel and de Buck, 1938; Fong, 1937; O'Leary and Welsh, 1933; James et al., 1936; Paulian, 1935; Nicol, 1932; Eldridge et al., 1925; Freeman et al., 1934) and up to 10–14% with the Madagascar strain (James, 1933) (reviewed in detail in Chapter 5). Early descriptions of severe disease in naturally acquired vivax malaria were confined to case reports and small case series (see Price et al., 2007a; Tan et al., 2008 for reviews), but more recently, larger series have clearly associated *P. vivax* infections with severe and fatal disease in both children and adults (Barcus et al., 2007; Rodriguez-Morales et al., 2008; Tjitra et al., 2008; Kochar et al., 2005, 2009, 2010; Andrade et al., 2010; Alexandre et al., 2010; Yadav et al., 2012; Lacerda et al., 2012; Nurleila et al., 2012; Manning et al., 2011; Lanca et al., 2012; Kaushik et al., 2012; Sharma et al., 2012; Naha et al., 2012). These series highlight substantial morbidity, particularly severe anaemia

(Price et al., 2009; Tjitra et al., 2008; Nurleila et al., 2012; Manning et al., 2011; Poespoprodjo et al., 2009; Genton et al., 2008) and low birth weight (Nosten et al., 1999; Poespoprodjo et al., 2008; Rijken et al., 2012a).

Severe vivax malaria has been reported in case series from Indonesia (Barcus et al., 2007; Tjitra et al., 2008; Nurleila et al., 2012; Lampah et al., 2011), Papua New Guinea (Manning et al., 2011; Genton et al., 2008), India (Kochar et al., 2005, 2009, 2010; Yadav et al., 2012; Kaushik et al., 2012; Sharma et al., 2012; Naha et al., 2012; Tanwar et al., 2011; Kapoor and Gupta, 2012; Jat et al., 2012), Thailand (Luxemburger et al., 1997), Venezuela (Rodriguez-Morales et al., 2008), Brazil (Andrade et al., 2010b; Alexandre et al., 2010; Lacerda et al., 2012; Lanca et al., 2012), Malaysia (Barber et al., 2012) and Sudan (Mahgoub et al., 2012), with most, but not all, series including fatal cases. These case series include a variety of severe manifestations associated with *P. vivax* infection, including severe anaemia, respiratory distress and acute lung injury (ALI), acute kidney injury (AKI), splenic rupture, jaundice, coma, multiorgan dysfunction and shock. Each of these syndromes is reviewed separately below.

Most recent series report PCR-exclusion of mixed infection with *P. falciparum*, however, in common with most series of falciparum malaria, investigation for concurrent infections or co-morbidities has been mostly incomplete or omitted. The most convincing evidence for *P. vivax* being a major contributory factor is apparent for severe anaemia (Price et al., 2009), acute respiratory distress syndrome (ARDS) (Tan et al., 2008; Valecha et al., 2009) and, increasingly, AKI (Chung et al., 2008; Kute et al., 2012; Sinha et al., 2012). For some of the other reported severe syndromes, the extent to which they are attributable to *P. vivax* is not yet clear. For these syndromes, acute and chronic infectious and non-infectious co-morbidities are likely to be important contributing factors or alternative causes (Price et al., 2009; Kitchen, 1949b; Lacerda et al., 2012; Lampah et al., 2011; Anstey et al., 2009; Rajahram et al., 2012). However, the presence of co-morbidities does not necessarily negate a crucial role that *P. vivax* may play in the pathophysiology of the disease. In *P. falciparum* infection, malnutrition, HIV and invasive bacterial infections are now recognized as being biologically associated with severe disease due to falciparum malaria rather than just being alternative diagnoses in co-incidentally parasitized patients in falciparum-endemic areas (Berkley et al., 2009). Similar relationships, including other co-morbidities, may apply with *P. vivax* but require more systematic study.

6. RISK OF SEVERE DISEASE AND DEATH

Plasmodium vivax infection in healthy travellers from non-endemic countries and healthy residents of low-endemicity temperate regions of Asia rarely causes severe disease (see Price et al., 2009 for review). These populations are predominantly adults without co-morbidities, with ready access to early diagnosis and effective treatment, and usually associated with infrequent recurrence due to low risk of reinfection and better anti-relapse therapy (Price et al., 2009). Deaths in travellers have been reported, but are generally associated with co-morbidities which may have contributed substantially to pathology (Stoppacher and Adams, 2003). In other endemic areas, the risk of severe disease rises with increasing transmission intensity of early and frequent relapse. Such regions often contain poorly resourced communities with less access to early diagnosis and treatment and greater prevalence of co-morbidities such as malnutrition and co-infections (Price et al., 2009).

Few studies have identified population-based risk of severe disease from *P. vivax* in endemic settings. In one series from Indonesian Papua, the risk of severe disease and death associated with *P. vivax* was estimated at one in 270 and one in 3959 clinical infections, respectively (Tjitra et al., 2008). This compared to estimates for *P. falciparum* of one in 185 and one in 1742, respectively (Tjitra et al., 2008). In Indonesian Papua, the hospital mortality rates in those hospitalised with *P. vivax* confirmed by microscopy were reported as 0.8–1.6%, similar to that of *P. falciparum* infection (1.6–2.2%) (Price et al., 2009; Barcus et al., 2007; Tjitra et al., 2008). Similarly, case-fatality rates in Indian children hospitalised with PCR-confirmed *P. vivax* monoinfection (3.9%) were comparable to that seen in PCR-confirmed *P. falciparum* infection (3.2%) (Kochar et al., 2010). A higher mortality (8.3%) was observed in a Brazilian study of 24 children admitted to an intensive care unit with severe disease associated with microscopy-diagnosed *P. vivax* (Lanca et al., 2012). In other settings, case-fatality rates are lower, and in children hospitalised in Thailand, were lower with *P. vivax* infection (0.22%) than with *P. falciparum* (1.2%) (Wattanagoon et al., 1994). The risk of death varies significantly between syndromes. For infants hospitalised with severe anaemia without respiratory distress, mortality was 0.4%, but increased to 10.2% in whom respiratory distress was also present (Tjitra et al., 2008).

Over the past few years, increasing reports of severe vivax has raised awareness of the potential adverse consequences of *P. vivax*, however, almost all have been limited by incomplete systematic investigation of concurrent infectious and non-infectious co-morbidities that may be either alternative causes of death or the major underlying cause of death. It is notable that in the 2012 Manaus autopsy series, at least four of 17 (23.5%) vivax-associated deaths were attributable to alternative causes (Lacerda et al., 2012), including yellow fever and meningitis (Lacerda et al., 2012). This proportion is identical to the 23% of *P. falciparum*-associated deaths in Malawi attributable to alternative non-falciparum causes (Taylor et al., 2004). A further 41% of the Manaus vivax malaria autopsy series had major underlying co-morbidities contributing to death (Lacerda et al., 2012) (Section 7.4). This has long been recognized in the malaria literature. In 1925, James noted that in southern England, untreated vivax malaria in those who were 'enfeebled from any cause, not infrequently results fatally' (James, 1925; Hutchinson and Lindsay, 2006). Attributable fractions for severe and fatal disease have been calculated for falciparum malaria (Bejon et al., 2007) but not yet for vivax malaria (Anstey and Price, 2007).

7. VULNERABLE GROUPS

7.1. Young Children

In co-endemic areas, morbidity from *P. vivax*, including both uncomplicated and severe disease, generally occurs at a younger age compared to *P. falciparum* (Tjitra et al., 2008; Kochar et al., 2010; Michon et al., 2007; Lin et al., 2010). Children under 5 years of age are at greatest risk of anaemia from *P. vivax*, especially severe vivax anaemia (Tjitra et al., 2008; Alexandre et al., 2010; Kochar et al., 2010; Lanca et al., 2012; Poespoprodjo et al., 2009; Genton et al., 2008; Douglas et al., 2012). The risk in the first year of life is particularly striking. In Indonesian Papua, more infants under 1 year of age are hospitalised with vivax malaria than falciparum malaria (Tjitra et al., 2008; Poespoprodjo et al., 2009); these patients having a significantly greater risk of severe anaemia than those hospitalised with falciparum malaria (odds ratio 2.4) (Tjitra et al., 2008). The risk to infants starts *in utero*, *P. vivax* infection during pregnancy being associated with congenital malaria (Poespoprodjo et al., 2011) and a greater risk of clinical disease and severe anaemia in the neonatal period (Lanca et al., 2012; Poespoprodjo et al., 2009).

7.2. Malnutrition

Malnutrition is a clear risk factor for both severe disease and mortality in falciparum malaria (Berkley et al., 2009) and is also likely to contribute to pathology and disease severity in *P. vivax* infections. In a large Indian series, malnutrition was present in 69% of severe vivax cases in children and 75% of deaths associated with severe vivax malaria (Kochar et al., 2010). In Brazil, 17% of children with *P. vivax* requiring intensive care admission had malnutrition (Lanca et al., 2012).

7.3. HIV Infection

HIV is a risk factor for both severe disease and death in children with falciparum malaria (Berkley et al., 2009; Bejon et al., 2007), but whether this also applies to vivax malaria is unclear. HIV prevalence is increasing in PNG and Indonesian Papua (Pontororing et al., 2010), where two large series of patients with severe vivax were reported (Tjitra et al., 2008; Genton et al., 2008) although neither study reported co-infection rates. In the Manaus autopsy series, one of the 17 vivax-associated deaths occurred in a patient with HIV infection listed as a contributing co-morbidity (Lacerda et al., 2012). Prospective studies of the role of HIV infection in increasing risk of severe disease are required, but in the interim, clinicians should consider testing for HIV infection in those with severe vivax malaria.

7.4. Other Co-morbidities

A range of acute and chronic infectious and non-infectious co-morbidities has also been proposed as contributory factors for severe and fatal disease in *P. vivax* infection (see Price et al., 2009; Lacerda et al., 2012; Anstey et al., 2009 for review). The haemodynamic effects of the systemic inflammatory response to *P. vivax* and anaemia are likely to contribute to decompensation of concurrent acute and chronic disease with potentially fatal consequences, whereas otherwise healthy children and adults may have made an uncomplicated recovery (Douglas et al., 2012). Concomitant pneumonia is particularly important since vivax-associated anaemia will exacerbate hypoxia and impair recovery.

Over 60 years ago, Kitchen described his experience of vivax malariotherapy in neurosyphilis, remarking on the importance of underlying co-morbidities in severe disease in adults: 'Our own experience suggests that serious trends which develop during vivax infections ought properly be classified as complications in practically all instances; events of this sort that we have observed would appear to have been based on primary deficiencies on the part of the patient, or due to intercurrent conditions' (Kitchen, 1949b). In the Manaus autopsy series, 76% of *P. vivax*-associated deaths had concurrent acute and chronic co-morbidities, including pneumonia, HIV, decompensated cirrhosis, emphysema, diabetes, haemorrhagic stroke and congestive heart failure (Lacerda et al., 2012). Similarly, a USA series reported two malaria deaths due to *P. vivax*, both of which were in patients with pre-existing cardiac disease (Stoppacher and Adams, 2003). Among the 24 Brazilian children with severe *P. vivax* infection requiring intensive care admission, 25% had concomitant acute gastroenteritis, and in total, 58% had a concurrent acute or chronic co-morbidity potentially contributing to severe illness (Lanca et al., 2012). All the co-morbidities above are common in malaria-endemic regions, and it is likely that their importance in contributing to severe and fatal outcomes in *P. vivax* infections is underestimated (see Price et al., 2009; Anstey et al., 2009 for review).

8. SPECIFIC CLINICAL MANIFESTATIONS OF SEVERE VIVAX MALARIA

8.1. Severe Malaria in Adults

A broad range of severe vivax syndromes has been described in adult series similar to that described in adult falciparum malaria. This spectrum could reflect the application of criteria for severity, which were developed explicitly for defining patients with *P. falciparum* infection at risk of dying (World Health Organization, 2010). However, two recent series from Brazil suggest that the criteria in current WHO guidelines for severe falciparum malaria are sensitive in identifying children requiring intensive care admission (Lanca et al., 2012) and identifying patients at risk of death (Lacerda et al., 2012). The latter report suggested adding splenic rupture as a severe malaria criterion, but it was not clear from this study whether antemortem application of the current severe malaria criterion of shock would have already identified patients subsequently dying of splenic rupture.

8.1.1. Severe anaemia

There is a continuum of anaemia associated with *P. vivax* malaria ranging from minor to profound. Severe anaemia secondary to *P. vivax* infection is typically defined as a haemoglobin concentration of <5 g/dl (World Health Organization, 2010) though forthcoming WHO guidelines will define a cut off of 7 g/dl in adults. The degree of anaemia in an individual is influenced by a wide range of factors including (but not limited to) premorbid haemoglobin concentration, level of immunity, duration of infection before treatment, the number of prior episodes of malaria, as well as the presence of comorbid conditions. Collins et al. (2003) analysed data from adult neurosyphilis patients with primary sporozoite-induced *P. vivax* infections. They showed that haemoglobin fell rapidly during the first 4–5 weeks of untreated infection and subsequently climbed gradually as immunity was acquired. Premorbid haemoglobin concentrations had still not been reached by week 11 of infection (Collins et al., 2003).

Anaemia associated with *P. vivax* infection is amongst the commonest manifestations of severe vivax disease and is more strongly skewed towards young children than the anaemia caused by *P. falciparum* infection (Tjitra et al., 2008; Poespoprodjo et al., 2009; Michon et al., 2007; Lin et al., 2010; Ladeia-Andrade et al., 2009; Douglas et al., 2012). Even in adults, severe anaemia can be a common manifestation of severe disease as highlighted in two series from Papua, Indonesia. In Timika, southern Papua, 10% of all adults hospitalised with *P. vivax* infections between 2005 and 2007 had severe anaemia (haemoglobin <5 g/dl) (Tjitra et al., 2008). In Jayapura, northern Papua, 6% of adults hospitalised with *P. vivax* malaria had a haemoglobin concentration of <6 g/dl (Barcus et al., 2007). In other series from India and Brazil, severe anaemia accounted for almost one-third of all cases of severe adult vivax malaria (Kochar et al., 2009; Andrade et al., 2010). Conversely, in a hypoendemic region of north western Thailand, none of the 1978 adults and children of age >5 year diagnosed with acute vivax malaria over a 3-year period required hospitalisation for severe anaemia or blood transfusion, though haemoglobin was not routinely quantitated in the community (Luxemburger et al., 1997).

Post-pubescent females are at greater risk than young males of being admitted to hospital with *P. vivax* malaria in Southern Papua (Tjitra et al., 2008) and in one large analysis from the same region, had more severe anaemia than males (unpublished data). This is likely to relate at least partially to lower premorbid haemoglobin concentrations. In addition, pregnant women with vivax malaria have at least a twofold higher risk of moderate anaemia than pregnant women without vivax infection (Section 8.3), a risk that has been demonstrated in several different geographical locations (Nosten et al., 1999; Poespoprodjo et al., 2008; Rodriguez-Morales et al., 2006).

As mentioned above, the anaemia of vivax malaria may be modulated by myriad co-morbidities including gastrointestinal helminth infections and malnutrition (Section 8.2.1), haemoglobin and red cell abnormalities, chronic inflammatory or myelosuppressive diseases and acute bacterial infections (Douglas et al., 2012). Population-based studies of the haematological morbidity associated with vivax malaria have yet to address adequately all of these co-factors.

8.1.2. Acute lung injury and respiratory distress

There have been over 20 reports of well-defined ALI and ARDS as a single complication of PCR-confirmed *P. vivax*. Initial reports were mostly managed in non-endemic country hospitals, with a very low case-fatality rate and with the majority occurring after commencement of antimalarial drug treatment (see Price et al., 2007a; Tan et al., 2008; Habib and Singh, 2004; Lawn et al., 2003; Pukrittayakamee et al., 1998; Torres et al., 1997; Kotwal et al., 2005; Maguire et al., 2005; Curlin et al., 1999; Carlini et al., 1999; Munteis et al., 1997; Saleri et al., 2006; Tanios et al., 2001; Kumar et al., 2007; Agarwal et al., 2007; Taylor et al., 2012 for review). One report from India of a 20-year-old woman with PCR- and autopsy-confirmed vivax malaria documents presentation with ARDS pre-treatment, resulting in a fatal outcome (Valecha et al., 2009).

More recently reported cases of ALI/ARDS in adults have mostly been part of larger series of severe vivax malaria from endemic settings, generally with less-clearly defined ARDS criteria, occurring as part of multiple organ dysfunction/failure (Kochar et al., 2005, 2009; Andrade et al., 2010; Alexandre et al., 2010). Clinical ARDS/'respiratory failure' occurred in 10–32% of patients with severe malaria and had a case-fatality rate of 50–67% (Kochar et al., 2005, 2009; Andrade et al., 2010; Alexandre et al., 2010), much higher than the very low fatality rate seen in reports of single-organ ARDS (Taylor et al., 2012).

In the Brazilian autopsy series, 15 of 17 (88%) had respiratory distress before death, again mostly commencing after the start of antimalarial drug treatment (Lacerda et al., 2012). ARDS and/or lung oedema was identified as the commonest complication contributing to death, occurring in six overall (35%) and in three of the four (75%), in whom vivax was considered the direct cause of death (Lacerda et al., 2012). In two of the latter cases, lung oedema occurred in conjunction with other major pathology (one each with coma/encephalitis and splenic rupture) (Lacerda et al., 2012).

8.1.3. Coma and other vivax-associated neurological complications

Coma is a common complication of *P. falciparum* infection comprising approximately 50% of adult severe falciparum malaria complications (Yeo et al., 2007; Dondorp et al., 2008; Hien et al., 1996). In contrast, coma is a less-common manifestation of vivax malaria. From 1921 to 2011, over 100 cases of *P. vivax* monoinfection in adults and children associated with coma were described in reports or series (Kochar et al., 2005, 2009, 2010; Lampah et al., 2011; Tanwar et al., 2011; Romanenko,

1936; Dhayagude and Purandare, 1943; Tarejev et al., 1944; Boshes, 1947; Hill et al., 1963; Gopinathan and Subramaniam, 1982; Sachdev and Man, 1985; Arora et al., 1988; Valecha et al., 1992; Tilluckdharry et al., 1996; Beg et al., 2002; Ozen et al., 2006; Thapa et al., 2007, 2009; Harish and Gupta, 2009; Sarkar and Bhattacharya, 2008; Kasliwal et al., 2009; Parakh et al., 2009). Only studies in the past 15 years have been able to exclude mixed species infection with *P. falciparum* by PCR methods (Kochar et al., 2005, 2009, 2010; Lampah et al., 2011; Tanwar et al., 2011; Beg et al., 2002; Thapa et al., 2007, 2009; Harish and Gupta, 2009; Sarkar and Bhattacharya, 2008; Kasliwal et al., 2009; Parakh et al., 2009). In Indonesian Papua, coma associated with PCR-confirmed *P. vivax* monoinfection (and without overt co-morbidities) occurred 23 times less frequently than that seen with falciparum malaria and was estimated as occurring in one in 29,500 infections (Lampah et al., 2011). In Thailand, the risk of hospitalisation with impaired consciousness with microscopy-diagnosed *P. vivax* (not-PCR-confirmed) was one in 858 infections, with the risk being 15.2-fold less than that with *P. falciparum* (Luxemburger et al., 1997). Most other reports of coma in association with PCR-confirmed *P. vivax* monoinfection have been from the Indian sub-continent (Kochar et al., 2005, 2009, 2010; Tanwar et al., 2011; Beg et al., 2002; Thapa et al., 2007, 2009; Harish and Gupta, 2009; Sarkar and Bhattacharya, 2008; Kasliwal et al., 2009; Parakh et al., 2009); however, denominators of the surveillance and population-based risks were not reported. Only one case with no identified alternative aetiology has been reported from South America, a PCR-confirmed *P. vivax* monoinfection (Lacerda et al., 2012).

No series, including those with PCR-confirmed *P. vivax* monoinfection, has reported complete, systematic microbiological and radiological investigations to exclude bacterial and viral infection and other causes of coma. Such studies are needed. In falciparum malaria, 23% of *P. falciparum* infection meeting the WHO clinical case definition for cerebral malaria had an alternative cause of coma at autopsy (Taylor et al., 2004). In a prospective series of 24 cases of coma associated with *P. vivax* infection diagnosed by microscopy, 75% had PCR evidence of *P. falciparum* (co)infection or other bacterial or non-infective causes of coma; almost all PCR-confirmed cases of coma in *P. vivax* monoinfection in this series were in young adults with low parasitaemia, no other organ dysfunction and good outcomes (Lampah et al., 2011). *Plasmodium vivax*-associated coma has been associated with thrombotic thrombocytopenic purpura in some (Sinha et al., 2012) but not other (Lampah et al., 2011) series. Other adult series with *P. vivax* monoinfection have described coma in association with multiorgan dysfunction (Kochar et al., 2005, 2009). A post-malarial neurological syndrome with tremor and myoclonus has also been reported after recovery from coma associated with PCR-confirmed *P. vivax* monoinfection (Lampah et al., 2011), similar to that occasionally seen following coma in falciparum malaria (Nguyen et al., 1996).

Other, rarer, neurological complications reported in association with *P. vivax* infection include facial diplegia, both during (Sim et al., 2010) and following (Kochar et al., 2007) recovery from febrile illness, acute inflammatory polyneuropathy (Chakravarty et al., 2004), acute disseminated encephalomyelitis (Koibuchi et al., 2003) and anterior ischemic optic neuropathy (Flower et al., 2011). Retinopathy is common in severe falciparum malaria, particularly in cerebral malaria (Beare et al., 2006). While not seen in a series of uncomplicated adult vivax malaria in Bangladesh (Abu Sayeed et al., 2011), cases of retinal haemorrhages in pure vivax malaria have been described elsewhere without central nervous system complications (Lee et al., 2010; Choi et al., 2004). A further case of retinal haemorrhages and coma attributed to *P. vivax* had a mixed infection with *P. falciparum*, which may have been responsible for the complications reported (Biswas et al., 1996).

8.1.4. Acute kidney injury

AKI has now been reported in a large number of series of adult vivax malaria (Kochar et al., 2005, 2009; Andrade et al., 2010; Alexandre et al., 2010; Chung et al., 2008; Kute et al., 2012; Sinha et al., 2012). There appear to be geographic differences in risk and severity, with little or no vivax-associated AKI being reported in returned travellers (Tan et al., 2008) or from some vivax-endemic areas, such as Thailand (Luxemburger et al., 1997; Piyaphanee et al., 2007) and Vietnam (Hien et al., 1996). AKI is reported but rarely severe in Korea (Chung et al., 2008). Across northern India, severe dialysis-requiring AKI and/or AKI-related death is increasingly reported, including Rajasthan (Kochar et al., 2005, 2009), Gujarat (Kute et al., 2012a, 2012b), Delhi (Sinha et al., 2012) and Uttar Pradesh (Prakash et al., 2003), with only the Rajasthan series being PCR-confirmed *P. vivax* monoinfection. In Gujarat, the proportion of all malaria-associated AKI due to microscopy-diagnosed *P. vivax* is reported to have risen from 2% (Kanodia et al., 2010) to 12% (Kute et al., 2012b) by 2010–2011. AKI in PCR-confirmed monoinfection has also been reported from Brazil (Andrade et al., 2010b; Alexandre et al., 2010; Lacerda et al., 2012). No population-based data on risk of AKI has been reported from any region.

In most series, AKI is commonly associated with multiorgan dysfunction (Kochar et al., 2005, 2009; Andrade et al., 2010; Kute et al., 2012b), and is a risk factor for fatal outcome (Kute et al., 2012b). Shock is also a common association (Kochar et al., 2005, 2009; Andrade et al., 2010; Alexandre et al., 2010; Kute et al., 2012b). Renal biopsies in four patients with vivax-associated AKI reported patchy cortical necrosis in three cases and acute tubular necrosis in the other (Kute et al., 2012b). The presentations reported frequently mimicked AKI seen with sepsis and/or shock, but since investigation for bacterial sepsis was not universal, the role of concomitant bacterial sepsis as a contributor to these syndromes cannot be excluded. Furthermore, in the largest autopsy series of deaths associated with *P. vivax* infection, all six cases with AKI were associated with pre-existing co-morbidities predisposing to AKI or multiorgan dysfunction (heart failure, sickle cell haemolytic crisis and chronic liver disease) or alternative aetiologies (primaquine-triggered haemolysis or yellow fever) (Lacerda et al., 2012).

Other *P. vivax* series have reported AKI associated with thrombotic microangiopathy (Sinha et al., 2012; Saharan et al., 2008) in both adults and children, and haemolytic uraemic syndrome (HUS) in children (Sharma et al., 1993). In the largest series to date (4 adults and 5 children), persistent vivax-associated AKI was associated with thrombocytopenia (range 8,000–82,000/μl) and anaemia (Hb 4.4–9.6), with schistocytes on peripheral film in 77% (Sinha et al., 2012). Bleeding manifestations (77%) and splenomegaly (44%) were common, with one of the patients with schistocytes having concomitant coma (Sinha et al., 2012). Biopsies showed universal ischaemia and endothelial injury and arteriolar thrombi in two cases, consistent with thrombotic microangiopathy (Sinha et al., 2012). Fatal AKI associated with crescentic glomerulonephritis has also been reported in *P. vivax* infection (Patel et al., 2012).

8.1.5. Shock and multiorgan dysfunction

Shock has been associated with adult *P. vivax* infection in reports and series from India, Brazil, Malaysia and Korea (Kochar et al., 2005, 2009; Alexandre et al., 2010; Song et al., 2007; Barber et al., 2012) usually, but not always (Barber et al., 2012), as part of multiple organ dysfunction (Kochar et al., 2005, 2009; Andrade et al., 2010; Alexandre et al., 2010). These clinical syndromes are also seen with severe bacterial sepsis, raising the possibility of bacterial co-infection in these patients. The aforementioned cases and series did not describe universal, systematic surveillance with culture of blood and other fluids.

8.1.6. Bacterial co-infection and bacteraemia

Gram-negative bacteraemia is a common complication in severe falciparum malaria, particularly in children (Berkley et al., 1999; Scott et al., 2011), and is a risk factor for death (Berkley et al., 1999). However, bacteraemia risk in *P. vivax* infection has not been studied systematically. A large prospective blood culture and malaria microscopy study in Kolkata, India showed that 2 of 90 (2.2%) cases of malaria (97% of which were *P. vivax*) had *Salmonella* bacteraemia, one *Salmonella typhi* and one non-typhoidal *Salmonella* (Sur et al., 2006). Of the two cases of *Salmonella* bacteraemia reported from Thailand in association with PCR-confirmed *P. vivax* monoinfection, one patient had multiorgan dysfunction (renal failure, jaundice and hypotension) meeting 2010 criteria for severe malaria (World Health Organization, 2010), with concurrent serogroup D *Salmonella* bacteraemia (Piyaphanee et al., 2007). The other otherwise-uncomplicated *P. vivax* case had non-typhoidal *Salmonella* bacteraemia (Piyaphanee et al., 2007). In Malaysia, five of 43 patients hospitalised with vivax malaria had hypotension and one had *Streptococcus pneumoniae* bacteremia (Barber et al., 2012).

8.1.7. Splenic rupture and infarction

Splenic rupture is a well-recognized, life-threatening but probably under-reported complication of *P. vivax* infection in adults (Lacerda et al., 2012; Kitchen, 1949a; Lubitz, 1949; Moses, 1945; Kapland et al., 1946; Imbert et al., 2009). In a systematic review of 55 cases of malaria-associated splenic rupture reported over a 50-year period (1958–2008), the mortality rate was 22%, with *P. vivax* accounting for approximately half of all cases (Imbert et al., 2009). In the 2012 Brazilian autopsy series of 17 patients with deaths associated with *P. vivax*, three patients (18%) had splenic rupture, all adults, two of whom also had lung oedema (Lacerda et al., 2012). Of the four deaths considered directly caused by *P. vivax*, splenic rupture was present in two (Lacerda et al., 2012).

Splenic infarction is thought to be rare in vivax malaria, though it is also likely to be under-recognised and under-reported (Gupta et al., 2010; Imbert et al., 2010). Splenic infarction is not commonly associated with splenic rupture, being found on histopathology in <2% of cases of splenic rupture (Imbert et al., 2009). Although not necessarily causally linked to symptoms, splenic infarction was identified in 38% of 34 Korean vivax patients undergoing CT scanning for abdominal pain (Kim et al., 2010). Splenic rupture or subcapsular haematoma was identified in a further 3 (9%) of these cases (Kim et al., 2010).

8.1.8. Thrombocytopenia

Although not a defining feature of severe malaria, vivax-associated thrombocytopenia occurs in 24–94% of patients with vivax malaria (see Tan et al., 2008; Lacerda et al., 2011 for reviews). It has been associated with other severe manifestations (Kochar et al., 2005, 2009; Andrade et al., 2010; Alexandre et al., 2010). In one series, severe epistaxis associated with thrombocytopenia requiring blood and platelet transfusions was reported in 5% (2/40) of adults with severe vivax malaria (Kochar et al., 2009).

8.1.9. Other complications

Jaundice is common, occurring in 36–57% of adults in severe vivax series, and has been associated with multiorgan dysfunction (Kochar et al., 2005, 2009; Andrade et al., 2010; Alexandre et al., 2010). Acalculous cholecystitis has been described in adults (Curley et al., 2011) with gall bladder wall oedema and periportal oedema described in 32% and 35%, respectively, of 34 vivax malaria patients undergoing CT scanning for abdominal pain (Kim et al., 2010). Series from the 1940s have reported patients with acute lower abdominal pain associated with vivax malaria, but not with splenomegaly, mimicking appendicitis and other acute surgical conditions (Kitchen, 1949b; Most and Hayman, 1946). Another rarely reported complication is rhabdomyolysis (Siqueira et al., 2010).

8.2. Severe Malaria in Children

Most of the severe manifestations seen in children with *P. falciparum* malaria have also been reported in children with *P. vivax*.

8.2.1. Severe anaemia

In regions with high transmission of *P. falciparum*, severe anaemia is the most common manifestation of severe falciparum malaria in children, with falciparum-associated severe anaemia being rare beyond childhood (Bloland et al., 1999). In Kilifi, Kenya, severe anaemia was the most common manifestation of severe falciparum malaria in children admitted to hospital between 1989 and 1991 (27.5%) and was associated with a 4.7% case fatality (Marsh et al., 1995). Similarly, in paediatric vivax malaria, the major severe manifestation in most series is severe anaemia (Rodriguez-Morales et al., 2008; Tjitra et al., 2008; Alexandre et al., 2010; Kochar et al., 2010; Poespoprodjo et al., 2009; Luxemburger et al., 1997). However, none of these reports describe the confounding effects of other co-factors. They are also almost all cross-sectional surveys, and thus are unable to quantify the decrement in haemoglobin with successive episodes of malaria caused by reinfection and relapses. Despite these limitations, the association between vivax infection and severe anaemia is strong, particularly in infancy (Tjitra et al., 2008; Poespoprodjo et al., 2009; Genton et al., 2008).

Severe anaemia associated with *P. vivax* infection tends to be even more strongly skewed to childhood than *P. falciparum*, and in regions with comparatively high endemicity, the first year of life in particular (Poespoprodjo et al., 2009). In Papua, Indonesia, 36% of all infants under the age of 3 months, who were admitted to hospital with vivax malaria, had severe anaemia (haemoglobin <5 g/dl) compared with 18% of infants with *P. falciparum* malaria (Poespoprodjo et al., 2009). Across the border in Papua, New Guinea, 1.6% of all children with vivax malaria presenting to community clinics had severe anaemia (Genton et al., 2008). At a hospital in Delhi, India, 20% (7/35) of children with vivax malaria diagnosed in the paediatric outpatient or emergency department had severe anaemia (Kaushik et al., 2012) and in Manaus, Brazil, 9% of children of 0–14-years old were admitted to an intensive care unit with vivax malaria fulfilled criteria for severe anaemia (Lanca et al., 2012). None of the seven children with severe anaemia in the latter study subsequently died.

Cross-sectional data show that the apparent severity of haematological morbidity associated with *P. vivax* in childhood is not simply due to selection bias. In the D'Entrecasteaux Islands, children between 0 and 6 years of age with vivax malaria had a significantly lower mean haemoglobin concentration (8.7 g/dl) than children with *P. falciparum* (9.0 g/dl) (Spencer, 1966).

One speculative reason for the significant haematological morbidity associated with vivax malaria in early childhood is that there is a greater pool of young red blood cells (RBCs) in circulation and thus more potential host cells susceptible to invasion (Douglas et al., 2012). A further potential explanation is that severe vivax-associated anaemia in childhood is a consequence of multiple additive haematological insults from frequent relapses. However, severe anaemia is commonly seen in children <2 months of age in Papua, Indonesia (Poespoprodjo et al., 2009). Multiple relapses cannot explain severe anaemia at such an early age. Sustained parasitaemia, initially subclinical, from congenital malaria may be an alternative explanation in this age group.

Infection with gastrointestinal helminths can cause anaemia through chronic blood loss. There is good evidence that mixed infection with hookworm and falciparum malaria causes an additive reduction in haemoglobin in preschool children (Brooker et al., 2007). To our knowledge, only one study has assessed the impact of co-infection with gastrointestinal helminths on the anaemia of vivax malaria in children. Melo et al. (2010) showed that in the Western Brazilian Amazon, co-infection with hookworm, *Ascaris lumbricoides* and *Trichuris trichiura*, actually attenuated the reduction in haemoglobin associated with vivax malaria.

8.2.2. Respiratory distress

Respiratory distress is a common severe manifestation in most series of severe paediatric vivax malaria (Tjitra et al., 2008; Kochar et al., 2010; Yadav et al., 2012; Genton et al., 2008). As in adults, the definitions used for respiratory distress vary considerably between series (Taylor et al., 2012), complicating comparison. Respiratory distress occurred more frequently in vivax malaria (60%) than in falciparum malaria (41%) (Genton et al., 2008) in Papua, New Guinean children. These rates were substantially higher than those reported in hospitalised children in Papua, Indonesia, a reflection of the broader inclusion criteria in PNG. In the Papua study, rates were similar between *P. vivax* (2.3%) and *P. falciparum* (2.5%) (Tjitra et al., 2008). In a study from Rajasthan, India, respiratory distress was more frequent in children with falciparum malaria (18%) than those with vivax malaria (11%), however, this ratio was reversed in young children of age <5 years (2% and 15% respectively) (Kochar et al., 2010).

Both hypoxemia and metabolic acidosis have been described in children with vivax-associated respiratory distress (Kochar et al., 2010). Metabolic acidosis (Marsh et al., 1995; Miller et al., 2002), community-acquired pneumonia (O'Dempsey et al., 1993), aspiration pneumonia, sepsis and severe anaemia are important factors in respiratory distress in children with falciparum malaria (Taylor et al., 2012), but their relative contribution to the respiratory distress associated with *P. vivax* is unclear. Two of 24 children in a Delhi series of severe vivax malaria had respiratory distress and two were reported to have lobar pneumonia, though whether these were the same patients was not specified (Kaushik et al., 2012). Of 24 children admitted to an intensive care unit in Brazil with severe disease associated with *P. vivax* infection, respiratory distress was the commonest complication (n = 16; 67%). Three of these (12.5% overall) children were reported to have strictly defined ARDS (Lanca et al., 2012); none died. One of the patients with respiratory distress (6%) had pneumonia and empyema identified as the cause of respiratory distress (Lanca et al., 2012). One of the 65 cases of PCR-confirmed severe vivax cases from Rajasthan had pulmonary oedema in association with multiorgan dysfunction (Kochar et al., 2010).

8.2.3. Coma

PCR-confirmed *P. vivax* monoinfection has also been associated with coma in children (Kochar et al., 2010; Lacerda et al., 2012; Lampah et al., 2011; Tanwar et al., 2011; Thapa et al., 2007, 2009; Harish and Gupta, 2009; Parakh et al., 2009), however, exclusion of alternative aetiologies was not possible. Two Brazilian children with vivax-associated coma underwent autopsy, both 8-year-old females reported in two separate series (Lacerda et al., 2012; Lanca et al., 2012). In the first, the 2012 Manaus autopsy series, one child with an inflammatory cerebrospinal fluid had no alternative cause evident at autopsy but encephalitis was not excluded (Lacerda et al., 2012). The other child was admitted to intensive care in Manaus with microscopic diagnosis of *P. vivax* with coma (Lanca et al., 2012). The outcome was fatal, however, autopsy revealed an underlying aetiology of viral encephalitis.

8.2.4. Renal dysfunction and acute kidney injury

AKI is less common and less severe among children with severe falciparum malaria than in adults (Dondorp et al., 2008; Hien et al., 1996; Anstey et al., 1996; Weber et al., 1999). AKI, with varying definitions, has been reported increasingly as a manifestation of severe vivax malaria in children (Kochar et al., 2010; Yadav et al., 2012; Manning et al., 2011; Kaushik et al., 2012; Jat et al., 2012) occurring in 3–15% of severe cases in these series, usually as part of multiorgan dysfunction. In paediatric series including both severe falciparum malaria and severe vivax malaria, renal impairment occurred with a lower (Kochar et al., 2010) or similar (Yadav et al., 2012; Manning et al., 2011) frequency in those with *P. vivax* infection. Mortality rates of up to 30% have been reported in children with PCR-confirmed *P. vivax* monoinfection-associated AKI, all of whom had associated multiorgan dysfunction (Kochar et al., 2010).

The majority (77%) of reports of AKI secondary to *P. vivax*-associated thrombotic microangiopathy (Sinha et al., 2012; Saharan et al., 2008) and HUS (Sharma et al., 1993) have occurred in children. Nephrotic syndrome, more commonly associated with mesangioproliferative and membranoproliferative glomerulonephritis from *Plasmodium malariae* (Gilles and Hendrickse, 1963, 1960; van Velthuysen and Florquin, 2000), has also been more recently described in association with *P. vivax* in children (Bircan et al., 1997; David et al., 2009).

8.2.5. Shock and multiorgan dysfunction

Shock has been described in several series of children with vivax malaria, often associated with multiorgan dysfunction (Kochar et al., 2010; Kaushik et al., 2012) and was present in 54% of 24 children requiring ICU admission in Brazil (Lanca et al., 2012). Many series are limited by incomplete systematic investigation for bacterial sepsis and other potential co-morbidities. Although one series described negative pre-antibiotic blood cultures among *P. vivax*-infected patients with shock (Lanca et al., 2012), 38% of these had an identified additional infectious co-morbidity potentially contributing to shock.

8.2.6. Other manifestations

Jaundice is common in severe vivax malaria (Alexandre et al., 2010; Kochar et al., 2010; Lanca et al., 2012; Jat et al., 2012), mostly in association with other severe manifestations (Lanca et al., 2012). In Brazilian children, jaundice did not predict need for ICU admission (Lanca et al., 2012). Hypoglycaemia (blood glucose <2.2 mmol/l or <40 mg/dl) (World Health Organization, 2010) was found in 12.5% of children with vivax malaria requiring admission to intensive care unit in Brazil (Lanca et al., 2012). Less-common manifestations reported in association with *P. vivax* also include haemoglobinuria in the absence of G6PD deficiency (Kochar et al., 2010), acalculous cholecystitis (Curley et al., 2011), gangrene (Raghunandan et al., 2012) and fatal pulmonary haemorrhage and haematemesis (Jat et al., 2012).

8.3. Vivax Malaria in Pregnancy

8.3.1. Effects on mother

In contrast to *P. falciparum*, *P. vivax* is not commonly associated with severe malaria in pregnant women (Nosten et al., 1999; Poespoprodjo et al., 2008; Rijken et al., 2012a; McGready et al., 2012a, 2012b) and in a large series from the Thai–Burma border, no maternal deaths were associated with vivax malaria over a 25-year period (McGready et al., 2012b). Despite its rarity in large series from Thailand and Indonesia (Nosten et al., 1999; Poespoprodjo et al., 2008; McGready et al., 2012a), severe maternal malaria, including severe anaemia, has been reported in association with *P. vivax* infection in small series from India (Kochar et al., 2005; Nayak et al., 2009) with poor pregnancy outcomes but again with no maternal deaths. In all endemic regions, vivax-associated maternal anaemia is common, being approximately twice as likely in pregnant women infected with *P. vivax* than without (Nosten et al., 1999; Poespoprodjo et al., 2008).

8.3.2. Effects on foetus and neonate

Plasmodium vivax infection in pregnancy causes a reduction in birth weight (median 108 g) (Nosten et al., 1999; Poespoprodjo et al., 2008; Rijken et al., 2012a), approximately 70% of that observed following maternal falciparum malaria (median 150–192 g) (Poespoprodjo et al., 2008; Rijken et al., 2012a). A single episode of *P. vivax* infection in the first trimester increases risk of miscarriage to 2.7-fold and fourfold with asymptomatic and symptomatic malaria, respectively, a rate similar to that with *P. falciparum* (McGready et al., 2012a). Other consequences of maternal vivax malaria (Nosten et al., 1999; Poespoprodjo et al., 2008; Rijken et al., 2012a; Rodriguez-Morales et al., 2006; McGready et al., 2012a; Singh et al., 1999), including maternal anaemia and low birth weight, contribute to greater infant mortality (Luxemburger et al., 2001). *Plasmodium vivax* in pregnancy is thus responsible for substantial indirect mortality in the first year of life.

8.3.3. Congenital malaria

Plasmodium vivax can cause congenital malaria (Poespoprodjo et al., 2011; Valecha et al., 2007; Del Punta et al., 2010; McGready et al., 2004; Liu et al., 2012; Vottier et al., 2008), through transplacental infection *in utero* or during delivery (Rijken et al., 2012a; Poespoprodjo et al., 2011). A Papuan study found that *P. vivax* congenital infection (alone or mixed) occurred in 1.6 per 1000 live births (Poespoprodjo et al., 2011). In Papua, congenital malaria was independently associated with low birth weight and was mostly asymptomatic at birth (Poespoprodjo et al., 2011). Similar to congenital falciparum malaria (Poespoprodjo et al., 2010), congenital vivax malaria can cause severe illness mimicking neonatal sepsis (Del Punta et al., 2010).

9. SEVERE AND FATAL DISEASE RESULTING FROM TREATMENT WITH PRIMAQUINE

Morbidity and mortality secondary to adverse effects of primaquine are likely to be underestimated in populations with unmonitored use of primaquine and high prevalence of G6PD deficiency (see Baird and Surjadjaja, 2010 and Chapter 4). Primaquine-induced haemolysis in patients with G6PD deficiency can cause life-threatening AKI and severe anaemia. In Manaus, G6PD-associated AKI and severe haemolytic anaemia accounted for two (8%) of the 24 cases of vivax malaria requiring intensive care admission (Lanca et al., 2012). Furthermore, of the 17 deaths in the Manaus autopsy series, two (12%) resulted from AKI and severe anaemia due to G6PD-related primaquine-induced haemolysis (Lacerda et al., 2012).

10. RISK FACTORS FOR UNCOMPLICATED AND SEVERE VIVAX MALARIA

As in falciparum malaria (Miller et al., 2002), host, parasite and socio-geographic factors may explain the variance in the overall risk of severe disease and death in vivax malaria, as well as specific severe manifestations.

10.1. Host Risk Factors

Young age is a risk factor for both uncomplicated and severe vivax malaria (Section 7) (Poespoprodjo et al., 2009), with infants and young children at increased risk of severe anaemia (see Douglas et al., 2012 for review). Gender may also influence risk of severe disease from *P. vivax*, with a higher risk of hospitalisation in females apparent in both adults and children (Tjitra et al., 2008; Kochar et al., 2010). This association needs confirmation in other large series. Adults and children with malnutrition and other co-morbidities are also at increased risk of severe and fatal disease (Section 7).

10.2. Haematinic Replacement

Iron deficiency protects against *P. falciparum* infection and severe disease (Gwamaka et al., 2012) and supplementation may increase the risk of high parasitaemia infections (Nyakeriga et al., 2004; Sazawal et al., 2006). Although there are fewer data for *P. vivax* infection, a link has been reported between iron supplementation and morbidity in vivax malaria, with haematinic treatment during pregnancy being associated with an increased risk of maternal vivax malaria (Nacher et al., 2003). In children, the evidence is less clear. Sixteen weeks of supplemental iron in prepubescent Papua, New Guinean school children provided an overall haematological benefit compared with placebo and did not affect the risk of vivax-associated morbidity (Harvey et al., 1989). A randomized trial from Peru demonstrated that iron plus zinc supplementation decreased morbidity (not otherwise specified) associated with vivax malaria in children under 5 years of age, however iron supplementation alone increased morbidity in those older than 5 years (Richard et al., 2006).

10.3. Host Genetics

Genetic polymorphisms have been associated with both increased risk (alpha and beta thalassaemia (Williams et al., 1996; O'Donnell et al., 2009)) and decreased risk (Duffy antigen negativity (Miller et al., 1976) and G6PD deficiency (Leslie et al., 2010)) of *P. vivax* parasitaemia. The rarity of morbidity from *P. vivax* in West Africa is attributed to the high prevalence of

Duffy antigen negativity (Miller et al., 1976; Ryan et al., 2006), however, the role of Duffy antigen in protecting against vivax anaemia and other severe disease is less well defined (Chapter 2, Part B). Alpha thalassaemia has been associated with reduced risk of severe anaemia from *P. falciparum* (Fowkes et al., 2008), attributed to a lesser reduction in haemoglobin from the loss of microcytic cells, and has been hypothesised to similarly protect against vivax anaemia (Fowkes et al., 2008). The role of ovalocytosis and other red-cell polymorphisms is reviewed by Zimmerman et al. in Chapter 2, Part B.

10.4. Parasite Factors

10.4.1. Differential virulence

Data from the malariotherapy era showed clear differences in risk of morbidity and mortality from different strains of *P. vivax*, with the greatest risk reported in patients infected with the Madagascar strain (case-fatality rate of 10–14%) (James, 1933). Whether naturally acquired *P. vivax* strains vary in virulence and risk of severe vivax malaria is unknown.

10.4.2. Chloroquine resistance

The high prevalence of chloroquine-resistant *P. vivax* found in Indonesian Papua (Sumawinata et al., 2003; Ratcliff et al., 2007) may be an important contributor to the high risk of severe vivax disease, particularly, severe anaemia reported in this region (Tjitra et al., 2008) compared to areas with less chloroquine resistance (Genton et al., 2008). In Africa, the rise of chloroquine resistance in *P. falciparum* coincided with an increase in severe malarial anaemia and mortality (Trape et al., 1998). The prevalence of chloroquine-resistant *P. vivax* is rising with reports of varying degrees of confidence and severity apparently across the vivax endemic world (Douglas et al., 2010). The role of emerging chloroquine resistance and vivax severe disease warrants further investigation (Price et al., 2009).

10.4.3. Mixed Plasmodium *infections*

Studies in areas of low malaria endemicity, such as Thailand, have shown that mixed infections with *P. vivax* appear to attenuate *P. falciparum* disease severity (Luxemburger et al., 1997; Snounou and White, 2004; Price et al., 2001; Mayxay et al., 2004). Conversely, in areas with higher endemicity of both species, mixed infections are associated with an increased risk of severe malaria (Tjitra et al., 2008; Genton et al., 2008), including severe anaemia (Tjitra et al., 2008; Genton et al., 2008) and coma (Manning et al., 2011). In some series from these regions, mixed *P. falciparum*–*P. vivax* infections have a higher reported case-fatality rate among patients with severe malaria than seen with either species alone (Yadav et al., 2012; Manning et al., 2011).

11. PATHOPHYSIOLOGY OF DISEASE IN VIVAX MALARIA

11.1. Comparative Pathobiology of *P. vivax*

There are a number of differences in pathobiology between *P. vivax* and *P. falciparum* that are important in understanding the pathophysiology of vivax malaria (see Kitchen, 1949b; Anstey et al., 2009 for reviews).

11.1.1. Parasite Biomass

Plasmodium falciparum invades RBC of all ages, and progresses to high parasite burdens if uninhibited by treatment or host immunity, with parasite biomass a major independent determinant of the risk of death (Dondorp et al., 2005; Yeo et al., 2008). In contrast, *P. vivax* has a very strong predilection for infecting RBCs that have emerged from the bone marrow within the last 14 days (Kitchen, 1949b; Simpson et al., 1999), particularly early in the course of infection (Kitchen, 1938). This property contributes to the lower parasite biomass seen in *P. vivax* infections. Unlike *P. falciparum* infections, parasitaemias in vivax malaria rarely exceed 2% of circulating RBCs (Ross and Thomson, 1911; Kitchen, 1949b; Field and Shute, 1956). Although a high *P. vivax* parasite burden (140,000/μl) compared to that usually observed has been reported in a fatal case of vivax ARDS (Valecha et al., 2009), this appears to be exceptional. Although the series reporting parasite counts in severe vivax malaria have documented relatively high parasitaemias for *P. vivax*, all have been <100,000/μl (about 2% parasitaemia), ranging from 6000 to 90,400 (n = 11) (Kochar et al., 2005), 680 to 28,847/μl (n = 17) (Alexandre et al., 2010), 8400 to 60,000/μl (n = 40) (Kochar et al., 2009) and 7600 to 60,000/μl (Kochar et al., 2010). None of these studies presented comparative data of parasitaemia between patients with uncomplicated and severe vivax malaria. The correlation of *P. vivax* parasitaemia and disease severity has been limited by the use of semi-quantitative measures (grading 1+–4+) frequently used in endemic areas, extensive pretreatment, lack of reporting and lack of comparison with uncomplicated vivax malaria. Nevertheless, two recent studies have demonstrated an association between disease severity and semi-quantitative parasitaemia in *P. vivax* infection (Nurleila et al., 2012; Lanca et al., 2012).

11.1.2. Relapse

A fundamental difference between *P. falciparum* and *P. vivax* is the ability of *P. vivax* to relapse from dormant hypnozoites and cause repeated episodes of clinical and subclinical infections. Differences in relapse patterns may be a major contributor to the geographic variation in vivax morbidity and disease severity. In tropical regions, *P. vivax* is characterized by frequent relapses 3–6 weeks apart (Goller et al., 2007 and reviewed by Coatney et al., 1971), resulting in recurrent infections often with heterologous strains to which there is little cross-immunity (Imwong et al., 2007; Chen et al., 2007). Frequent recurrent infections prevent adequate time for the patient to achieve haematological recovery from each bout of haemolysis. In contrast, in temperate areas, relapses are fewer and delayed, and the associated impact of each recurrence is less (see Price et al., 2009; Goller et al., 2007; Song et al., 2003 for review).

11.1.3. Greater inflammatory response in P. vivax than P. falciparum

Plasmodium vivax has a lower pyrogenic threshold than *P. falciparum* (Ross and Thomson, 1911; Kitchen, 1949a). Cytokine production (Hemmer et al., 2006; Karunaweera et al., 1992; Yeo et al., 2010a), endothelial activation (Yeo et al., 2010a) and pulmonary inflammatory responses (Anstey et al., 2007) are higher during and after *P. vivax* infections than in *P. falciparum* infections of similar parasite biomass. Although the inflammatory correlates of the lower pyrogenic threshold have been reported, the underlying mechanism(s) have not. Hypothesised reasons include differences between the two species in candidate 'malaria toxin(s)', including glycosylphosphatidylinositol (GPI) (Boutlis et al., 2005) and *Plasmodium* DNA associated with haemozoin (Parroche et al., 2007). Because *P. vivax* has a greater GC content, approximately twice that of *P. falciparum* (Carlton et al., 2008), greater concentrations of TLR9-stimulating CpG motifs within *P. vivax* haemozoin may account for greater inflammatory responses (Anstey et al., 2009). It is possible that *Plasmodium* spp. priming of the innate immune response to bacterial products (Franklin et al., 2009) may be greater in *P. vivax* than that shown in *P. falciparum*, though this remains to be demonstrated. If so, this may explain the septic shock-like presentations increasingly reported in severe vivax disease (Leoratti et al., 2012) (Section 11.1.6).

In falciparum malaria, plasma concentrations of the pro-inflammatory cytokine TNF (Kern et al., 1989; Grau et al., 1989; Day et al., 1999) and the anti-inflammatory cytokine IL-10 (Anstey et al., 1996; Day et al., 1999) are directly related to disease severity. Although *P. vivax* is capable of eliciting greater concentrations of both pro- and anti-inflammatory cytokines than *P. falciparum* (Yeo et al., 2010a; Hemmer et al., 2006), the relationships to disease severity are different in vivax malaria. While plasma concentrations of the pro-inflammatory cytokines TNF and IFNg are directly related to disease severity, plasma concentrations of IL-10 are *inversely* related to vivax disease severity (Andrade et al., 2010b), suggesting a deficiency in the anti-inflammatory response in severe vivax malaria, with an unopposed pro-inflammatory response. Plasma concentrations of superoxide dismutase, an enzyme produced in response to oxidative stress, have also been associated with vivax disease severity (Andrade et al., 2010a), but a role in pathogenesis has not been determined.

Another putative toxin that appears unique to *P. vivax* is a lipid found in the cholesterol/triglyceride fraction of plasma at the time of paroxysmal fever (Karunaweera et al., 2003, 2007). This lipid has greater activity than GPI-like phospholipids (Karunaweera et al., 2007), and together with host cytokines, is known to mediate *in vitro* aggregation of leucocytes, mostly neutrophils (Karunaweera et al., 2003, 2007). This lipid may also contribute to the greater pyrogenicity of *P. vivax*.

11.1.4. Cytoadherence and rosetting

A central mechanism in the pathophysiology of severe falciparum malaria is the cytoadherence of late stages of *P. falciparum* to activated microvascular endothelium resulting in sequestration and microvascular obstruction (Marchiafava and Bignami, 1894; MacPherson et al., 1985; Pongponratn et al., 2003; Turner et al., 1994). Since all stages of *P. vivax* are visible in peripheral blood, albeit with partial depletion of mature stages (Field and Shute, 1956; Rudolf and Ramsay, 1927), sequestration is not thought to occur to a significant degree in vivax malaria or cause end-organ dysfunction in the same manner as *P. falciparum* (Anstey et al., 2009). Recent *in vitro* data show that *P. vivax*-infected RBCs do cytoadhere to endothelial cells, via ICAM-1 and chondroitin sulphate-A (CSA), with a similar strength but a 10-fold lower frequency than *P. falciparum*-infected RBCs (Carvalho et al., 2010). Another study showed no cytoadherence to ICAM-1 but did confirm cytoadherence to the glycosaminoglycans, CSA and hyaluronic acid (Chotivanich et al., 2012).

Despite some cytoadherence *in vitro*, evidence for sequestration-mediated pathology in vivax malaria *in vivo* is at best modest. Indirect physiological studies partitioning pulmonary gas transfer in adults with vivax malaria showed impairment of the pulmonary capillary vascular component, suggesting encroachment and potential sequestration by parasitized red cells in the lung in vivax malaria (Anstey et al., 2007). The few published twentieth century reports of autopsy in vivax malaria showed overall little evidence for microvascular accumulation of vivax-infected RBCs (see Anstey et al., 2009; Ewing, 1901; Billings and Post, 1915; Bruetsch, 1932 for review), though one described isolated 'intra-capillary masses of swollen, infected erythrocytes and pigment'(Billings and Post, 1915) and another 'frequent...mature malaria parasite(s) within a red blood cell...taking up the entire lumen in immediate contact with the endothelial cell'(Bruetsch, 1932b) in at least some vessels in the brain, and 'an unusually large number of infected red cells, young plasmodia and pigment' in intestinal submucosal vessels (Bruetsch, 1932b). These findings were not seen in other early autopsy studies (Clark and Tomlinson, 1949). Ocular histopathology in a later fatal case of vivax-associated coma reported accumulation of *P. vivax*-parasitized erythrocytes, including schizonts within retinal and choroidal blood vessels, however, antemortem peripheral blood showed a mixed infection with *P. falciparum*, with the latter likely to explain the histopathology reported (Biswas et al., 1996).

In contrast, a recent autopsy of fatal ARDS in vivax malaria before any antimalarial treatment showed no evidence for pulmonary or other organ sequestration (Valecha et al., 2009). In the largest autopsy study to date, from Manaus Brazil, the majority of patients had received antimalarial treatment before death (Lacerda et al., 2012). In one of the two cases with ARDS (one of six with lung edema), scanty parasitized red cells were seen in alveolar capillaries despite antimalarial drug clearance of parasites from peripheral blood (Lacerda et al., 2012). Published placental histology is even more limited, but has not shown placental accumulation of *P. vivax*-infected red cells to date (McGready et al., 2004). Taken together, the autopsy findings suggest that significant microvascular obstruction from sequestration of parasitized red cells does not appear to occur in vivax malaria, though it is possible that in some circumstances, more limited cytoadherence to endothelial cells may occur, and may amplify local inflammatory responses in affected organs, such as the lung or placenta.

Rosetting, adherence of non-infected to infected RBCs, has been linked to the pathophysiology of severe falciparum malaria (Carlson et al., 1990). Rosetting has been described *ex vivo* in vivax malaria (Udomsanpetch et al., 1995), however, its role in vivax pathophysiology is unknown.

11.1.5. Deformability and fragility of parasitized erythrocytes

In *P. falciparum*, deformability of both infected and non-infected RBCs is reduced and is thought to contribute to impaired organ perfusion (Dondorp et al., 1999). In contrast, the deformability of vivax-infected RBCs is increased (Suwanarusk et al., 2004; Handayani et al., 2009). This may enable *P. vivax* to pass through the narrow interendothelial slits of the splenic sinusoids (mean dimensions 1.89 × 0.65 μm) (Handayani et al., 2009; Deplaine et al., 2011) and makes it an unlikely mechanism for impaired organ perfusion in severe vivax malaria. Such deformability may, however, be accompanied by increased fragility of both *P. vivax*-infected and -non-infected RBCs (Handayani et al., 2009), though the extent to which this is an artefact of the *ex vivo* microfluidic system used is unknown.

11.1.6. Endothelial activation and altered thrombostasis

Concentrations of circulating endothelial activation markers are as high (ICAM-1 and E-selectin) or higher (angiopoietin-2) in uncomplicated vivax malaria than in falciparum malaria (Yeo et al., 2010a; Jakobsen et al., 1994). Endothelial dysfunction and impaired NO bioavailability are significant contributors to severe falciparum malaria (Yeo et al., 2007, 2009, 2010b), but their importance in severe vivax malaria is not known. Autopsies in fatal cases of *P. vivax* malariotherapy of neurosyphilis demonstrated endothelial 'stimulation' (Bruetsch, 1932a; Clark and Tomlinson, 1949), and endothelial activation and damage has been described in fatal vivax-associated ARDS (Valecha et al., 2009). With the reduced ability of *P. vivax* to cytoadhere, pathogenic consequences of endothelial activation in promoting sequestration of parasitized red cells will be much less in vivax malaria than in falciparum malaria.

However, other consequences of endothelial activation and altered thrombostasis may be more important in *P. vivax* infection. Elevated von Willebrand factor (VWF) and reduced ADAMTS-13 have been associated with disease severity in falciparum malaria (Larkin et al., 2009; Lowenberg et al., 2010). Increased procoagulant activity (Hemmer et al., 2006), VWF (de Mast et al., 2008) and ADAMTS-13 deficiency (de Mast et al., 2008) have now been reported in uncomplicated vivax malaria. The role of altered haemostatic pathways, intravascular coagulation and endothelial inflammation through increased formation of ultra-large VWF and platelet aggregates in severe vivax malaria is not known. However, with the reports of thrombotic microangiopathy in some cases, vivax-associated kidney disease (Sinha et al., 2012; Saharan et al., 2008; Sharma et al., 1993), these processes may contribute to at least a proportion of vivax associated AKI, coma, anaemia and/or thrombocytopenia.

11.1.7. Susceptibility to bacterial co-infection?

It remains to be determined whether, as with falciparum malaria (Berkley et al., 1999), the risk of bacteraemia is increased in vivax malaria. *Plasmodium falciparum* is thought to increase the risk of concurrent bacteraemia through multiple mechanisms, including impairment of granulocyte oxidative burst through induction of hemoxygenase-1 (Cunnington et al., 2012), impaired phagocyte function through ingestion of haemozoin (Schwarzer et al., 2008) and quenching of monocyte nitric oxide (Yeo et al., 2009). Similar pathophysiological processes are plausible in vivax malaria but have not been studied specifically with this species. However, impairment of neutrophil chemotaxis has been shown in *P. vivax* infection and hypothesised to contribute to an increased risk of bacterial co-infection in vivax malaria (Leoratti et al., 2012).

Whether or not *P. vivax* increases *susceptibility* to bacterial co-infection, it is likely to exacerbate the *consequences* of bacterial co-infection. *Plasmodium falciparum* is capable of priming the innate immune response due to interferon-γ-induced enhancement of toll-like receptor expression and function (Franklin et al., 2009). This has been proposed to prime the host to a greater systemic inflammatory response to bacterial and other TLR agonists than would otherwise occur, e.g. during bacterial, viral and other co-infections (Franklin et al., 2009; Leoratti et al., 2012). While data in *P. vivax* are awaited, given the greater inflammatory response in *P. vivax* than *P. falciparum*, it is possible that marked priming of innate immune responses in *P. vivax* infections may account for the relatively high proportion of severe vivax reports occurring with septic shock and multiorgan dysfunction, particularly in endemic regions with a high prevalence of bacterial co-infections. With such TLR priming, excessive sepsis-like responses to circulating bacterial products may well occur with bacterial burdens less than those detectable using routine blood culture systems, which, along with widespread community availability of antibiotics, may be a plausible explanation for septic-shock presentations in *P. vivax* despite a low rate of reported blood culture-positivity.

11.2. Pathophysiology of Specific Syndromes of Severe Vivax Malaria

11.2.1. Severe vivax anaemia

The aetiology of vivax-associated anaemia is complex and confounded by intercurrent infection with *P. falciparum*. Likely mechanisms include loss of both vivax-infected and uninfected red cells from the circulation and impaired RBC production (see Anstey et al., 2009; Douglas et al., 2012 for reviews). Malariotherapy data show that removal of red cells from the circulation is particularly pronounced during acute infection (Collins et al., 2003; Kitchen, 1938), over and above the rate modelled from reticulocyte loss and impaired supply of mature red cells (McQueen and McKenzie, 2004; Antia et al., 2008). The removal of red cells occurs from both extravascular loss in the spleen and from intravascular red cell loss (Douglas et al., 2012). Despite the lower parasite biomass of *P. vivax* relative to *P. falciparum*, red-cell removal is comparable because of the greater proportional removal of uninfected red cells following *P. vivax* infection. In vivax malaria, approximately 34 uninfected red cells are removed for every infected red cell (Collins et al., 2003; Douglas et al., 2012) compared to approximately eight uninfected red cells for every infected red cell in falciparum malaria (Price et al., 2001; Jakeman et al., 1999). Mechanisms for this difference are not clear, but may relate to the greater proportionate inflammatory response in vivax relative to falciparum malaria (Hemmer et al., 2006; Yeo et al., 2010a), both systemically and possibly in the spleen where the majority of extravascular haemolysis occurs (Douglas et al., 2012). Littman showed the importance of the spleen as a site of extravascular RBC removal when he found that a relapsed *P. vivax* infection in a splenectomised patient did not cause anaemia, whereas *P. vivax* infection in the same patient before splenectomy caused severe anaemia (Littman, 1974). The greater proportionate inflammatory response to *P. vivax* has been hypothesized to be paraleled by greater red-cell oxidative stress (Douglas et al., 2012; Bhattacharya and Swarup-Mitra, 1987; Meera et al., 1999; Erel et al., 1997). Although removal of both infected and uninfected erythrocytes is greatest during the early stages of infection, enhanced removal of uninfected RBCs has been shown to persist for at least 5 weeks after antimalarial treatment (Looareesuwan et al, 1987; Woodruff et al., 1979).

An additive factor in the anaemia of *P. vivax* is relapse. In equatorial latitudes, over 80% of *P. vivax* infections relapse at 3–4-week intervals and progressive anaemia is associated with recurrent episodes of haemolysis and dyserythropoiesis prior to haematological recovery from the preceding infection (Price et al., 2009). In areas of chloroquine resistance, this is exacerbated further by delayed parasite clearance, and recrudescent and chronic infections (Price et al., 2007b).

Impaired production of red cells is also likely to contribute to vivax-related anaemia, particularly with chronic and repeated infections, but the exact mechanisms are unclear. Cytokine-related dyserythropoiesis has been demonstrated in *P. vivax* malaria in adults (Wickramasinghe et al., 1989) but not yet in children. Wickramasinghe et al. proposed that this might relate to direct toxicity of *P. vivax* on erythroblasts or enhanced bone marrow phagocyte activity with associated release of locally cytotoxic molecules (Wickramasinghe and Abdalla, 2000). Erythroblasts can be infected and lysed by *P. vivax in vitro*, and recent bone marrow studies have shown erythroblast infection *in vivo* in adult *P. vivax* infection (Ru et al., 2009). Bone marrow hypoxia may contribute to impaired erythropoiesis in *P. falciparum* malaria as a result of marrow sinusoid obstruction by cytoadherent schizonts (Gandapur et al., 1997; Chang and Stevenson, 2004). Sequestration of schizonts in the marrow of patients with *P. vivax* infection has not been clearly demonstrated and, therefore, hypoxia is unlikely to be an important mechanism of impaired erythrocyte production in this disease. Impaired erythropoietin production or response is also an important mechanism of anaemia in *P. falciparum* malaria (Vedovato et al., 1999; Burgmann et al., 1996; el Hassan et al., 1997) but erythropoietin metabolism is yet to be studied in vivax infection.

11.2.2. Acute lung injury

Even in uncomplicated vivax malaria, clinical pulmonary function testing shows subclinical evidence of small airways obstruction, impaired ventilation and reduced gas transfer at the alveolar-capillary membrane (Anstey et al., 2002, 2007). Non-invasive imaging shows that these changes are associated with increased pulmonary phagocytic cell activity (Anstey et al., 2002). Autopsy findings in a fatal Indian case of ARDS from PCR-confirmed *P. vivax* prior to antimalarial treatment showed heavy intravascular monocytic infiltrates with diffuse endothelial and alveolar damage (Valecha et al., 2009). In contrast, in the Manaus autopsy series, mostly post-treatment, the predominant interstitial inflammatory cells were neutrophils (Lacerda et al., 2012). While indirect *in vivo* gas transfer studies infer possible accumulation of *P. vivax*-infected RBCs in the lung (Anstey et al., 2007) and parasites were seen in alveolar capillaries after peripheral blood clearance in one of the Manaus autopsy cases of ALI (Lacerda et al., 2012), the Indian severe vivax ARDS autopsy did not show sequestration of parasitized RBCs in the pulmonary vasculature (Valecha et al., 2009). As in ALI in other disease settings, ARDS in vivax malaria probably results from soluble mediators, diffuse alveolar-endothelial capillary damage, exacerbated by shock, with increases in alveolar permeability and altered alveolar fluid clearance (Tan et al., 2008; Valecha et al., 2009; Suratt and Parsons, 2006).

The majority of cases of vivax-associated ARDS develop after the start of antimalarial chemotherapy (see Taylor et al., 2012; Anstey et al., 2009 for reviews) and gas transfer studies show progressive deterioration in alveolar-capillary function following treatment (Anstey et al., 2007). This may reflect exacerbation of an inflammatory response associated with parasite-killing and/or monocyte/pigment accumulation (Tan et al., 2008; Anstey et al., 2007), though vivax-associated ARDS can occur prior to the start of antimalarial chemotherapy (Valecha et al., 2009; Lomar et al., 2005). As chloroquine-resistance spreads, and other treatment regimens without the anti-inflammatory effects of chloroquine become necessary (Douglas et al., 2010), it is possible that the incidence of ARDS post-treatment may rise (Anstey et al., 2007).

The majority of respiratory distress, in both falciparum and vivax malaria, occurs in young children (Tjitra et al., 2008; Genton et al., 2008; Marsh et al., 1995). ALI is reportedly rare in young children, but is described in recent paediatric severe vivax malaria series (Lanca et al., 2012). An increase in mortality in children with severe falciparum malaria receiving fluid boluses (Maitland et al., 2011) suggests that pulmonary oedema in children with severe malaria may be more prevalent than previously realized. Nevertheless, while ALI appears to be a major cause of vivax-associated respiratory distress in adults, other causes are likely to be important in causing respiratory distress in children. As in falciparum malaria, these may include metabolic acidosis, reported in paediatric vivax-associated respiratory distress (Kochar et al., 2010), as well as severe anaemia (Tjitra et al., 2008), pneumonia and/or sepsis. Concomitant pneumonia was seen in 8% of children with severe vivax malaria in a recent series (Kaushik et al., 2012).

11.2.3. Acute kidney injury

Mechanisms underlying AKI in vivax malaria are not clear. With its frequent association with shock and multiorgan dysfunction, a sepsis-like syndrome either from *P. vivax* itself (see below) or concomitant bacterial sepsis may be a significant contributor to AKI. The extent to which acute tubular necrosis underlies AKI in the setting of multiorgan dysfunction, as seen frequently in falciparum-associated AKI, or acute cortical necrosis, as reported in one series (Kute et al., 2012b), requires further study.

With the increasing reports of thrombotic microangiopathy in vivax-associated AKI (Sinha et al., 2012; Saharan et al., 2008; Sharma et al., 1993), the extent to which thrombotic microangiopathy underlies vivax-associated AKI also requires further prospective studies. Elevated VWF and low levels of the VWF-cleaving protein ADAMTS-13 are known to occur in vivax malaria (de Mast et al., 2009) and are linked to disease severity in falciparum malaria (Larkin et al., 2009); these processes may underlie thrombotic microangiopathy in these patients. Autopsy studies will also be useful. Although six of the 17 fatal Manaus autopsy cases had AKI (as part of multiorgan dysfunction), renal histopathology was not included in the initial histopathology report from this series (Lacerda et al., 2012).

11.2.4. Coma

The aetiology of the coma associated with *P. vivax* is not known and the role of co-infections remains unclear (see Lampah et al., 2011; Anstey et al., 2009 for review). One case of coma associated with microscopy-diagnosed *P. vivax–P. falciparum* mixed infection reported accumulation of '*P. vivax*-infected' red cells within retinal and choroidal blood vessels (Biswas et al., 1996), however, these are likely to have been *P. falciparum*. The rarity of coma in *P. vivax* relative to *P. falciparum* and its absence in *Plasmodium knowlesi* (William et al., 2011; Daneshvar et al., 2009) make it unlikely that the cerebrovascular sequestration of parasitized red cells characteristic of *P. falciparum* occurs to a significant degree with these other parasites. This contention is supported by the absence of *P. vivax* DNA, albeit post-treatment, in both the single Brazilian autopsy case of coma attributed to *P. vivax* (Lacerda et al., 2012) and the post-mortem brain aspirates from three PNG children with coma associated with mixed *P. falciparum–P. vivax* infection (Manning et al., 2012).

While coma has been described in one of nine patients with thrombotic microangiopathy-associated AKI (Sinha et al., 2012), none of the cases of PCR-confirmed vivax-associated coma in Papua had renal impairment, or significant anaemia or thrombocytopenia to suggest a systemic thrombotic microangiopathy in these patients (Lampah et al., 2011). Nevertheless, prospective evaluation of these parameters in other series of vivax-associated coma, as well as schistocytes, plasma concentrations of VWF and ADAMTS-13, are needed to investigate the potential contribution of thrombotic microangiopathy in vivax-associated coma.

11.2.5. Multiorgan dysfunction and shock

The extent to which multiorgan dysfunction and shock are attributable to sepsis-like inflammatory responses to *P. vivax*, the gram-negative bacterial co-infection known to occur during *P. vivax* infection (Piyaphanee et al., 2007; Sur et al., 2006), or both, requires further study. Negative blood cultures reported in some series are limited by widespread pre-hospitalisation use of antibiotics in the community, and the known insensitivity of pre-antibiotic blood cultures in bacterial sepsis (Davis et al., 2011), with concomitant sepsis not being excluded by negative blood cultures.

Nevertheless, plausible hypotheses have been advanced to explain septic shock-like presentations in *P. vivax* infections, particularly in endemic settings with a high prevalence of bacterial co-infections, including *Plasmodium*-induced priming of host inflammatory responses to bacterial TLR agonists (Franklin et al., 2009; Leoratti et al., 2012) (Section 11.1.6).

11.2.6. Pregnancy-associated malaria and low birth weight

P. vivax does not appear to sequester in the placenta although published reports are limited (McGready et al., 2004). Placental histopathology shows haemozoin deposition, but inflammatory and pathological changes are otherwise modest (McGready et al., 2004). Systemic and placental inflammatory responses and microvascular dysfunction from vivax malaria may cause deleterious utero-placental haemodynamic effects and foetal growth restriction (McGready et al., 2004; Rogerson et al., 2007; Umbers et al., 2011; Rijken et al., 2012b). Maternal anaemia, common in vivax malaria (Poespoprodjo et al., 2008; Rijken et al., 2012a), is likely to be an additive cause of low birth weight (Rogerson et al., 2007).

12. CONCLUSION

It is clear that *P. vivax* is responsible for major morbidity and significant mortality in vivax-endemic areas. A single infection causing febrile illness in otherwise healthy individuals rarely progresses to severe disease, particularly with early treatment and anti-relapse therapy. Nevertheless, it has been long recognized that *P. vivax* can cause severe illness and fatal outcome in the presence of acute and chronic co-morbidities. Such co-morbidities, including malnutrition, anaemia, concurrent bacterial infections and chronic renal, liver and lung disease, are common in vivax-endemic regions. While some of these co-morbidities are alternative causes of severe and fatal disease in co-incidentally parasitized patients, others are likely to be major co-factors in the genesis of severe pathology in *P. vivax* infections that may have otherwise caused uncomplicated disease.

Severe and fatal disease also results from recurrent and/or chronic infections in endemic areas, including severe anaemia and malnutrition, particularly in early childhood. Other severe manifestations include ALI, AKI and, uncommonly, coma. The role of bacterial and other co-infections in the increasing reports of shock and multiorgan failure requires further investigation. In pregnancy, *P. vivax* infection is clearly recognized as a major cause of maternal anaemia, miscarriage, preterm delivery, low birth weight and congenital malaria, with significant impact on maternal and infant health.

Mechanisms underlying the pathogenesis of all of the severe vivax syndromes, particularly severe anaemia, AKI, coma, shock and multiorgan failure, remain poorly understood and require further investigation. Similarly the role of risk factors and co-morbidities in contributing to severe and fatal outcomes needs to be elucidated. Although recent advances have built upon the large and neglected knowledge base of the nineteenth and early twentieth centuries, a continued focus on filling these knowledge gaps is needed to target appropriate strategies for treatment and prevention of disease and death from *P. vivax*.

ACKNOWLEDGMENTS

NMA is supported by a National Health and Medical Research Council Practitioner Fellowship; RNP is supported by a Wellcome Trust Senior Fellowship.

REFERENCES

Abu Sayeed, A., et al., 2011. Malarial retinopathy in Bangladeshi adults. Am. J. Trop. Med. Hyg. 84 (1), 141–147.

Adams, F., 1849. The Genuine Works of Hippocrates. Sydenham Society, London.

Agarwal, R., Nath, A., Gupta, D., 2007. Noninvasive ventilation in *Plasmodium vivax* related ALI/ARDS. Intern. Med. 46 (24), 2007–2011.

Alexandre, M.A., et al., 2010. Severe *Plasmodium vivax* malaria, Brazilian Amazon. Emerg. Infect. Dis. 16 (10), 1611–1614.

Andrade, B.B., et al., 2010a. Plasma superoxide dismutase-1 as a surrogate marker of vivax malaria severity. PloS Negl. Trop. Dis. 4 (4), e650.

Andrade, B.B., et al., 2010b. Severe *Plasmodium vivax* malaria exhibits marked inflammatory imbalance. Malar. J. 9, 13.

Anstey, N.M., et al., 1996. Nitric oxide in Tanzanian children with malaria: inverse relationship between malaria severity and nitric oxide production/nitric oxide synthase type 2 expression. J. Exp. Med. 184 (2), 557–567.

Anstey, N.M., et al., 2002. Pulmonary manifestations of uncomplicated falciparum and vivax malaria: cough, small airways obstruction, impaired gas transfer, and increased pulmonary phagocytic activity. J. Infect. Dis. 185 (9), 1326–1334.

Anstey, N.M., et al., 2007. Lung injury in vivax malaria: pathophysiological evidence for pulmonary vascular sequestration and posttreatment alveolar-capillary inflammation. J. Infect. Dis. 195 (4), 589–596.

Anstey, N.M., et al., 2009. The pathophysiology of vivax malaria. Trends Parasitol. 25, 220–227.

Anstey, N.M., Price, R.N., 2007. Improving case definitions for severe malaria. PLoS Med. 4 (8), e267.

Antia, R., Yates, A., de Roode, J.C., 2008. The dynamics of acute malaria infections. I. Effect of the parasite's red blood cell preference. Proc. Biol. Sci. 275 (1641), 1449–1458.

Arora, R.C., et al., 1988. Cerebral malaria caused by *Plasmodium vivax*. J. Assoc. Physicians India 36, 564.

Baird, J.K., Surjadjaja, C., 2010. Consideration of ethics in primaquine therapy against malaria transmission. Trends Parasitol..

Baird, J.K., 2007. Neglect of *Plasmodium vivax* malaria. Trends Parasitol. 23 (11), 533–539.

Baird, J.K., 2009. Severe and fatal vivax malaria challenges 'benign tertian malaria' dogma. Ann. Trop. Paediatr. 29, 251–252.

Barber, B.E., William, T., Grigg, M.J., Menon, J., Auburn, S., Marfurt, J., Anstey, N.M., Yeo, T.W., 2012. A prospective comparative study of knowlesi, falciparum and vivax malaria in Sabah, Malaysia: high proportion with severe disease from Plasmodium knowlesi and P. vivax but no mortality with early referral and artesunate therapy. Clin. Infect. Dis. doi: 10.1093/cid/cis902.

Barcus, M.J., et al., 2007. Demographic risk factors for severe and fatal vivax and falciparum malaria among hospital admissions in northeastern Indonesian Papua. Am. J. Trop. Med. Hyg. 77 (5), 984–991.

Bassat, Q., Alonso, P.L., 2011. Defying malaria: fathoming severe *Plasmodium vivax* disease. Nat. Med. 17 (1), 48–49.

Beare, N.A., et al., 2006. Malarial retinopathy: a newly established diagnostic sign in severe malaria. Am. J. Trop. Med. Hyg. 75 (5), 790–797.

Beg, M., et al., 2002. Cerebral involvement in benign tertian malaria. Am. J. Trop. Med. Hyg. 67, 230–232.

Bejon, P., et al., 2007. Defining childhood severe falciparum malaria for intervention studies. PLoS Med. 4 (8), e251.

Berkley, J., et al., 1999. Bacteraemia complicating severe malaria in children. Trans. R Soc. Trop. Med. Hyg. 93, 283–286.

Berkley, J.A., et al., 2009. HIV infection, malnutrition, and invasive bacterial infection among children with severe malaria. Clin. Infect. Dis. 49 (3), 336–343.

Bhattacharya, J., Swarup-Mitra, S., 1987. Reduction in erythrocytic GSH level and stability in *Plasmodium vivax* malaria. Trans. R Soc. Trop. Med. Hyg. 81 (1), 64–66.

Billings, F., Post, W., 1915. Fatal malaria due to the tertian parasite. Trans. Chicago Path Soc. 9, 209–215.

Bircan, Z., et al., 1997. Two cases of nephrotic syndrome and tertian malaria in south-eastern Anatolia. Pediatr. Nephrol. 11 (1), 78–79.

Biswas, J., et al., 1996. Ocular malaria. A clinical and histopathologic study. Ophthalmology 103 (9), 1471–1475.

Bloland, P.B., et al., 1999. Longitudinal cohort study of the epidemiology of malaria infections in an area of intense malaria transmission II. Descriptive epidemiology of malaria infection and disease among children. Am. J. Trop. Med. Hyg. 60 (4), 641–648.

Boshes, B., 1947. Neuropyschiatric manifestations during the course of malaria. Arch. Neurol. Pyschiatr. 58, 14–27.

Boutlis, C.S., et al., 2005. Glycosylphosphatidylinositols in malaria pathogenesis and immunity: potential for therapeutic inhibition and vaccination. Curr. Top. Microbiol. Immunol. 297, 145–185.

Boyd, M.F., 1938. The threshold of parasite density in relation to clinical activity in primary infections with *Plasmodium vivax*. Am. J. Trop. Med. s1-18 (5), 497–503.
Boyd, M.F., 1941. Historical Introduction. A Symposium on Human Malaria, Am Assoc Advancement Sci, Washington DC, p. 163.
Brooker, S., et al., 2007. Epidemiology of *Plasmodium*-helminth co-infection in Africa: populations at risk, potential impact on anemia, and prospects for combining control. Am. J. Trop. Med. Hyg. 77 (Suppl. 6), 88–98.
Bruetsch, W., 1932a. The histopathology of therapeutic (tertian) malaria. Am. J. Psychiatry 12, 19–65.
Bruetsch, W.L., 1932b. The histopathology of therapeutic (tertian) malaria. Am. J. Psychiatry 89 (1), 19–65.
Buck, R.L., et al., 1983. Malaria at San Lazaro hospital, Manila, Philippines, 1979–1981. Am. J. Trop. Med. Hyg. 32 (2), 212–216.
Burgmann, H., et al., 1996. Serum levels of erythropoietin in acute *Plasmodium falciparum* malaria. Am. J. Trop. Med. Hyg. 54 (3), 280–283.
Carlini, M.E., White Jr., A.C., Atmar, R.L., 1999. Vivax malaria complicated by adult respiratory distress syndrome. Clin. Infect. Dis. 28 (5), 1182–1183.
Carlson, J., et al., 1990. Human cerebral malaria: association with erythrocyte rosetting and lack of anti-rosetting antibodies. Lancet 336 (8729), 1457–1460.
Carlton, J.M., et al., 2008. Comparative genomics of the neglected human malaria parasite *Plasmodium vivax*. Nature 455 (7214), 757–763.
Carrara, V.I., et al., 2006. Deployment of early diagnosis and mefloquine-artesunate treatment of falciparum malaria in Thailand: the Tak Malaria Initiative. PLoS Med. 3 (6), e183.
Carvalho, B.O., et al., 2010. On the Cytoadhesion of *Plasmodium vivax*-infected erythrocytes. J. Infect. Dis.
Chakravarty, A., et al., 2004. Acute inflammatory demyelinating polyneuropathy following *plasmodium vivax* malaria. Neurol. India 52 (1), 130–131.
Chang, K.-H., Stevenson, M.M., 2004. Malarial anaemia: mechanisms and implications of insufficient erythropoiesis during blood-stage malaria. Int. J. Parasitol. 34 (13–14), 1501–1516.
Chen, N., et al., 2007. Relapses of *Plasmodium vivax* infection result from clonal hypnozoites activated at predetermined intervals. J. Infect. Dis. 195, 934–941.
Choi, H.J., et al., 2004. Retinal haemorrhage in vivax malaria. Trans. R Soc. Trop. Med. Hyg. 98 (6), 387–389.
Chotivanich, K., et al., 2012. *Plasmodium vivax* adherence to placental glycosaminoglycans. PLoS ONE 7 (4), e34509.
Chung, B.H., et al., 2008. Predictors of *Plasmodium vivax* malaria-induced nephropathy in young Korean men. Nephron Clin. Pract. 110 (3), c172–c177.
Clark, H., Tomlinson, W., 1949. The pathologic anatomy of malaria. In: Boyd, M. (Ed.), Malariology, WB Saunders, Philadelphia, pp. 874–903.
Coatney, G.R., et al., 1971. The Primate Malarias. National Institutes of Health, Bethesda, Md p. 1–366.
Collins, W.E., Jeffery, G.M., Roberts, J.M., 2003. A retrospective examination of anemia during infection of humans with *Plasmodium vivax*. Am. J. Trop. Med. Hyg. 68 (4), 410–412.
Cunnington, A.J., et al., 2012. Malaria impairs resistance to Salmonella through heme- and heme oxygenase-dependent dysfunctional granulocyte mobilization. Nat. Med. 18 (1), 120–127.
Curley, J.M., Mody, R.M., Gasser Jr., R.A., 2011. Malaria caused by *Plasmodium vivax* complicated by acalculous cholecystitis. Am. J. Trop. Med. Hyg. 85 (1), 42–49.
Curlin, M.E., et al., 1999. Noncardiogenic pulmonary edema during vivax malaria. Clin. Infect. Dis. 28 (5), 1166–1167.
Daneshvar, C., et al., 2009. Clinical and laboratory features of human *Plasmodium knowlesi* infection. Clin. Infect. Dis. 49 (6), 852–860.

David, J., Shanbag, P., More, V., 2009. *Plasmodium vivax* malaria presenting as the nephrotic syndrome in an infant. Trop. Doct 39 (2), 127–128.
Davis, J.S., et al., 2011. Sepsis in the tropical Top end of Australia's northern Territory: disease burden and impact on indigenous Australians. Med. J. Aust. 194 (10), 519–524.
Day, N.P., et al., 1999. The prognostic and pathophysiologic role of pro- and antiinflammatory cytokines in severe malaria. J. Infect. Dis. 180 (4), 1288–1297.
de Mast, Q., et al., 2008. ADAMTS13 deficiency with elevated levels of ultra-large and active von Willebrand factor in malaria (abstract). Am. J. Trop. Med. Hyg. 79, s348.
de Mast, Q., et al., 2009. ADAMTS13 deficiency with elevated levels of ultra-large and active von Willebrand factor in *P. falciparum* and *P. vivax* malaria. Am. J. Trop. Med. Hyg. 80 (3), 492–498.
Del Punta, V., et al., 2010. Congenital *Plasmodium vivax* malaria mimicking neonatal sepsis: a case report. Malar. J. 9, 63.
Deplaine, G., et al., 2011. The sensing of poorly deformable red blood cells by the human spleen can be mimicked in vitro. Blood 117 (8), e88–95.
Dhayagude, R.G., Purandare, N.M., 1943. Autopsy study of cerebral malaria with reference to malarial granulomas. Arch. Pathol. 36, 550–558.
Dobson, M., 1997. Contours of Death and Disease in Early Modern England. Cambridge University Press, , Cambridge.
Dondorp, A.M., et al., 1999. Red blood cell deformability as a predictor of anemia in severe falciparum malaria. Am. J. Trop. Med. Hyg. 60 (5), 733–737.
Dondorp, A.M., et al., 2005. Estimation of the total parasite biomass in acute falciparum malaria from plasma PfHRP2. PLoS Med. 2, e204.
Dondorp, A.M., et al., 2008. The relationship between age and the manifestations of and mortality associated with severe malaria. Clin. Infect. Dis. 47 (2), 151–157.
Douglas, N.M., et al., 2010. Artemisinin combination therapy for vivax malaria. Lancet Infect. Dis. 10, 405–416.
Douglas, N.M., et al., 2012. The anaemia of *Plasmodium vivax* malaria. Malar. J. 11 (1), 135.
el Hassan, A.M., et al., 1997. Decreased erythropoietin response in *Plasmodium falciparum* malaria-associated anaemia. Eur. J. Haematol. 59 (5), 299–304.
Eldridge, W., et al., 1925. Treatment of paresis: results of inoculation with the organism of benign tertian malaria. J. Am. Med. Assoc. 84, 1097–1101.
Erel, O., et al., 1997. Oxidative stress and antioxidative status of plasma and erythrocytes in patients with vivax malaria. Clin. Biochem. 30 (8), 631–639.
Ewing, J., 1901. Malarial parasitology. J. Exp. Med. 5, 429–491.
Falconer, A., 1919. The pulmonary manifestations in malaria. QJM 13, 25–34.
Fernando, S.D., et al., 2003a. The impact of repeated malaria attacks on the school performance of children. Am. J. Trop. Med. Hyg. 69 (6), 582–588.
Fernando, D., de Silva, D., Wickremasinghe, R., 2003b. Short-term impact of an acute attack of malaria on the cognitive performance of schoolchildren living in a malaria-endemic area of Sri Lanka. Trans. R Soc. Trop. Med. Hyg. 97 (6), 633–639.
Field, J., Shute, P., 1956. Plasmodium vivax. The Microscopic Diagnosis of Human Malaria. A Morphological Study of the Erythrocytic Parasites, vol. 2 Institute for Medical Research, Kuala Lumpur.
Flower, B., et al., 2011. Blind, breathless, and paralysed from benign malaria. Lancet 377 (9763), 438.
Fong, T., 1937. A study of the mortality rate and complications following therapeutic malaria. South. Med. J. 30, 1084–1088.
Fowkes, F.J., et al., 2008. Increased microerythrocyte count in homozygous alpha(+)-thalassaemia contributes to protection against severe malarial anaemia. PLoS Med. 5 (3), e56.
Franklin, B.S., et al., 2009. Malaria primes the innate immune response due to interferon-gamma induced enhancement of toll-like receptor expression and function. Proc. Natl. Acad. Sci. U S A 106 (14), 5789–5794.

Freeman, W., Eldridge, W., Hall, R., 1934. Malaria treatment of dementia paralytica: results in 205 cases after five to eleven years. South. Med. J. 27, 122–126.
Gamage-Mendis, A.C., et al., 1991. Clustering of malaria infections within an endemic population: risk of malaria associated with the type of housing construction. Am. J. Trop. Med. Hyg. 45 (1), 77–85.
Gandapur, A.S., Malik, S.A., Raziq, F., 1997. Bone marrow changes in human malaria: a retrospective study. J. Pak. Med. Assoc. 47 (5), 137–139.
Genton, B., et al., 2008. *Plasmodium vivax* and mixed infections are associated with severe malaria in children: a prospective cohort study from Papua New Guinea. PLoS Med. 5 (6), e127.
Gilles, H.M., Hendrickse, R.G., 1960. Possible aetiological role of *Plasmodium malariae* in "nephrotic syndrome" in Nigerian children. Lancet 1 (7128), 806–807.
Gilles, H.M., Hendrickse, R.G., 1963. Nephrosis in Nigerian children. Role of *Plasmodium malariae*, and effect of antimalarial treatment. Br. Med. J. 2 (5348), 27–31.
Golgi, C., 1886. Sull"infezione malarica. Archivo Per Le Scienze Mediche 10, 109–135.
Golgi, C., 1889. On the cycle of development of malarial parasites in tertian fever: differential diagnosis between the intracellular malarial parasites of tertian and quartan fever. Archivo Per Le Scienza Mediche 13, 173–196.
Goller, J.L., et al., 2007. Regional differences in the response of *Plasmodium vivax* malaria to primaquine as anti-relapse therapy. Am. J. Trop. Med. Hyg. 76 (2), 203–207.
Gopinathan, V.P., Subramaniam, A.R., 1982. Pernicious syndromes in *Plasmodium* infections. Med. J. Aust. 2, 568–572.
Gopinathan, V.P., Subramanian, A.R., 1986. Vivax and falciparum malaria seen at an Indian service hospital. J. Trop. Med. Hyg. 89 (2), 51–55.
Grau, G.E., et al., 1989. Tumor necrosis factor and disease severity in children with falciparum malaria. N. Engl. J. Med. 320 (24), 1586–1591.
Gupta, B.K., et al., 2010. A case series of splenic infarction during acute malaria in northwest Rajasthan, India. Trans. R Soc. Trop. Med. Hyg. 104 (1), 81–83.
Gwamaka, M., et al., 2012. Iron deficiency protects against severe *Plasmodium falciparum* malaria and death in young children. Clin. Infect. Dis. 54 (8), 1137–1144.
Habib, A.G., Singh, K.S., 2004. Respiratory distress in nonimmune adults with imported malaria. Infection 32 (6), 356–359.
Handayani, S., et al., 2009. High deformability of *Plasmodium vivax*-infected red blood cells under microfluidic conditions. J. Infect. Dis. 199 (3), 445–450.
Hankey, D.D., et al., 1953. Korean vivax malaria. I. Natural history and response to chloroquine. Am. J. Trop. Med. Hyg. 2 (6), 958–969.
Harish, R., Gupta, S., 2009. *Plasmodium vivax* malaria presenting with severe thrombocytopenia, cerebral complications and hydrocephalus. Indian J. Pediatr. 76 (5), 551–552.
Harris, I., et al., 2010. A large proportion of asymptomatic *Plasmodium* infections with low and sub-microscopic parasite densities in the low transmission setting of Temotu Province, Solomon Islands: challenges for malaria diagnostics in an elimination setting. Malar. J. 9, 254.
Harvey, P.W., et al., 1989. The effect of iron therapy on malarial infection in Papua New Guinean schoolchildren. Am. J. Trop. Med. Hyg. 40 (1), 12–18.
Hemmer, C.J., et al., 2006. Stronger host response per parasitized erythrocyte in *Plasmodium vivax* or ovale than in *Plasmodium falciparum* malaria. Trop. Med. Int. Health 11 (6), 817–823.
Hill, G.J., et al., 1963. Vivax malaria complicated by aphasia and hemiparesis. Arch. Intern. Med. 12, 863–868.
Hutchinson, R.A., Lindsay, S.W., 2006. Malaria and deaths in the English marshes. Lancet 367 (9526), 1947–1951.
Hyman, A., 1945. Clinical masquerades of malaria. Nav. Med. Bull. 45, 287–303.
Imbert, P., et al., 2010. Left upper quadrant abdominal pain in malaria: suspect pathological splenic rupture first. Trans. R Soc. Trop. Med. Hyg. 104 (9), 628.

Imbert, P., Rapp, C., Buffet, P.A., 2009. Pathological rupture of the spleen in malaria: analysis of 55 cases (1958–2008). Travel Med. Infect. Dis. 7 (3), 147–159.

Imwong, M., et al., 2007. Relapses of *Plasmodium vivax* infection usually result from activation of heterologous hypnozoites. J. Infect. Dis. 195 (7), 927–933.

Jakeman, G.N., et al., 1999. Anaemia of acute malaria infections in non-immune patients primarily results from destruction of uninfected erythrocytes. Parasitology 119 (Pt 2), 127–133.

Jakobsen, P.H., et al., 1994. Increased plasma concentrations of sICAM-1, sVCAM-1 and sELAM-1 in patients with *Plasmodium falciparum* or *Plasmodium vivax* malaria and association with disease severity. Immunology 83, 665–669.

James, S., Nicol, W., Shute, P., 1936. Clinical and parasitological observations on induced malaria. Proc. R Soc. Med. 29, 879–894.

James, S.P., 1925. The disappearance of malaria from England. Proc. R Soc. Med. 23, 71–85.

James, S.P., 1926. Report on the First Laboratory Work in Malaria in England. League of Nations Health Organisation, Geneva.

James, S.P., 1931. Some general results of a study of induced malaria in England. Trans. R Soc. Trop. Med. Hyg. 24, 477.

James, S., 1933. Third general report of the malaria Commission of the health Organization of the League of Nations. Quart Bull. Health Org. League Nation. 2, 217.

Jat, K.R., Guglani, V., Khairwa, A., 2012. Severe and complicated *Plasmodium vivax* malaria in children. Trop. Doct.

Kanodia, K.V., et al., 2010. Malaria induced acute renal failure: a single center experience. Saudi J. Kidney Dis. Transpl. 21 (6), 1088–1091.

Kapland, L.I., Read, H.S., Mullins, D.F., 1946. Spontaneous rupture of the spleen during malaria therapy; report of a case. Ann. Intern. Med. 24, 707–714.

Kapoor, K., Gupta, S., 2012. Malarial acute kidney injury in a paediatric intensive care unit. Trop. Doct. (p. epub ahead of print).

Karunaweera, N.D., et al., 1992. Tumour necrosis factor-dependent parasite-killing effects during paroxysms in non-immune *Plasmodium vivax* malaria patients. Clin. Exp. Immunol. 88 (3), 499–505.

Karunaweera, N.D., et al., 2003. The paroxysm of *Plasmodium vivax* malaria. Trends Parasitol. 19 (4), 188–193.

Karunaweera, N., et al., 2007. *Plasmodium vivax*: paroxysm-associated lipids mediate leukocyte aggregation. Malar. J. 6, 62.

Karyana, M., et al., 2008. Malaria morbidity in Papua Indonesia, an area with multidrug resistant *Plasmodium vivax* and *Plasmodium falciparum*. Malar. J. 7 (1), 148.

Kasliwal, P., Rao, M.S., Kujur, R., 2009. *Plasmodium vivax* malaria: an unusual presentation. Indian J. Crit. Care Med. 13 (2), 103–105.

Kaushik, J.S., Gomber, S., Dewan, P., 2012. Clinical and epidemiological profiles of severe malaria in children from Delhi, India. J. Health Popul. Nutr. 30 (1), 113–116.

Kern, P., et al., 1989. Elevated tumor necrosis factor alpha and interleukin-6 serum levels as markers for complicated *Plasmodium falciparum* malaria. Am. J. Med. 87 (2), 139–143.

Kim, E.M., et al., 2010. Abdominal computed tomography findings of malaria infection with *Plasmodium vivax*. Am. J. Trop. Med. Hyg. 83 (6), 1202–1205.

Kitchen, S.F., 1938. The infection of reticulocytes by *Plasmodium vivax*. Am. J. Trop. Med. Hyg. s1-18 (4), 347–359.

Kitchen, S., 1949a. Symptomatology: general Considerations. In: Boyd, M.F. (Ed.), Malariology, W.B. Saunders, Philadelphia, pp. 966–994.

Kitchen, S.F., 1949b. Vivax Malaria. In: Boyd, M.F. (Ed.), Malariology, W.B. Saunders, Philadelphia, pp. 1027–1045.

Kochar, D.K., et al., 2005. Plasmodium vivax malaria. Emerg. Infect. Dis. 11, 132–134.

Kochar, D.K., et al., 2007. Post-malaria neurological syndrome–a case of bilateral facial palsy after *Plasmodium vivax* malaria. J. Vector Borne Dis. 44 (3), 227–229.
Kochar, D.K., et al., 2009. Severe *Plasmodium vivax* malaria: a report on serial cases from Bikaner in northwestern India. Am. J. Trop. Med. Hyg. 80 (2), 194–198.
Kochar, D.K., et al., 2010. Clinical features of children hospitalized with malaria–a study from Bikaner, northwest India. Am. J. Trop. Med. Hyg. 83 (5), 981–989.
Koibuchi, T., et al., 2003. Acute disseminated encephalomyelitis following *Plasmodium vivax* malaria. J. Infect. Chemother. 9 (3), 254–256.
Kotwal, R.S., et al., 2005. An outbreak of malaria in US Army Rangers returning from Afghanistan. JAMA 293 (2), 212–216.
Kumar, S., et al., 2007. *P. vivax* malaria complicated by shock and ARDS. Scand. J. Infect. Dis. 39 (3), 255–256.
Kute, V.B., et al., 2012a. Outcome and prognostic factors of malaria-associated acute kidney injury requiring hemodialysis: a single center experience. Indian J. Nephrol. 22 (1), 33–38.
Kute, V.B., et al., 2012b. *Plasmodium vivax* malaria-associated acute kidney injury, India, 2010–2011. Emerg. Infect. Dis. 18 (5), 842–845.
Lacerda, M.V., et al., 2011. Thrombocytopenia in malaria: who cares? Mem. Inst. Oswaldo Cruz 106 (Suppl. 1), 52–63.
Lacerda, M.V., et al., 2012. Postmortem characterization of patients with clinical diagnosis of *Plasmodium vivax* malaria: to what extent does this parasite Kill? Clin. Infect. Dis. (p. epub ahead of print).
Ladeia-Andrade, S., et al., 2009. Age-dependent acquisition of protective immunity to malaria in riverine populations of the Amazon Basin of Brazil. Am. J. Trop. Med. Hyg. 80 (3), 452–459.
Lampah, D.A., et al., 2011. Coma associated with microscopy-diagnosed *Plasmodium vivax*: a prospective study in Papua, Indonesia. PloS Negl. Trop. Dis. 5 (6), e1032.
Lanca, E.F., et al., 2012. Risk factors and characterization of *Plasmodium vivax*-associated admissions to pediatric intensive care units in the Brazilian Amazon. PLoS ONE 7 (4), e35406.
Larkin, D., et al., 2009. Severe *Plasmodium falciparum* malaria is associated with circulating ultra-large von Willebrand multimers and ADAMTS13 inhibition. PLoS Pathog. 5 (3), e1000349.
Lawn, S.D., et al., 2003. Case reports: pernicious complications of benign tertian malaria. Trans. R. Soc. Trop. Med. Hyg. 97 (5), 551–553.
Lee, J.H., et al., 2010. Retinal hemorrhage in *Plasmodium vivax* malaria. Am. J. Trop. Med. Hyg. 82 (2), 219–222.
Lee, G., et al., 2012. Comparative effects of vivax malaria, fever and diarrhoea on child growth. Int. J. Epidemiol. 41 (2), 531–539.
Leoratti, F.M., et al., 2012. Neutrophil Paralysis in *Plasmodium vivax* malaria. PloS Negl. Trop. Dis. 6 (6), e1710.
Leslie, T., et al., 2010. The impact of phenotypic and genotypic G6PD deficiency on risk of *Plasmodium vivax* infection: a case-control study amongst Afghan refugees in Pakistan. PLoS Med. 7 (5), e1000283.
Lin, E., et al., 2010. Differential patterns of infection and disease with *P. falciparum* and *P. vivax* in young Papua New Guinean children. PLoS ONE 5 (2).
Littman, E., 1974. Splenectomy in hereditary spherocytosis: effect on course of relapsing vivax malaria. Am. J. Med. Sci. 267 (1), 53–56.
Liu, X., et al., 2012. A case of congenital *Plasmodium vivax* malaria from a temperate region in central China. Malar. J. 11 (1), 182.
Lomar, A.V., et al., 2005. Acute respiratory distress syndrome due to vivax malaria: case report and literature review. Braz. J. Infect. Dis. 9 (5), 425–430.
Looareesuwan, S., et al., 1987. Reduced erythrocyte survival following clearance of malarial parasitaemia in Thai patients. Br. J. Haematol. 67 (4), 473–478.

Lowenberg, E.C., et al., 2010. Severe malaria is associated with a deficiency of von Willebrand factor cleaving protease, ADAMTS13. Thromb. Haemost. 103 (1), 181–187.

Lubitz, J., 1949. Pathology of ruptured spleen in acute vivax malaria. Blood 4, 1168–1176.

Luxemburger, C., et al., 1996. The epidemiology of malaria in a Karen population on the western border of Thailand. Trans. R Soc. Trop. Med. Hyg. 90 (2), 105–111.

Luxemburger, C., et al., 1997. The epidemiology of severe malaria in an area of low transmission in Thailand. Trans. R Soc. Trop. Med. Hyg. 91 (3), 256–262.

Luxemburger, C., et al., 1998. Clinical features cannot predict a diagnosis of malaria or differentiate the infecting species in children living in an area of low transmission. Trans. R Soc. Trop. Med. Hyg. 92 (1), 45–49.

Luxemburger, C., et al., 2001. Effects of malaria during pregnancy on infant mortality in an area of low malaria transmission. Am. J. Epidemiol. 154 (5), 459–465.

MacPherson, G.G., et al., 1985. Human cerebral malaria. A quantitative ultrastructural analysis of parasitized erythrocyte sequestration. Am. J. Pathol. 119 (3), 385–401.

Maguire, G.P., et al., 2005. Lung injury in uncomplicated and severe falciparum malaria: a longitudinal study in papua, Indonesia. J. Infect. Dis. 192 (11), 1966–1974.

Mahgoub, H., et al., 2012. Severe *Plasmodium vivax* malaria among Sudanese children at New Halfa hospital, eastern Sudan. Parasite Vectors 5 (1), 154.

Maitland, K., et al., 1997. Absence of malaria-specific mortality in children in an area of hyperendemic malaria. Trans. R Soc. Trop. Med. Hyg. 91 (5), 562–566.

Maitland, K., et al., 2011. Mortality after fluid bolus in African children with severe infection. N. Engl. J. Med..

Manning, L., et al., 2011. Features and prognosis of severe malaria caused by *Plasmodium falciparum*, *Plasmodium vivax* and mixed *Plasmodium* species in Papua New Guinean children. PLoS ONE 6 (12), e29203.

Manning, L., et al., 2012. A histopathologic study of fatal paediatric cerebral malaria caused by mixed *Plasmodium falciparum*/*Plasmodium vivax* infections. Malar. J. 11, 107.

Manson, P., 1900. Experimental Proof of the Mosquitomalaria Theory. Br. Med. J. 2 (2074), 949–951.

Marchiafava, E., Bignami, A., 1894a. On Summer-autumnal Malaria Fevers. The New Sydenham Society, London.

Marchiafava, E., Bignami, A., 1894b. On Summer-Autumnal Fever. New Sydenham Society, London.

Marsh, K., et al., 1995. Indicators of life-threatening malaria in African children. N. Engl. J. Med. 332 (21), 1399–1404.

Mayxay, M., et al., 2004. Mixed-species malaria infections in humans. Trends Parasitol. 20 (5), 233–240.

McGready, R., et al., 2004. The effects of *Plasmodium falciparum* and *P. vivax* infections on placental histopathology in an area of low malaria transmission. Am. J. Trop. Med. Hyg. 70 (4), 398–407.

McGready, R., et al., 2012a. Adverse effects of falciparum and vivax malaria and the safety of antimalarial treatment in early pregnancy: a population-based study. Lancet Infect. Dis. 12 (5), 388–396.

McGready, R., et al., 2012b. Effect of early detection and treatment on malaria related maternal mortality on the north-western border of Thailand 1986–2010. PLoS ONE 7 (7), e40244.

McQueen, P.G., McKenzie, F.E., 2004. Age-structured red blood cell susceptibility and the dynamics of malaria infections. Proc. Natl. Acad. Sci. U S A 101 (24), 9161–9166.

Meera, S., et al., 1999. In vitro studies on peroxidative changes leading to hemolysis of erythrocytes infested with malarial parasite *Plasmodium vivax*. Indian J. Exp. Biol. 37 (7), 729–730.

Melo, G.C., et al., 2010. Concurrent helminthic infection protects schoolchildren with *Plasmodium vivax* from anemia. PLoS ONE 5 (6), e11206.

Michon, P., et al., 2007. The risk of malarial infections and disease in Papua New Guinean children. Am. J. Trop. Med. Hyg. 76 (6), 997–1008.
Miller, L.H., et al., 1976. The resistance factor to *Plasmodium vivax* in blacks. The Duffy-blood-group genotype, FyFy. N. Engl. J. Med. 295 (6), 302–304.
Miller, L., et al., 2002. The pathogenic basis of malaria. Nature 415, 673–679.
Moses, W.R., 1945. Splenic rupture in therapeutic malaria. South. Med. J. 38, 745.
Most, H., Hayman Jr., J.M., 1946. Uncommon clinical manifestations of vivax malaria. J. Am. Med. Assoc. 130, 480–485.
Munteis, E., et al., 1997. Pulmonary involvement in a case of *Plasmodium vivax* malaria. Chest 111 (3), 834–835.
Nacher, M., et al., 2003. Haematinic treatment of anaemia increases the risk of *Plasmodium vivax* malaria in pregnancy. Trans. R Soc. Trop. Med. Hyg. 97 (3), 273–276.
Naha, K., Dasari, S., Prabhu, M., 2012. Spectrum of complications associated with *Plasmodium vivax* infection in a tertiary hospital in South-Western India. Asian Pac. J. Trop. Med. 5 (1), 79–82.
Nayak, K.C., et al., 2009. Spectrum of vivax malaria in pregnancy and its outcome: a hospital-based study. J. Vector Borne Dis. 46 (4), 299–302.
Nguyen, T.H., et al., 1996. Post-malaria neurological syndrome. Lancet 348 (9032), 917–921.
Nicol, W., 1932. A review of seven years malaria therapy in general paresis. J. Ment. Sci. 78, 843–866.
Nicolajev, B., 1939. The duration and course of malaria infection in man. Med. Parasitol. Parasit Dis. 8, 191.
Nosten, F., et al., 1999. Effects of *Plasmodium vivax* malaria in pregnancy. Lancet 354 (9178), 546–549.
Nurleila, S., et al., 2012. Serious and fatal illness associated with falciparum and vivax malaria among patients admitted to hospital at West Sumba in eastern Indonesia. Am. J. Trop. Med. Hyg. 87 (1), 41–49.
Nyakeriga, A.M., et al., 2004. Iron deficiency and malaria among children living on the coast of Kenya. J. Infect. Dis. 190 (3), 439–447.
O'Dempsey, T.J., et al., 1993. Overlap in the clinical features of pneumonia and malaria in African children. Trans. R Soc. Trop. Med. Hyg. 87 (6), 662–665.
O'Donnell, A., et al., 2009. Interaction of malaria with a common form of severe thalassemia in an Asian population. Proc. Natl. Acad. Sci. U S A 106 (44), 18716–18721.
O'Leary, P., Welsh, A., 1933. Treatment of neurosyphilis with malaria: observations on nine hundred and eighty-four cases in the last nine years. JAMA 101, 495–501.
Ozen, M., et al., 2006. Cerebral malaria owing to *Plasmodium vivax*: case report. Ann. Trop. Paediatr. 26, 141–144.
Parakh, A., et al., 2009. *Plasmodium vivax* malaria in children: uncommon manifestations. Ann. Trop. Paediatr. 29 (4), 253–256.
Parroche, P., et al., 2007. Malaria hemozoin is immunologically inert but radically enhances innate responses by presenting malaria DNA to Toll-like receptor 9. Proc. Natl. Acad. Sci. U S A 104 (6), 1919–1924.
Patel, M.P., et al., 2012. An unusual case of *Plasmodium vivax* malaria monoinfection associated with crescentic glomerulonephritis: a need for vigilance. Parasitol. Res..
Paulian, D., 1935. Statistique surd ix ans de malaria therapie. Rev. Neurol. 63, 963–965.
Piyaphanee, W., et al., 2007. Concurrent salmonella bacteremia in P. vivax infection–a report of 2 cases at the Hospital for Tropical Diseases, Thailand. Southeast. Asian J. Trop. Med. Public Health 38 (4), 616–618.
Poespoprodjo, J.R., et al., 2008. Adverse pregnancy outcomes in an area where multidrug-resistant *plasmodium vivax* and *Plasmodium falciparum* infections are endemic. Clin. Infect. Dis. 46 (9), 1374–1381.
Poespoprodjo, J.R., et al., 2009. Vivax malaria: a major cause of morbidity in early infancy. Clin. Infect. Dis. 48 (12), 1704–1712.

Poespoprodjo, J.R., et al., 2010. Severe congenital malaria acquired in utero. Am. J. Trop. Med. Hyg. 82 (4), 563–565.
Poespoprodjo, J.R., et al., 2011. Highly effective therapy for maternal malaria associated with a lower risk ofVertical transmission. J. Infect. Dis. 204, 1613–1619.
Pongponratn, E., et al., 2003. An ultrastructural study of the brain in fatal *Plasmodium falciparum* malaria. Am. J. Trop. Med. Hyg. 69 (4), 345–359.
Pontororing, G., et al., 2010. The burden and treatment of HIV in tuberculosis patients in Papua Province, Indonesia: a prospective observational study. BMC Infect. Dis. 10 (1), 362.
Prakash, J., et al., 2003. Acute renal failure in *Plasmodium vivax* malaria. J. Assoc. Physicians India 51, 265–267.
Price, R.N., et al., 2001. Factors contributing to anemia after uncomplicated falciparum malaria. Am. J. Trop. Med. Hyg. 65 (5), 614–622.
Price, L., et al., 2007a. Acute respiratory distress syndrome in *Plasmodium vivax* malaria: case report and review of the literature. Trans. R Soc. Trop. Med. Hyg. 101 (7), 655–659.
Price, R.N., et al., 2007b. Vivax malaria: neglected and not benign. Am. J. Trop. Med. Hyg. 77 (Suppl. 6), 79–87.
Price, R.N., Douglas, N.M., Anstey, N.M., 2009. New developments in *Plasmodium vivax* malaria: severe disease and the rise of chloroquine resistance. Curr. Opin. Infect. Dis. 22, 430–435.
Pukrittayakamee, S., et al., 1998. Pulmonary oedema in vivax malaria. Trans. R Soc. Trop. Med. Hyg. 92 (4), 421–422.
Raghunandan, J., et al., 2012. Peripheral gangrene in an 18-month-old boy with *Plasmodium vivax* malaria. Paediatr. Int. Child. Health 32 (3), 164–166.
Rajahram, G.S., et al., 2012. Deaths due to *Plasmodium knowlesi* malaria in Sabah, Malaysia: association with reporting as *Plasmodium malariae* and delayed parenteral artesunate. Malar. J. 11 (1), 284.
Ratcliff, A., et al., 2007. Therapeutic response of multidrug-resistant *Plasmodium falciparum* and P. vivax to chloroquine and sulfadoxine-pyrimethamine in southern Papua, Indonesia. Trans. R Soc. Trop. Med. Hyg. 101 (4), 351–359.
Richard, S.A., et al., 2006. Zinc and iron supplementation and malaria, diarrhea, and respiratory infections in children in the Peruvian Amazon. Am. J. Trop. Med. Hyg. 75 (1), 126–132.
Rijken, M.J., et al., 2012a. Malaria in pregnancy in the Asia-Pacific region. Lancet Infect. Dis. 12 (1), 75–88.
Rijken, M.J., et al., 2012b. Ultrasound evidence of early fetal growth restriction after maternal malaria infection. PLoS ONE 7 (2), e31411.
Rodriguez-Morales, A.J., et al., 2006. Pregnancy outcomes associated with *Plasmodium vivax* malaria in Northeastern Venezuela. Am. J. Trop. Med. Hyg. 74 (5), 755–757.
Rodriguez-Morales, A.J., Benitez, J.A., Arria, M., 2008. Malaria mortality in Venezuela: focus on deaths due to *Plasmodium vivax* in children. J. Trop. Pediatr. 54 (2), 94–101.
Rogerson, S.J., et al., 2007. Malaria in pregnancy: pathogenesis and immunity. Lancet Infect. Dis. 7 (2), 105–117.
Romanenko, A.P., 1936. Two cases of comatose form of tertian malaria treated by subcutaneous injection of quinine and methylene blue. Med. Parasitol. Parasit Dis. 5, 925.
Ross, R., Thomson, D., 1911. Some enumerative studies on malaria fever. Proc. R Soc. B 83, 159–173.
Ru, Y.X., et al., 2009. Invasion of erythroblasts by *Plasmodium vivax*: a new mechanism contributing to malarial anemia. Ultrastruct. Pathol. 33 (5), 236–242.
Rudolf, G., Ramsay, J., 1927. Enumeration of parasites in therapeutic malaria. J. Trop. Med. Hyg. 30, 1–8.

Ryan, J.R., et al., 2006. Evidence for transmission of *Plasmodium vivax* among a duffy antigen negative population in Western Kenya. Am. J. Trop. Med. Hyg. 75 (4), 575–581.
Sachdev, H.P.S., Man, M., 1985. Vivax cerebral malaria. J. Trop. Pediatr. 31, 213–215.
Saharan, S., et al., 2008. Thrombotic microangiopathy associated with *Plasmodium vivax* malaria. Pediatr. Nephrol..
Saleri, N., et al., 2006. Acute respiratory distress syndrome in *Plasmodium vivax* malaria in traveler returning from Venezuela. J. Travel Med. 13 (2), 112–113.
Sarkar, S., Bhattacharya, P., 2008. Cerebral malaria caused by *Plasmodium vivax* in adult subjects. Indian J. Crit. Care Med. 12 (4), 204–205.
Sazawal, S., et al., 2006. Effects of routine prophylactic supplementation with iron and folic acid on admission to hospital and mortality in preschool children in a high malaria transmission setting: community-based, randomised, placebo-controlled trial. Lancet 367 (9505), 133–143.
Schwarzer, E., et al., 2008. Hemozoin and the human monocyte–a brief review of their interactions. Parassitologia 50 (1–2), 143–145.
Scott, J.A., et al., 2011. Relation between falciparum malaria and bacteraemia in Kenyan children: a population-based, case-control study and a longitudinal study. Lancet 378 (9799), 1316–1323.
Sharma, J., et al., 1993. *Plasmodium vivax* malaria presenting as hemolytic–uremic syndrome. Indian Pediatr. 30 (3), 369–371.
Sharma, R., et al., 2012. *Plasmodium vivax* malaria admissions and risk of mortality in a tertiary-care children's hospital in North India. Paediatr. Int. Child. Health 32 (3), 152–157.
Shute, P., 1939. Protracted incubation periods in indigenous cases of malaria in England. J. Trop. Med. 46 (5), 57.
Sim, J.E., Choi, Y.C., Kim, W.J., 2010. Facial diplegia in *Plasmodium vivax* malaria. J. Clin. Neurol. 6 (2), 102–103.
Simpson, J.A., et al., 1999. Red cell selectivity in malaria: a study of multiple-infected erythrocytes. Trans. R Soc. Trop. Med. Hyg. 93 (2), 165–168.
Singh, N., Shukla, M.M., Sharma, V.P., 1999. Epidemiology of malaria in pregnancy in central India. Bull. World Health Organ 77 (7), 567–572.
Sinha, A., et al., 2012. Thrombotic microangiopathy and acute kidney injury following vivax malaria. Clin. Exp. Nephrol..
Siqueira, A.M., et al., 2010. Severe rhabdomyolysis caused by *Plasmodium vivax* malaria in the Brazilian Amazon. Am. J. Trop. Med. Hyg. 83 (2), 271–273.
Snounou, G., White, N.J., 2004. The co-existence of *Plasmodium*: sidelights from falciparum and vivax malaria in Thailand. Trends Parasitol. 20 (7), 333–339.
Song, H.H., et al., 2003. Clinical features of *Plasmodium vivax* malaria. Korean J. Intern. Med. 18 (4), 220–224.
Song, J.Y., et al., 2007. Two cases of *Plasmodium vivax* Malaria with the clinical picture resembling toxic shock. Am. J. Trop. Med. Hyg. 77 (4), 609–611.
Spencer, T.E., 1966. Haemoglobin levels in the D'Entrecasteaux Islands in relation to malaria and nutrition. Med. J. Aust. 2 (23), 1093–1097.
Stirk, E., 1943. Pulmonary signs in malaria. J. R. Nav. Med. Serv. 29, 272–274.
Stoppacher, R., Adams, S.P., 2003. Malaria deaths in the United States: case report and review of deaths, 1979–1998. J. Forensic Sci. 48 (2), 404–408.
Sumawinata, I.W., et al., 2003. Very high risk of therapeutic failure with chloroquine for uncomplicated *Plasmodium falciparum* and *P. vivax* malaria in Indonesian Papua. Am. J. Trop. Med. Hyg. 68 (4), 416–420.
Sur, D., et al., 2006. The malaria and typhoid fever burden in the slums of Kolkata, India: data from a prospective community-based study. Trans. R Soc. Trop. Med. Hyg. 100 (8), 725–733.
Suratt, B.T., Parsons, P.E., 2006. Mechanisms of acute lung injury/acute respiratory distress syndrome. Clin. Chest Med. 27 (4), 579–589 (abstract viii).

Suwanarusk, R., et al., 2004. The deformability of red blood cells parasitized by *Plasmodium falciparum* and *P. vivax*. J. Infect. Dis. 189 (2), 190–194.
Swellengrebel, N.H., de Buck, A., 1938. Malaria in the Netherlands. Scheltema & Holkema Ltd, Amsterdam.
Tan, L.K., et al., 2008. Acute lung injury and other serious complications of *Plasmodium vivax* malaria. Lancet Infect. Dis. 8 (7), 449–454.
Tanios, M.A., et al., 2001. Acute respiratory distress syndrome complicating *Plasmodium vivax* malaria. Crit. Care Med. 29 (3), 665–667.
Tanwar, G.S., et al., 2011. Clinical profiles of 13 children with *Plasmodium vivax* cerebral malaria. Ann. Trop. Paediatr. 31 (4), 351–356.
Tarejev, E.M., Gontayava, A.A., Rotenburg, S.S., 1944. Fulminating type of tertian malaria. Sov. Med. 4, 12.
Taylor, T.E., et al., 2004. Differentiating the pathologies of cerebral malaria by postmortem parasite counts. Nat. Med. 10, 143–145.
Taylor, W., et al., 2012. Respiratory manifestations of malaria. Chest 142, 492–505.
Thapa, R., et al., 2009. Childhood cerebral vivax malaria with pancytopenia. J. Pediatr. Hematol. Oncol. 31 (2), 116–117.
Thapa, R., Patra, V., Kundu, R., 2007. *Plasmodium vivax* cerebral malaria. Indian Pediatr. 44, 433–434.
Tilluckdharry, C.C., et al., 1996. A case of vivax malaria presenting with pyschosis. West. Indian Med. J. 45, 39–40.
Tjitra, E., et al., 2008. Multidrug-resistant *Plasmodium vivax* associated with severe and fatal malaria: a prospective study in Papua, Indonesia. PLoS Med. 5 (6), e128.
Torres, J., et al., 1997. Acute non-cardiogenic lung injury in benign tertian malaria. Lancet 350, 31–32.
Tran, T.H., Phu, N.H., Mai, N.T.H., Chau, T.T.H., Loc, P.P., Sinh, D.X., Chuong, L.V., Vinh, H., Waller, D., Peto, T.E.A., White, N.J., 1996. Controlled trial of artemether or quinine in Vietnamese adults with severe falciparum malaria. New Engl. J. Med. 335 (2), 76–83.
Trape, J.F., et al., 1998. Impact of chloroquine resistance on malaria mortality. C R Acad. Sci. 321 (8), 689–697.
Turner, G.D., et al., 1994. An immunohistochemical study of the pathology of fatal malaria. Evidence for widespread endothelial activation and a potential role for intercellular adhesion molecule-1 in cerebral sequestration. Am. J. Pathol. 145, 1057–1069.
Udomsanpetch, R., et al., 1995. Rosette formation by *Plasmodium vivax*. Trans. R Soc. Trop. Med. Hyg. 89 (6), 635–637.
Umbers, A.J., Aitken, E.H., Rogerson, S.J., 2011. Malaria in pregnancy: small babies, big problem. Trends Parasitol. 27 (4), 168–175.
Valecha, N., et al., 1992. Cerebral symptoms with *P. vivax* malaria. Indian Pediatr. 29, 1176–1178.
Valecha, N., et al., 2007. Congenital malaria with atypical presentation: a case report from low transmission area in India. Malar. J. 6, 43.
Valecha, N., et al., 2009. Histopathology of fatal respiratory distress caused by *Plasmodium vivax* malaria. Am. J. Trop. Med. Hyg. 81 (5), 758–762.
van Velthuysen, M.L., Florquin, S., 2000. Glomerulopathy associated with parasitic infections. Clin. Microbiol. Rev. 13 (1), 55–66 (table of contents).
Vedovato, M., et al., 1999. Defective erythropoietin production in the anaemia of malaria. Nephrol. Dial. Transplant. 14 (4), 1043–1044.
Vitor-Silva, S., et al., 2009. Malaria is associated with poor school performance in an endemic area of the Brazilian Amazon. Malar. J. 8, 230.
Vottier, G., et al., 2008. Congenital malaria in neonates: two case reports and review of the literature. Acta Paediatr. 97 (4), 505–508.
Wattanagoon, Y., et al., 1994. Convulsions in childhood malaria. Trans. R Soc. Trop. Med. Hyg. 88 (4), 426–428.

Weber, M.W., et al., 1999. Renal involvement in Gambian children with cerebral or mild malaria. Trop. Med. Int. Health 4 (5), 390–394.

Wickramasinghe, S.N., Abdalla, S.H., 2000. Blood and bone marrow changes in malaria. Baillieres Best Pract. Res. Clin. Haematol. 13 (2), 277–299.

Wickramasinghe, S.N., et al., 1989. Dyserythropoiesis and ineffective erythropoiesis in *Plasmodium vivax* malaria. Br. J. Haematol. 72 (1), 91–99.

William, T., et al., 2011. Severe *Plasmodium knowlesi* malaria in a tertiary care hospital, Sabah, Malaysia. Emerg. Infect. Dis. 17 (7), 1248–1255.

Williams, T.N., et al., 1996. High incidence of malaria in alpha-thalassaemic children. Nature 383 (6600), 522–525.

Williams, T.N., et al., 1997. *Plasmodium vivax*: a cause of malnutrition in young children. QJM 90 (12), 751–757.

Woodruff, A.W., Ansdell, V.E., Pettitt, L.E., 1979. Cause of anaemia in malaria. Lancet 1 (8125), 1055–1057.

World Health Organization, 2010. Guidelines for the Treatment of Malaria. World Health Organization, .

Yadav, D., et al., 2012. Changing profile of severe malaria in north Indian children. Indian J. Pediatr. 79 (4), 483–487.

Yeo, T.W., et al., 2007. Impaired nitric oxide bioavailability and L-arginine reversible endothelial dysfunction in adults with falciparum malaria. J. Exp. Med. 204 (11), 2693–2704.

Yeo, T.W., et al., 2008. Angiopoietin-2 is associated with decreased endothelial nitric oxide and poor clinical outcome in severe falciparum malaria. Proc. Natl. Acad. Sci. U S A 105, 17097–17102.

Yeo, T.W., et al., 2009. Relationship of cell-free hemoglobin to impaired endothelial nitric oxide bioavailability and perfusion in severe falciparum malaria. J. Infect. Dis. 200, 1522–1529.

Yeo, T.W., et al., 2010a. Greater endothelial activation, Weibel-Palade body release and host inflammatory response to *Plasmodium vivax*, compared with *Plasmodium falciparum*: a prospective study in Papua, Indonesia. J. Infect. Dis. 202 (1), 109–112.

Yeo, T.W., et al., 2010b. Increased Asymmetric Dimethylarginine in severe falciparum malaria: association with impaired nitric oxide bioavailability and fatal outcome. PLoS Pathog. 6 (4), e1000868.

CHAPTER FOUR

Diagnosis and Treatment of *Plasmodium vivax* Malaria

J. Kevin Baird*, Jason D. Maguire†, Ric N. Price‡
*Eijkman-Oxford Clinical Research Unit, Jalan Diponegoro No. 69, Jakarta, Indonesia
†Portsmouth Naval Hospital, Portsmouth, Virginia, USA
‡Global Health Division, Menzies School of Health Research, Charles Darwin University, Darwin, Australia

Contents

Advances in Parasitology, Volume 80
ISSN 0065-308X, http://dx.doi.org/10.1016/B978-0-12-397900-1.00004-9

Abstract

Infection by *Plasmodium vivax* poses unique challenges for diagnosis and treatment. Relatively low numbers of parasites in peripheral circulation may be difficult to confirm, and patients infected by dormant liver stages cannot be diagnosed before activation and the ensuing relapse. Radical cure thus requires therapy aimed at both the blood stages of the parasite (blood schizontocidal) and prevention of subsequent relapses (hypnozoitocidal). Chloroquine and primaquine have been the companion therapies of choice for the treatment of vivax malaria since the 1950s. Confirmed resistance to chloroquine occurs in much of the vivax endemic world and demands the investigation of alternative blood schizontocidal companions in radical cure. Such a shift in practice necessitates investigation of the safety and efficacy of primaquine when administered with those therapies, and the toxicity profile of such combination treatments, particularly in patients with glucose-6-phosphate dehydrogenase deficiency. These clinical studies are confounded by the frequency and timing of relapse among strains of *P. vivax*, and potentially by differing susceptibilities to primaquine. The inability to maintain this parasite in continuous *in vitro* culture greatly hinders new drug discovery. Development of safe and effective chemotherapies for vivax malaria for the coming decades requires overcoming these challenges.

1. INTRODUCTION

Accurate diagnosis and targeted treatment of vivax malaria constitutes the front line of defence against this infection. Indeed in the absence of a vaccine and the relative ineffectiveness of vector control to prevent it, diagnosis and treatment may be the only effective response available. Given the burdens of risk and disease detailed in Chapter 1, and the daunting challenges to effective diagnosis and treatment detailed in this chapter, *Plasmodium vivax* can be appreciated as very serious threat to human health.

Treatment of infection by *P. vivax* imposes far greater complexity, nuance, and risk to the patient relative to treatment of uncomplicated falciparum malaria. Treating both liver and blood stages, as opposed to blood stages alone, greatly amplifies the complexity of the chemotherapeutic problem. The only drug available for attacking the liver stages, primaquine, causes mild to severe haemolysis in patients with glucose-6-phosphate dehydrogenase deficiency (G6PDd) – a risk exacerbated by both a higher frequency of this trait where malaria is most common, and the absence of the laboratory capacities required to identify those at risk of harm where most patients live. Those patients will thus very often not have access to safe and effective therapy (Baird, 2008).

The poor state of vivax malaria chemotherapeutics derives from its historic neglect (Mendis et al., 2001; Baird, 2007; Price et al., 2007b). The erroneous presumption of a benign course outlined in this chapter and Chapter 5 helps to explain such neglect. This is compounded by an inability to maintain discreet culture-adapted lines of blood (or liver) stages of *P. vivax* in continuous *ex vivo* culture (as achieved with *Plasmodium falciparum*), a consequence of a disregard of essential research, especially in chemotherapeutics. Indeed a mere 3.1% of global funding in malaria research was committed to *P. vivax* between 2007 and 2009 (PATH, 2011). Unless this is rectified, the prospects for progress against *P. vivax* will remain poor.

Blood schizontocidal therapeutics in use against *P. vivax* today were first processed through pre-clinical and clinical development programs targeting *P. falciparum*. Assessment of the efficacy of these drugs against vivax malaria typically occurs either in late phase III or post-registration for treatment of falciparum malaria. Treatment of the acute attack of vivax malaria has been and continues to be through the chemotherapeutics surrogacy of *P. falciparum*. This approach has nonetheless yielded excellent therapies for acute vivax blood stage infections, but not hypnozoites. *Plasmodium falciparum* lacks a hypnozoite stage and such surrogacy ceases there. Hypnozoitocidal therapies must be directed specifically at *P. vivax* (with surrogacy for *Plasmodium ovale*). Primaquine was developed as anti-relapse therapy in the 1940s only after events in the Pacific War stirred scientific action directed at tackling the very serious strategic problems imposed by relapse. Those efforts occurred before hypnozoites were known (although assumed to exist), and before identification of G6PDd as the basis for haemolysis caused by 8-aminoquinolines-like primaquine. In other words, anti-relapse therapy was never optimized for activity against the responsible parasite stage, or for safe use in G6PDd patients.

The current front-line therapy for radical cure of vivax malaria in most nations – chloroquine (for the acute attack) and primaquine (presumptive anti-relapse therapy (PART)) – has been used, or at least recommended, globally since the 1950s. Evidence of resistance to chloroquine appeared during the late 1980s from the island of New Guinea (Whitby et al., 1989), has spread across the Indonesian archipelago, and now reaches into the Mekong region where it threatens the Indian subcontinent (Baird, 2009; Price et al., 2009). Newer blood schizontocides developed for falciparum malaria have shown very good activity against the acute attack of vivax malaria (Pukrittayakamee et al., 2000; Price and Douglas, 2010), but simply substituting any one of those for chloroquine as a new partner with primaquine in radical cure risks safety and efficacy of the treatment (Baird, 2011). A substantial body of evidence (reviewed here) shows that the activity of primaquine against relapse varies substantially with blood schizontocidal companion drug. New blood schizontocide(s) against the acute attack of vivax malaria also requires evidence of primaquine safety and efficacy against relapse for any given drug partnership for radical cure.

Chapter 2 explains the biology behind relapse, highlighting the technical and ethical challenges of measuring the efficacy of hypnozoitocidal therapies. Relapse rates vary dramatically across regions and parasitemia originating from them cannot be distinguished from those caused by recrudescence or re-infection. This problem renders clinical trials of hypnozoitocides in endemic zones subject to serious confounding. Indeed, chloroquine and primaquine were developed in clinical trials employing prisoner volunteers in the United States, effectively ruling out confounding by reinfection. Alternative approaches to gauging anti-relapse therapy in endemic settings needs to be explored and validated.

The greatest challenge for vivax chemotherapeutics is the safe delivery of primaquine. This risk of primaquine-induced haemolysis in patients with G6PDd is detailed in Chapter 4 of Volume B of this series and explains the impracticability of prolonged dosing of standard therapy – daily for 14 days – that leads to serious adherence problems. Finally, primaquine is contraindicated in pregnancy, a target population who would benefit most from radical cure, due to risk of haemolysis in a foetus with unknown G6PD status. The emerging threat of chloroquine-resistant *P. vivax* demands newer blood schizontocidal agents which, in turn, impose the necessity of demonstrating the safety and efficacy of primaquine with each new partner therapy in radical cure. Ultimately, the effectiveness of primaquine in endemic zones will require the development of safer and more practical dosing regimens. As currently prescribed, primaquine is not fit for purpose in endemic zones. These challenges define the direction of immediate work aimed at improving vivax malaria chemotherapeutics, while also highlighting the need for research and development of a new generation of therapies, especially hypnozoitocides that can be used in G6PDd or pregnant patients.

2. CHEMOTHERAPEUTICS TERMINOLOGY

Malariologists invented a lexicon in coping with the complexity of the parasites and the diseases they cause. Figure 4.1 illustrates the classes of antimalarial drugs divided according to targeted segment of the life cycle: *blood schizontocides* kill asexual blood stages; *tissue schizontocides* kill active asexual forms in the liver; *hypnozoitocides* kill dormant forms in the liver and *gametocytocides* kill sexual forms in the blood. *Sporontocides* kill forms in the mosquito, but no antimalarial drug is used clinically in this way. Drugs are classified according to therapeutic intent, or the nature of activity against the parasite at therapeutic doses.

Class identity has no specific bearing on chemical structure among members of the class. Some drugs belong to several classes, but most belong to just one. Class identity may change between species of plasmodia; for example, primaquine is blood schizontocidal, gametocytocidal, and hypnozoitocidal in vivax malaria, but is only gametocytocidal in falciparum malaria. Although blood schizontocidal for vivax malaria, primaquine is never administered with that therapeutic intent and few would refer to it as a member of that class. Likewise, chloroquine is biologically gametocytocidal in vivax malaria and not falciparum malaria, but the therapeutic intent of chloroquine is never gametocytocidal and is thus not thought of as such.

The aim in malaria chemotherapeutics is to achieve *radical cure* – elimination of all parasites from the body. What constitutes radical cure depends on the infecting species and the chemotherapeutic targets it presents. No recommended therapy aims to tackle all species and stages although such an approach has pragmatic advantages for deployment (Douglas et al., 2011; Baird, 2012a). For much of the endemic world, the radical cure of *P. vivax* includes therapy directed at hypnozoites. Not all patients with a diagnosis of vivax malaria have a relapse, hence the decision to prescribe radical cure is presumptive, based on knowledge of the clinical consequences of recurrent episodes, treatment efficacy and tolerability. The term presumptive anti-relapse therapy (previously called 'terminal

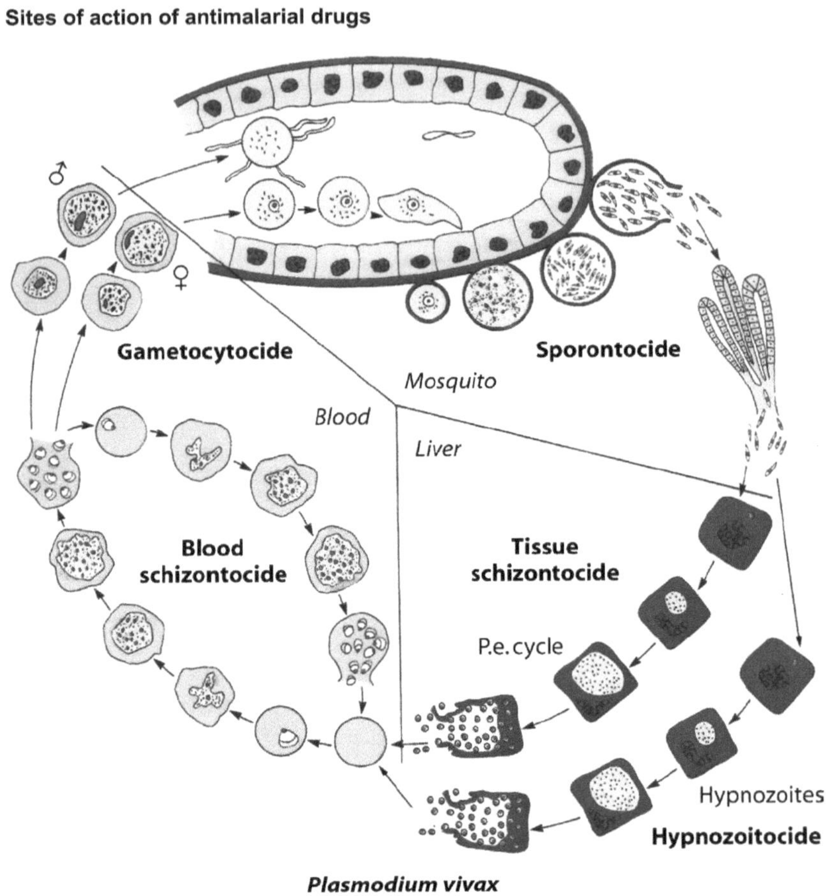

Figure 4.1 The chemotherapeutic compartments in vivax malaria. (*Created by Prof. Wallace Peters with art by Ms. Andrea Darlow, reprinted with their consent and the permission of the publishers of Clinical Microbiology Review where it was originally published* (Baird, 2009)). For colour version of this figure, the reader is referred to the online version of this book.

prophylaxis' in connection with ending travel without a primary attack) is used specifically for this application (Hill et al., 2006). We recommend 'PART' for radical cure of ill patients and 'post-travel PART' for healthy travellers. *Causal prophylaxis* involves drugs active against early developmental forms in hepatocytes, thus preventing both the acute attack and the seeding of the liver with hypnozoites negating the need for post-travel PART. Daily primaquine during exposure achieves this and is termed 'primaquine primary prophylaxis'.

In evaluating chemotherapeutics and clinical and parasitological responses, malariologists developed borrowed terms that carry specific meaning in this context. Parasites appearing in the blood, or *parasitemia*, typically derive from only three possible sources: (1) a new infection after therapy is called a *reinfection* and derives from recent inoculation with sporozoites from biting mosquitoes; (2) a new blood infection deriving from activated hypnozoites is called a *relapse* and (3) an old blood infection that had become sub-patent (below thresholds of diagnosis) but rallies to patency is called a *recrudescence*. The origin of a new parasitemia following therapy of the primary attack often defies certainty of its origin and is referred to with deliberate ambiguity as a *recurrence*.

3. HISTORICAL PERSPECTIVES

The history of the treatment of vivax malaria offers key insights today as the research community again faces the task of fielding new chemotherapeutics for vivax malaria after a 60-year hiatus. The contemporary malariologist dealing with chloroquine-resistant vivax malaria may recognize the strategic quandary provoked by the war-spurred loss of access to quinine in the 1940s. The medical and scientific response to that problem, with its technical insights and hard lessons learned, may give greater traction to today's response to what is an essentially similar problem.

This history offers other key perspectives, especially underlying biases that have profoundly shaped malaria chemotherapeutics over the past century. Chemotherapeutics research has been focused principally to the malaria problem as it occurs in travellers, especially soldiers at war. Chemotherapeutic strategies applied to such populations presume diagnosis and screening capacities that are poorly suited to malaria as it occurs in residents of endemic zones (Baird, 2012a, 2012b). A historical perspective provides a deeper understanding of this significant and persistent problem.

3.1. Quinine

Treatment for vivax malaria with so-called Jesuit's bark (from trees of the genus *Cinchona*, particularly the species *Cinchona ledgeriana*) began several hundred years before Laveran incriminated plasmodia as the cause of malaria in 1880, and the 1896 description of *P. vivax* as a species (Grassi and Feletti, 1890). Although the Inca of Peru presumably did not encounter vivax malaria before the arrival of Europeans in the New World, they apparently chewed the bitter bark of the cinchona tree (they called it quina–quina) to mitigate shivering at high Andean altitudes (Rocco, 2004). Spaniard Jesuits encountering vivax malaria patients likely reasoned the bark could also mitigate the shaking chills of that infection. That the bark proved curative rather than merely palliative may be considered a most improbable good fortune. Pelletier and Caventou in France extracted and identified quinine as the active ingredient of Jesuit's bark against malaria in 1820. Their discovery unleashed efforts that empowered economically successful cultivation of cinchona trees. That success, where it occurred, and geopolitical events would later give direct rise to modern antimalarials still in wide use.

Traders in cinchona bark began seeking out species and strains of trees of the highest quinine yield. In 1864, a British adventurer, Philip Ledger, smuggled 14 pounds of cinchona seeds out of Peru. The authorities captured and killed his Peruvian accomplice, Manuel Incra Mamani, who had obtained the seeds from somewhere in Bolivia (Taylor, 1945). Humanity owes a huge debt to Mamani because naturally occurring cinchona trees were not sustainable as a source of quinine. Even in his day, the trees were being exploited to an extent that risked extinction. Only commercial cultivation of superior trees using sophisticated, optimized techniques could meet global demand and prove economically sustainable (Taylor, 1945). Mamani and Ledger made this possible.

Although no one yet understood the extraordinary nature of Ledger's seeds, they would go on to produce trees having a bark containing up to 14% quinine sulfate content, compared to the more typical 2–4% among almost all other strains. Ledger's seeds were destined to make quinine production a profitable and sustainable enterprise. The British authorities showed no interest in the seeds, already having cinchona plantations in India using trees obtained by Markham – not yet appreciating that the economic difficulties on those plantations were the product of relatively inferior trees. Ledger sold a pound of the seeds to the Dutch authorities for 100 francs up front, followed by 24 British pound sterling after they proved viable by germinating. The remaining 13 pounds Ledger managed to sell to a British cinchona planter on home leave from India, but the planter lost confidence in the seeds and traded Ledger's for the standard strain being used in India. Ledger's seeds in India became lost.

The one remaining pound of Ledger's seeds arrived in Java in late 1865, and the Dutch used these to consolidate by 1918 a global cartel in the trade of quinine, controlling 95% of global production. According to Taylor (1945), in 1939, the Netherlands East Indies had 17,000 ha under cinchona cultivation in 110 estates that produced 12.6 million kilograms of cinchona bark – capable of yielding 1.3 million kilograms of purified quinine. As late as 1941, most malaria in the world was being treated with quinine from those plantations. The earliest cinchona plantations were established on a high plateau near Bandung, West Java (Taylor, 1945). In 1896, the Dutch established a Pasteur Institute adjacent to a cinchona plantation at Bandung, largely aimed at improving the efficiency (and profitability) of quinine extraction and purification. Bandung in contemporary Indonesia is the hub of a vibrant pharmaceutical manufacturing industry derived from the technology of quinine production.

3.2. Modern Antimalarials

In 1856, the chemist Perkins attempted to synthesize quinine in a dye works in Perth, Australia. That unsuccessful effort nonetheless resulted in the discovery of coal tar dyes and gave birth to the aniline dye industry (Field, 1938). In 1891, Erlich and Guttman, noting the affinity for parasites shown by the aniline dye methylene blue in stained blood films, successfully treated two malaria patients (Schlitzer, 2007). But methylene blue was never licensed or marketed as a therapy for malaria because it was considered inferior to quinine. Only the absence of quinine would inspire alternatives to it, and the consolidation of production at a single location and the consequences of war would provide such motivation.

According to Greenwood (1995), quinine was widely used in troops in many theatres of World War I (1914–1918). Governments having access to Java's Quinine Bureau, a co-op of planters controlled by the Dutch government with annual output of 102,000 kg quinine, seized available stocks for prioritizing military use. Germany, at war with its European neighbours, may have had limited access. This may have spurred the immediate efforts by German industry to replace quinine with synthetic antimalarials, but no documentation affirms this as a strategic imperative.

During the 1920s and 1930s, scientists at the Elberfeld laboratory of Bayer (one of six firms in the I.G. Farbenindustrie chemical conglomerate) used methylene blue as the starting point for rational systematic molecular assessments of the antimalarial activity of synthetic compounds. They used bird malaria models, principally *Plasmodium relictum* in Javanese finches, the export of which the Dutch authorities subsequently forbade as a potential threat to their hard-earned monopoly (Greenwood, 1995). The company strategized to break the Dutch commercial hold on treatment of malaria by offering alternatives to quinine. Bayer brought three important drugs to clinical trials and commercial distribution: pamaquine (an 8-aminoquinoline), atabrine (a 9-aminoacridine structurally similar to chloroquine) and sontochin (methylated chloroquine, sold today as Nivaquine). These drugs became the templates for modern antimalarial therapies used for most of the second half of the twentieth century. During this critical period of discovery (1920–1940), neither American nor British science brought any important technical advances or drugs to bear on the malaria problem (Greenwood, 1995). These governments showed little interest in improving upon quinine therapy, a view that would radically change with their loss of access to Java and its cinchona plantations.

Pamaquine was licensed and commercially distributed during the 1920s (as plasmochin or plasmoquine) for treatment of acute malaria. It became known as relatively toxic and never threatened quinine's commercial primacy. Later practitioners realized they could give lower doses of pamaquine with quinine co-therapy and achieve superior results more safely – unwittingly inventing radical cure with blood schizontocide and hypnozoitocide before the existence of that liver stage was known. Pamaquine tablets contained 20 mg base drug, and treatment of acute malaria with this drug alone typically consisted of three tablets daily for 5–10 days. The therapy was later formulated in tablets containing 10 mg pamaquine and 125 mg quinine sulfate, again three daily doses for as many days (5–10) (Brosius, 1927). Practitioners noted, even at the lower dose of pamaquine, gastric upset, cyanosis, and, in a minority of patients who tended to be African in origin, black urine as the principal toxic effects. In 1937, the Malaria Commission of the League of Nations recommended against using pamaquine (despite a 1933 position endorsing its wide use) in writing, '*...no chemotherapeutic remedy should be used for general purposes unless it will comply with the requirement that the dose which kills the parasite is very much smaller than the dose which the human host will tolerate without injury, and unfortunately, plasmoquine cannot yet be classified as a drug which fulfills this requirement*'. (quoted in (Field, 1938)). Pamaquine was effectively abandoned as a viable therapy.

Field (1938) describes efforts by French and Russian scientists to mitigate pamaquine toxicity by modifying the length of the alkyl chain – safer compounds, such as Russian plasmocide, seemed less effective, however. A French version of pamaquine (having one fewer carbon in its alkyl chain), rhodaquine, was apparently widely used in French colonies (Field, 1938).

Atabrine (also called atebrin or mepacrine) was synthesized in 1931 at Elberfeld. The French and Russians produced replicas dubbed quinacrine (exact) and acriquine-8 (lacking a methyl group in the alkyl side chain), respectively. The French authorities forbid the purchase of atabrine in lieu of quinacrine in their colonies (Field, 1938). Atabrine, however, also did not seriously challenge the quinine monopoly because it caused the skin and eyes of patients to turn distinctly yellow. Most practitioners of that era typically considered atabrine an option only in the absence of quinine. The onset of war in Germany in 1939 interrupted the development of sontochin but it had already been evaluated successfully in over 1000 malaria patients in Germany and elsewhere. The onset of war in the Pacific in late 1941 also forced the Allies to turn to both atabrine and pamaquine as their primary chemotherapeutics against malaria.

The occupation of the Netherlands by the Nazis and of Netherlands East Indies by the Imperial Japanese denied the Allies access to quinine. This prompted an urgent scale-up of the production of atabrine in the USA, and seizure of available pamaquine and quinine stocks by the authorities (Shannon, 1946). During mid- to late 1942, the Allies conducted their first offensive actions against the Japanese at Guadalcanal in the Solomon Islands and experienced the grave challenge of malaria (Downs et al., 1947). The inadequacy of atabrine for treatment and prophylaxis became apparent with attack rates of 1.7/person-year. The poor performance of atabrine was later found to be a consequence of variable and poor prescribing practice, in addition to poor adherence spurred by gastric upset and skin discoloration (Shannon, 1946). Divisions of soldiers evacuated from Guadalcanal for recuperation in non-malarious Fiji suffered relapse of vivax malaria at rates reaching 3.7 per person-year (Downs et al., 1947). The Allies faced a very serious threat with very poor chemotherapeutic options.

Even as war loomed in mid-1941, the US government organized the Conference on Chemotherapy of Malaria, which became the Board for the Coordination of Malaria Studies (BMCS) under the aegis of the National Research Council in November 1943 (Coatney, 1963). The vast network of laboratories and clinics it oversaw (including work with allied laboratories in the UK and Australia) managed the evaluation and progress of over 14,000 compounds from drug discovery to clinical trials (Clark, 1946). However, it would be the I.G. Farbenindustrie patents held in the USA by their cartel partner, Winthrop Chemical Company of New York that would yield the greatest chemotherapeutics prize – chloroquine.

Coatney (1963), a former BMCS board member, gives a detailed account of the discovery and development of chloroquine, the synthetic antimalarial that demolished quinine's centuries-long hold as the front line for treatment of the acute attack of malaria. In 1934, Andersag at Elberfeld synthesized a drug he named resochin. Kikuth evaluated the drug in birds with passing marks, and then Sioli evaluated it in four neurosyphilis patients treated with vivax malaria at a Dusseldorf clinic in 1935–1936. According to Coatney (1963), who cites post-war Allied intelligence reports from occupied Germany, Sioli had dismissed resochin as 'too toxic for practical use in humans'. Resochin was abandoned and Andersag quickly produced a methylated version of resochin, which he called sontochin in 1936. By 1939, sontochin had been evaluated in 1100 patients in Germany and Cameroon, and against all four species of human plasmodia with good results. In July 1941 Bayer sent sontochin tablets to the French firm Specia owned by Rhone-Poulenc. Some 5000 of these tablets were in the hands of French clinical investigators in May 1943 when the Allies liberated Tunis in North Africa.

According to Coatney (1963), American scientists had already taken notice of the 4-aminoquinolines and had independently taken special interest in those chlorinated at the 7-position. Later they discovered that Winthrop Chemical held a patent on the most active compound among those, the one Bayer called resochin and the Americans had dubbed chloroquine. Sontochin and chloroquine went to clinical trials under BCMS supervision and chloroquine became therapy of choice for acute malaria of any species in 1946 (Most et al., 1946b). An accounting of this history in a book by German authors, one in the employ of Bayer, agrees on all essential points (Jensen and Mehlhorn, 2009).

A higher priority for the BMCS was the replacement of pamaquine against relapse with less-toxic drugs. *Plasmodium vivax* caused five of six cases of malaria in Allied troops, and a drug–drug interaction with atabrine had forced them to stop using pamaquine against relapse (Baird, 2011; Office of the Surgeon General, 1943). Allied troops were thus fully exposed to that very serious threat. Moreover, relapse especially concerned the BMCS as a means of reintroducing the then vanishing endemic vivax malaria into the US via repatriation of several million service men and women. It is relevant to understand that the one drug available against relapse today was borne of serious concerns about not only the certainty and speed of military victory in the Pacific War but also a direct domestic public health threat to the US (Shannon, 1946; Clark, 1946). The federal government resorted to the use of prison inmate volunteers as experimental subjects in clinical trials of antimalarials (Alving et al., 1948; Coatney et al., 1948), a practice that continued well into the 1960s. The development of radical cure against vivax malaria required in that era, and still, does the mobilization of substantial technical, financial and social capital. The drug they developed and licensed – primaquine – required 8 years of heroic effort. Nonetheless, primaquine may be characterized as a deeply flawed drug. Even more remarkable, that product remains the only therapeutic option against a serious disease that threatens 40% of humanity.

3.3. Primaquine

Historical perspectives on the development of primaquine provide crucial insights today as momentum builds to apply it more aggressively in reaching the goals of elimination of malaria transmission (Baird, 2010; Marsh, 2010; Feachem et al., 2010). Knowledge of the activity of 8-aminoquinolines against relapse predated compelling evidence of hypnozoites (although the likely existence of such had been broadly accepted by the 1930s). This remained true through the BMCS efforts to replace pamaquine, and preclinical screening of compounds for this activity was not possible. The BMCS instead focused screening on toxicity in rats, dogs and monkeys, and in so doing, wholly missed the G6PDd key to that problem. Clinical trials on humans involving only a few dozen compounds assessed safety and efficacy. Those trials led to discovery of G6PDd as the basis of the most important toxicity associated with these compounds (Beutler, 2008). That understanding, however, occurred only after the selection and licensing of primaquine (in 1956). In effect, there has yet to be a systematic exploration of chemical classes for activity against hypnozoites and superior safety in G6PDd patients.

The research conducted principally by Alf S. Alving et al. at the University of Chicago involved the clinical vetting of the 8-aminoquinoline candidates in experimentally challenged prisoner volunteers at the Stateville Penitentiary at nearby Joliet, Illinois (Comfort, 2009). In addition to singling out primaquine as the superior compound, those investigators described three important therapeutic principles with 8-aminoquinolines against relapse:

(1) Total dose effect: whether administered as a single dose or multiple doses over as much as 8 weeks, primaquine provides good efficacy against relapse.

(2) Co-drug synergy: the efficacy of primaquine hinges upon the blood schizontocidal partner drug, even if that drug alone exerts no apparent effect on relapse.

(3) Safety: the haemolytic toxicity of the 8-aminoquinolines is attributable to an inborn X-linked deficiency of G6PD.

These principles guided Alving et al. in the development of chloroquine and primaquine for radical cure of vivax malaria. They demonstrated good efficacy of primaquine when administered with chloroquine, and then they exploited the total dose effect in developing relatively safe dosing regimens for undiagnosed G6PDd patients. That safety, however, was demonstrated only in healthy adult volunteers, relatively few of them, and only in African-Americans with the A-variant of relatively very mild sensitivity to primaquine haemolytic toxicity (Baird and Surjadjaja, 2011). Their work nonetheless met the needs of American soldiers, the intended beneficiaries of this Army-sponsored work. Shannon sat on the BMCS and in 1946 expressed their efforts like this (Shannon, 1946), '*The primary purpose of the investigations has been, at all times, to satisfy specific needs of the armed services*'.

Primaquine was never adapted for use by poorly resourced nations with endemic malaria. The currently prescribed doses derived directly from US Army development programs not intended or optimized for residents of endemic zones. The threat of haemolytic anaemia in undiagnosed G6PDd patients sharply restricts the effectiveness of this otherwise extraordinarily useful and versatile drug. Chemotherapeutic explorations aimed at finally adapting primaquine for use in poorly resourced systems must consider Alving's chemotherapeutic principles in doing so. The first two principles are considered below, and Chapter 4 of Volume B of this series provides a full consideration of the singularly important problem of G6PDd and primaquine therapy.

3.4. Total Dose Effect of Primaquine

In 1928 and 1930, Sinton and colleagues (Sinton and Bird, 1928; Sinton et al., 1930) described the successful treatment of vivax malaria delivering the same total dose of pamaquine over 21 days rather than the recommended 5–10 days. They applied smaller doses of pamaquine over longer periods as a means of mitigating pamaquine toxicity while incurring higher risk of poor adherence. Higher doses over shorter duration improved adherence but increased the risk of toxicity, and Sinton et al. (1930) warned against the approach. As is explained later, striking the same balance persists today in strategizing primaquine therapy for elimination. In 1946, Most et al. (1946a) chose 14 days as the appropriate balance between risk of harm and risk of poor adherence for pamaquine and quinine therapy of vivax malaria. Because the therapeutic indices of the candidate 8-aminoquinolines were not substantially higher than pamaquine, this 14-day regimen was applied as the dosing standard in almost all of the clinical trials of Alving et al.

Later studies of primaquine demonstrated essentially similar independence of dosing schedule from efficacy. Cooper et al. (1953) treated Chesson-challenged volunteers with 7- or 14-days of daily primaquine, but nearly all subjects relapsed. Subsequent work (Jones et al., 1953; Di Lorenzo et al., 1953) effectively repeated that experiment in soldiers naturally infected in Korea and demonstrated equal efficacy of the two regimens. After the discovery of G6PDd as the basis of primaquine sensitivity, Alving et al. (1960) applied weekly dosing for 8 weeks to safely accommodate G6PDd patients needing radical cure and demonstrated good efficacy against relapse. In the 1970s, Clyde and McCarthy (1977) reported good efficacy of 60 mg daily for 7 days compared to 30 mg daily for 14 days. Schmidt et al. (1977) reported conclusive affirmation of the total dose effect using primaquine and other 8-aminoquinolines (with chloroquine or quinine) in the *Plasmodium cynomolgi* relapse model. Figure 4.2 summarizes those findings. Across total doses ranging from completely ineffective to effective for primaquine and a methylated derivative, efficacy with a one, three, seven, or 14 doses (daily doses) varied little. A study in Colombia, however, reported 15%

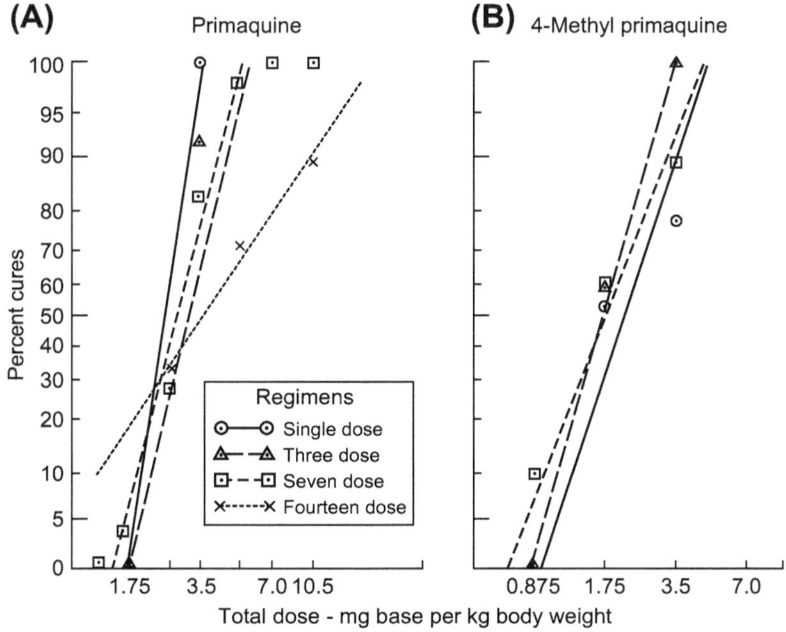

Figure 4.2 The total dose effect of 8-aminoquinolines against relapse of *Plasmodium cynomolgi* in rhesus macaques. (*Reprinted from Schmidt* (Schmidt et al., 1977) *with permission of the American Journal of Tropical Medicine & Hygiene.*)

recurrence rate (including reinfection) among patients given the standard 14-day regimen vs. 57% in patients given the same total dose over just 3 days (Carmona-Fonseca and Maestre, 2009). Controlled studies are needed to affirm short-duration, high-dose primaquine regimens, along with better G6PD diagnostics to protect patients from risk of harm from this approach.

3.5. Hypnozoitocidal Co-drug Synergy

Co-drug synergy represents a vital consideration in the safety, efficacy and effectiveness of primaquine, and in seeking new means of radical cure. The data demonstrating co-drug synergy thus merits deeper examination, especially because many workers today presume primaquine acts alone in killing hypnozoites. The effect of the blood schizontocidal partner on the efficacy of primaquine becomes decisively important with the encroaching loss of chloroquine efficacy against the acute attack. Replacing chloroquine in radical cure is not simply exchange of blood schizontocide for use with primaquine, but development of a pair of drugs, each impacting the safety and efficacy of the other.

In their approach to assessing 8-aminoquinolines, Alving et al. set out to standardise the prisoner volunteer challenge model (Alving et al., 1948; Craige et al., 1947a, 1947b). They selected the Chesson strain of *P. vivax*, a rapid and frequent relapsing strain isolated from an American soldier named Chesson who was infected in New Guinea during 1944 (Ehrman et al., 1945). In these challenge studies, they controlled for possible confounding by recrudescence by demonstrating complete efficacy of quinine alone against blood stage challenge with Chesson strain. They also carefully characterized pamaquine efficacy against relapse as the benchmark for discovery of drugs superior to it. In this work, they noted, similar to Sinton and colleagues had 20 years before them (Sinton and Bird, 1928; Sinton et al., 1930), distinct differences in efficacy when quinine and pamaquine were administered together.

Tables 4.2–4.4 list the data from Alving's group (Ruhe et al., 1949; Berliner et al., 1948b). Alving et al. (1955) referred to these findings in explaining the rationale for a clinical trial designed to test the hypothesis of potentiation of primaquine activity against relapse by drugs having no known effect on relapse when administered without primaquine. They wrote, '*These experiments, however, cannot be considered definitive because the individual groups treated were either too small or the different test groups lacked adequate controls. The possibility of potentiation of the action of any 8-aminoquinoline against*

Table 4.1 Evolution of Therapies for Vivax Malaria

Treatment	Developed	Discontinued	Advantage	Disadvantage
QN	1600s	1940s	Cures acute attack	- No effect on relapse - Side effects - Lengthy dosing
QN + PM	1924	1942	Achieves radical cure	- Dangerous toxicity of PM
AB + PM	1942	1943	Achieves radical cure	- AB–PM interaction exacerbates already dangerous PM toxicity
CQ + PQ	1950	Still in use	Achieves radical cure	- PQ toxicity in G6PDd - Lengthy PQ dosing - Resistance to CQ
ACT + PQ	2000s	In limited use	Cures CQ-resistant acute attack	- PQ safety/efficacy - Lengthy PQ dosing - Most ACTs unproven with PQ
CQ + TQ	In phase IIb trials in 2012	Not yet in use	Achieves radical cure with brief dosing	- TQ also toxic in G6PDd - CQ resistance

QN = quinine; PM = Pamaquine; AB = atabrine; CQ = chloroquine; PQ = primaquine; TQ tafenoquine.

tissue stages by concurrent administration of quinine, therefore, still remains a matter of uncertainty'. They designed and executed a clinical trial to address that uncertainty.

After following an elaborate protocol aimed at minimizing variance in the numbers of sporozoites inoculated via mosquito biting, 57 volunteers were randomized to three treatment groups of 19 men each. Two groups received identical 14-day doses of quinine and primaquine, one concurrently and the other quinine first followed by primaquine after a two-day pause. The third treatment group received chloroquine rather than quinine,

Table 4.2 Early Evidence for Quinine Synergizing Pamaquine against Relapse of Chesson Strain Vivax Malaria Reported in 1947

Total dose pamaquine (mg)	QN administration	Number of subjects	% relapse
240	Consecutive	1	100
420	Consecutive	5	100
434	Consecutive	7	100
882	Consecutive	5	100
240	Concurrent	5	80
420	Concurrent	5	60
630	Concurrent	5	60
882	Concurrent	5	40

Adapted from Craige et al. (1947a)

Table 4.3 Early Evidence for Quinine Synergizing Pamaquine against Relapse of Chesson Strain Vivax Malaria Reported in 1948

Total dose pamaquine	QN administration	Number of subjects	% relapse
None	N/A	5	80
420	Concurrent	4	50
840	Concurrent	18	11
1260	Concurrent	4	0
1260	Consecutive	5	40

Adapted from Berliner et al. (1948b)

Table 4.4 Early Evidence of Quinine Synergizing Pamaquine against Relapse of St. Elizabeth Strain Vivax Malaria Reported in 1949

Total dose pamaquine	QN administration	Number of subjects	% relapse
360	Consecutive	6	33
360	Concurrent	6	0
720	Concurrent	7	0
1080	Concurrent	6	0

Adapted from Ruhe et al. (1949)

and the same dose of primaquine. Table 4.5 summarizes the findings. Primaquine therapy concurrent with quinine achieved good efficacy (1 in 19 relapsed), whereas when administered consecutively, most of the volunteers (15 of 19) relapsed. Quinine synergized the activity of primaquine against relapse as did chloroquine but to a lesser degree (5 of 19 relapsed). Alving

Table 4.5 Definitive Evidence for Quinine Synergizing Primaquine against Relapse of Chesson Strain Vivax Malaria Reported in 1955

Total dose primaquine	Drug administration	Number of subjects	% relapse
240	Quinine consecutive	19	79
240	Quinine concurrent	19	5
240	Chloroquine concurrent	19	26

Adapted from Alving et al. (1955)

et al. (1955) concluded, '*Evidence is presented which indicates that either quinine or chloroquine, when administered concurrently with primaquine during the treatment of clinical attacks of mosquito-induced Chesson strain vivax malaria, potentiates the curative action of primaquine*'. This critically important observation and conclusion seems to have been lost to malariology during the 60-year hiatus from technical consideration of the problem.

Unpublished data from Alving et al. (1949) were described in a 1949 semi-annual report to their US NIH sponsors (the BMCS dissolved in 1946) showing synergistic effects against relapse between isopentaquine, quinine, and, remarkably, methylene blue. Nine of 10 subjects given isopentaquine alone (60 mg daily for 14 days) relapsed (or possibly recrudesced). Among 15 other subjects given the same dose of isopentaquine but with quinine, only five relapsed. Adding 0.5 g methylene blue daily in lieu of quinine in nine subjects resulted in just three relapses. And three subjects given methylene blue, a 60 mg regimen of isopentaquine (30 mg doses of this drug were also used but all relapsed regardless of adjunctive therapy), and quinine did not relapse. In the same report, Alving also describes one subject who haemolysed severely when given the 60 mg regimen of pentaquine, and again did so with a 30 mg daily regimen of pamaquine. When that subject was again treated with the 60 mg pentaquine regimen in combination with methylene blue, he suffered no haemolysis during the 14 days of dosing. This observation, viewed in the context of the severe limitations imposed by primaquine therapy in unscreened G6PDd patients, commands attention.

Later studies of the treatment of *P. cynomolgi* infections of rhesus monkeys (a superb model for *P. vivax* in humans) documented distinct therapeutic responses to primaquine when administered with quinine or chloroquine. Tables 4.6 and 4.7 summarize those findings for the M and B strains of that parasite, i.e. combining chloroquine with primaquine provided consistently superior efficacy to primaquine against relapse

Table 4.6 Variable Cure Rates of Primaquine against Relapse of M strain of *P. cynomolgi* in Rhesus Macaques when Co-administered with Quinine vs. Chloroquine

Daily primaquine dose (mg/kg) × 7 days	Blood schizontocide	Number of monkeys	% relapse
None	Quinine	83	98
None	Chloroquine	46	100
0.75	None	38	61
0.75	Quinine	55	45
0.75	Chloroquine	16	25
1.5	None	20	35
1.5	Quinine	5	13
1.5	Chloroquine	5	0

Adapted from Schmidt (1981)

Table 4.7 Variable Cure Rates of Primaquine against Relapse of B strain of *P. cynomolgi* in Rhesus Macaques when Co-administered with Quinine versus Chloroquine

Daily primaquine dose (mg/kg) × 7 days	Blood schizontocide	Number of monkeys	% relapse
None	Quinine	2	100
None	Chloroquine	2	100
0.38	Quinine	4	100
0.38	Chloroquine	4	75
0.75	Quinine	8	38
0.75	Chloroquine	8	0
1.5	Quinine	4	0
1.5	Chloroquine	4	0

Adapted from Schmidt (1981)

(Schmidt, 1981). Similarly, quinine synergized primaquine against relapse in this model (Schmidt et al., 1982). That laboratory also later reported screening 265 chemically diverse compounds, finding broad activity at tolerable doses only among 8-aminoquinolines (Schmidt, 1983). That screening of compounds and another sponsored by the US Army led to survey of 200 8-aminoquinoline compounds for anti-relapse activity in that model. That work led to discovery of the drug tafenoquine (30 years later in phase IIb trials in 2012) (Davidson et al., 1981). Schmidt later reported enhancement of primaquine activity against *P. cynomolgi* relapse by the concurrent administration of mirinamycin (a lincosamide class antibiotic) (Schmidt, 1985). This work aimed at reducing the effective doses of 8-aminoquinoline therapy against hypnozoites, so that '*...radical cure might* *be reduced to levels tolerated by all recipients*'. However, the Army program developing tafenoquine turned their energies to its use as an agent of prophylaxis rather than radical cure. No further exploration of synergy of the 8-aminoquinolines against relapse occurred.

Through a fitful drug development program stretching over more than three decades, the US Army abandoned tafenoquine for prophylaxis and today GlaxoSmithKline (GSK) and the Medicines for Malaria Venture (MMV) are developing the drug for radical cure of vivax malaria. In a series of experiments using *P. cynomolgi* challenge in rhesus monkeys not designed to detect the phenomenon, US Army scientists noted that the minimally effective dose of tafenoquine could be reduced 10-fold when administered with blood schizontocidal therapies (Dow et al., 2011). Exploring co-drug synergies with primaquine or tafenoquine may indeed yield safe 8-aminoquinoline therapies against relapse, should drug developers at last commit to the therapeutic principles laid down by Alving et al. over 60 years ago.

3.6. Chloroquine-Resistant Vivax Malaria

The emergence of resistance to chloroquine by the asexual blood stages of *P. vivax* represents another important historical event that provides key perspectives in managing the problem today. Although resistance to chloroquine in *P. falciparum* appeared as early as the late 1950s, the first report of this in *P. vivax* did not appear until 1989 (Whitby et al., 1989). There are several likely explanations for that seemingly long delay, one or all of which may have played a role. These are described later in this chapter.

The onset of resistance to chloroquine, and its continuing emergence today, profoundly impacts the treatment of vivax malaria and therefore its management as a global problem. This single problem effectively unhinges chemotherapeutics of the infection for the simple reason that there is only a single proven means of radical cure, i.e. chloroquine and primaquine. That 60-year-old treatment has failed in much of malarious Southeast Asia and now threatens South Asia, these areas accounting for over 80% of the global burden of *P. vivax*.

The emergence of chloroquine-resistant vivax malaria impels clinical development of new ways of achieving radical cure. A concerted effort to address this problem has not been undertaken since the 1940s, when large-scale studies in American prisoner volunteers were conducted. The choice of those subjects should not be construed as having been a matter of convenience – freedom from confounding by reinfection was essential to the successful development of chloroquine–primaquine. Today the use of prisoner volunteers is not an option, and yet the task of developing new radical cures is essential. Sections to follow in this chapter propose strategies for detecting and coping with the far-reaching problem of resistance to chloroquine in vivax malaria chemotherapeutics.

4. CHEMOPROPHYLAXIS AGAINST VIVAX MALARIA

Travellers from non-endemic zones may use antimalarial drugs to prevent acute malaria during and after the visit. The vast majority of these drugs provide suppressive rather than causal prophylaxis, i.e. they kill blood stages rather than liver stages. These drugs have no effect on latent hypnozoites and travellers may be directed to consume PART upon completion of travel in order to avoid relapse in the following weeks, months and years. In reality, however, few begin or finish this treatment and late relapse following exposure to endemic vivax malaria may be the rule with exclusive use of suppressive chemoprophylactic drugs (Swartz et al., 2003).

Early clinical trials with primaquine (and a Russian primaquine-like drug called quinocide) demonstrated causal prophylaxis against falciparum and vivax malarias (Baird et al., 2003; Lysenko, 1960). In falciparum malaria, a single 30 mg dose given within 48 h of sporozoite inoculation sufficed. In other words, just 30 mg of primaquine was sufficient to kill the early developing forms of *P. falciparum* in hepatocytes. The same single-dose experiment was not repeated for *P. vivax*, but 30 mg daily for 5 days also proved effective against experimental challenge (Baird et al., 2003; Arnold et al., 1954, 1955; Powell and Brewer, 1967; Alving et al., 1959). As will be seen later, the same regimen for preventing relapse after acute illness is almost completely inefficacious: killing developed hypnozoites requires higher doses than preventing their formation, which is presumably what causal prophylaxis accomplishes.

During the era of drug development in the 1950s, chloroquine became the universal choice for chemoprophylaxis by virtue of its superior safety, tolerability and convenience compared to primaquine. Primaquine for prophylaxis was not further pursued. Further, as chloroquine efficacy and utility in chemoprophylaxis ebbed during the 1970s and 1980s, drug developers naturally sought to replace it with other blood schizontocides. Suppressive prophylaxis had come to be viewed as the favoured approach.

The US Navy conducted a series of clinical trials during the 1990s and early 2000s that re-examined primaquine for causal prophylaxis. An early trial, as a result of initial safety and tolerability concerns, applied a regimen of three doses of 30 mg primaquine per week. That resulted in relatively poor efficacy (74% against *P. falciparum*; 90% against *P. vivax*): suggesting that parasites developing for more than 48 h before exposure to primaquine seemed more likely to survive (Baird et al., 1995b). Daily dosing proved more efficacious and equally well tolerated as weekly chloroquine over 16 weeks, and in one trial, over 50 weeks of daily dosing (Fryauff et al., 1995; Baird et al., 2001). These studies, conducted in endemic zones, could only gauge efficacy against the primary attack. The possibility of relapses could not be ruled out because of constant reinfection of subjects, but, at least in concept, prevention of primary attack by killing very early developmental forms would also kill parasites attempting development to hypnozoites. In 2012, primaquine is recommended by many authorities as safe and effective for primary prophylaxis against vivax malaria, despite lacking regulatory approval for this indication because of a lack of commercial incentives to obtain such for a drug without patent protection.

5. DIAGNOSIS OF VIVAX MALARIA

5.1. Clinical Diagnosis

Clinical diagnosis of vivax malaria is difficult because signs and symptoms are non-specific and impacted by endemicity and host immunity. Common symptoms include malaise, fever, chills, diaphoresis, headache, arthralgia, myalgia, cough, abdominal pain, nausea, vomiting, and diarrhoea, which can also occur with many common systemic febrile illnesses such as meningitis, pneumonia or gastroenteritis. Splenomegaly, anaemia, leukopenia and thrombocytopenia are also common, but non-specific. Depending on where infection is acquired, the differential diagnosis should also include dengue fever, typhoid fever (particularly in south central Asia), infectious mononucleosis, rickettsiosis, influenza, bacteremia and region-specific arboviral infections (Freedman et al., 2006; Wilson et al., 2007). The periodicity of fever, classically described as 48 h paroxysms, is a poor predictor of diagnosis since most patients, particularly non-immunes, present with asynchronous erythrocytic schizogony with 24-h cycles of paroxysm.

A multitude of clinical algorithms have been evaluated for utilization in resource-poor areas where microscopic diagnosis of malaria may be limited. Such efforts are designed to reduce inappropriate prescription of antimalarials for non-malarial illness, which contributes to emergence of antimalarial resistance (White and Olliaro, 1996; Gbotosho et al., 2009).

Clinical diagnostic algorithms are generally limited to falciparum malaria and are site-specific (Chandramohan et al., 2002; Hozhabri et al., 2002; Perisse and Strickland, 2008). However, clinical diagnostic algorithms do not reduce drug wastage, and in higher prevalence areas, risks patients with symptomatic malaria being left untreated (Chandramohan et al., 2002). In a low malaria prevalence region of Pakistan, for example, presence of fever for >3 days in the absence of cough was highly sensitive (100%), but poorly specific (63%) (Hozhabri et al., 2002). The problem with such an algorithm with respect to timely treatment of malaria is inherent. Sensitivity also declines with increasing age in moderate malaria transmission areas (Mwangi et al., 2005). Because a reliable, universally applicable clinical diagnostic algorithm has not, and will most likely not be established, timely laboratory diagnosis remains a critical component of management of malaria illness.

5.2. Microscopic Diagnosis

Diagnosis of vivax malaria should be established by microscopic examination of Giemsa-stained thick and thin smear blood samples obtained every 8–12 h over a 24–48 h period. Unfortunately, the quality of microscopic diagnosis of malaria is highly variable because of lack of standards or means of validating proficiency among microscopists. Slide preparation and staining methodology, microscopist skill and quality of microscope all contribute to the risk of diagnostic error. Even in the hands of presumed expert microscopists, results can vary significantly depending on the species, parasite density and method of enumeration (Milne et al., 1994; Maguire et al., 2006b; O'Meara et al., 2006). In a study of United Kingdom clinical laboratories, microscopists correctly identified only 63% of confirmed *P. vivax*-positive blood films (Milne et al., 1994). Twenty-eight expert microscopists from 13 countries who reviewed polymerase chain reaction (PCR)-confirmed Giemsa smears from *P. vivax*-infected individuals correctly diagnosed only 86% of vivax malaria, and agreed on only 47% of mixed infection species (Maguire et al., 2006b). Even in the best hands, diagnosis to species of early ring stage parasites is extremely difficult and hence in the context of a busy service laboratory, the first sighting of parasites is often assumed to be *P. falciparum*, classically regarded as the most threatening and clinically important species, i.e. the safest diagnosis for the patient. This also results in underestimates of mixed infections.

In recent years, the World Health Organization has made significant progress in establishing standards for microscopic diagnosis of malaria as well as guidelines for measuring proficiency (World Health Organization, 2009a). Depending on local resources, implementation of quality assurance programs to meet those standards may remain limited.

5.3. Molecular Diagnosis

PCR-based diagnosis of vivax malaria has been well established for almost two decades. The most common method utilizes species-specific PCR amplification of the small subunit rRNA DNA polymerase gene sequence (Kimura et al., 1995). However, this technique requires resources and expertise, which is not generally available in developing countries where diagnostic needs are greatest. The development of loop-mediated isothermal amplification (LAMP) methodology, which does not require use of a thermocycler, has positioned molecular diagnosis of malaria within reach for low resource settings where cost, accuracy, ease of performance and time are important considerations. In recent studies in China, comparing LAMP methodology against a standard nested PCR assay for the diagnosis of microscopically confirmed vivax malaria, sensitivity and specificity for LAMP was 98% and 100% and for nested PCR, 99% and 100%, respectively (Lu et al., 2011; Tao et al., 2011). Similar results have been reported in Korea (Chen et al., 2010). Deployment of LAMP-based diagnostics to low resource settings and further evaluation of sensitivity and specificity outside of the research setting is necessary before widespread use or replacement of microscopic diagnosis with LAMP can be recommended.

5.4. Antigen-Capture Rapid Diagnostic Testing

If available, an antigen-capture rapid diagnostic test (RDT), which can distinguish falciparum from non-falciparum malaria, may be helpful for establishing early diagnosis when microscopic diagnosis capabilities are not available. There are currently more than 200 monoclonal antibody-based RDT products currently available on the worldwide market (World Health Organization, 2012). A 2009 WHO report on the performance of RDTs for diagnosis of *P. vivax* identified CareStart Malaria (pan), Advantage (pan), SD BIOLINE Pf/Pv and OptiMAL as best performers with sensitivities >95% for both high and low parasite density infections. Since then, the WHO has developed a web-based interactive guide to help select suitable RDTs based on specific criteria such as targeted *Plasmodium* species, minimum panel detection score for that species at 200 and 2000 parasites/μl, maximum acceptable false-positive rate, test format and heat stability (WHO Global Malaria Programme, 2009). Currently, the only

US Food and Drug Administration (FDA) approved RDT is the Binax NOW Malaria Test, a card-based diffusion immunochromatographic assay which detects the presence of *P. falciparum*-specific histidine-rich protein II (HRPII) and a pan-malarial antigen common to all four human species. When used in endemic locations where prevalence of malaria is high, the overall sensitivities and specificities for *P. falciparum* exceeded 95%. Performance for diagnosis of *P. vivax* is much lower, sensitivity 69% and specificity 100%, and does not meet the WHO recommended minimum panel detection score. Because RDT sensitivity is directly related to parasite density, health care providers cannot rely on a negative RDT result to rule out malaria, particularly for vivax malaria. However, a positive result may be helpful in making management decisions before a microscopic diagnosis can be made.

6. TREATMENT OF UNCOMPLICATED VIVAX MALARIA

The goals of antimalarial chemotherapy are to reduce the immediate risk to the host, eradicate peripheral asexual parasitaemia, prevent the recurrent infection and interrupt the cycle of transmission. No drug on its own has the ability to achieve all of these aims. Of particular importance for the chemotherapy of *P. vivax* is its ability to form dormant liver stages (hypnozoites) capable of causing relapsing infections weeks to months after the initial blood stage infection. The chemotherapy of *P. vivax* is therefore complex and requires a strategy targeting a variety of specific key elements of the parasite life cycle (Fig. 4.1).

6.1. Treatment of *P. vivax* Erythrocytic Stages

6.1.1. Chloroquine

Chloroquine has been the treatment of choice for *P. vivax* malaria for almost 70 years. By the 1940s, clinical studies confirmed chloroquine to be an extremely potent antimalarial agent, effective against all plasmodia species, and its slow elimination from the blood allowing curative regimens to be achieved with short-course treatment regimens. Clinical experience in a huge number of patients demonstrated the drug to be safe and well tolerated, including its use in pregnant women and young infants. By the 1950s, its wide availability, familiarity amongst health care staff and low cost, resulted in it becoming the first-line treatment for both *P. falciparum* and *P. vivax*, with almost 90 metric tonnes of the drug being administered annually during its heyday.

Chloroquine is rapidly absorbed after oral administration, its bioavailability exceeding 90%. It has an extremely long and complex terminal elimination phase because of its extensive distribution in tissue, with high concentrations occurring in the liver, spleen, kidney and lung. This tissue penetration accounts for its enormous apparent volume of distribution (~200–300 l/kg), and it is this rather than the elimination process that determines the blood concentration profile of this drug. Its reported plasma half-life varies from 70 to 300 h, resulting in therapeutic drug concentrations being detected in the blood for between 3 weeks and 3 months after administration in some patients.

Against sensitive strains of *P. vivax*, chloroquine produces a very rapid clinical and parasitological response, patients usually clearing their fever and peripheral parasitaemia within 48 h. Chloroquine, similar to all blood schizontocidal agents effective against *P. vivax*, is gametocytocidal with rapid clearance of sexual stages during acute infection (Pukrittayakamee et al., 2000). Its slow elimination provides a period during which blood concentrations exceed the minimally effective concentration of the parasites, ensuring complete clearance of all erythrocytic stages of the parasite and providing a prolonged period of post-treatment prophylaxis (PTP) during which the drug can suppress new invasion of the blood stream by merozoites from reinfection or relapse from hypnozoites.

Although chloroquine has dominated antimalarial treatment of *P. vivax* for over half a century, this position is under threat from the emergence and spread of chloroquine-resistant (CQR) strains of the parasite. The first cases of chloroquine-resistant *P. vivax* were documented in 1989 (Rieckmann et al., 1989), almost 30 years after reports of the emergence of chloroquine-resistant *P. falciparum*. There are several likely explanations for that seemingly long delay.

First, chloroquine was evaluated and recommended for both of these species at the same time, and with the same recommended dosing, i.e. 1.5 g base total adult dose over 48 h. That dose, sufficient for *P. falciparum*, far exceeded the minimally effective dose of 0.3 g for *P. vivax* (Berliner et al., 1948a). Chemotherapeutic 'overkill' may have delayed onset of resistance to chloroquine in *P. vivax*.

Second, unlike *P. falciparum*, the gametocytes of *P. vivax* appear to be exquisitely sensitive to chloroquine (Jeffery, 1958). Drug-resistance phenotype does not necessarily carry over to non-target parasite stages. Thus, if chloroquine-resistant asexual blood stages emerged and survived exposure to that drug, the gametocytes they gave rise to would likely have been killed. Onset and transmission of chloroquine-resistant *P. vivax* likely required resistance in both asexual and sexual blood stages. The intrinsically chloroquine-tolerant gametocytes of falciparum malaria, on the other hand, would have greatly accelerated propagation of the drug-resistant genotype/phenotype.

A crucial determinant of *de novo* selection is the degree of selective drug pressure exerted on a parasite population (White, 1999). This is, at least in theory, significantly lower in *P. vivax* infections, a consequence of higher rates of asymptomatic parasitaemia and lower parasite biomass. In *P. falciparum* infections, gametocytogenesis occurs after the appearance of symptoms and is refractory to schizontocidal drugs. Carriage of *P. falciparum* gametocytes increases greatly with delayed patient clearance favouring the transmission of resistant genotypes (Price et al., 1999). In contrast, *P. vivax* gametocytes appear earlier, usually before presentation to the clinic, and are susceptible to schizontocidal agents ensuring transmission occurs before drug selection.

In any event, resistance to chloroquine did emerge, probably on the island of New Guinea at the eastern edge of the Indonesian archipelago. The treatment has been abandoned in Indonesia, the Solomon Islands, and Vanuatu. The same decision is under review in Cambodia, Thailand, and Vietnam. The problem has been documented at relatively low frequencies in Burma and India, and those nations account for the majority of vivax malaria cases globally. The loss of chloroquine to resistance unhinges vivax malaria therapeutics as a whole, that is, it incites loss of the only practical and proven means of radical cure – its pairing with primaquine. The reality of this problem and its serious consequences are already mobilizing the scientific community, as it did in the wake of the onset of World War II and the loss of quinine, to commit to the daunting challenge of developing and licensing new radical cures for vivax malaria. The three therapeutic principles for 8-aminoquinolines against relapse laid down by Alving et al. should guide renewed efforts in this arena.

The epicentre of *P. vivax* chloroquine resistance is the island of New Guinea. In both Indonesian Papua and Papua New Guinea, clinical trials of *P. vivax* have documented high rates of early clinical deterioration requiring hospitalization, delayed parasite clearance and recurrent parasitaemia within 28 days in over 65% of patients (Rieckmann et al., 1989; Baird et al., 1991; Sumawinata et al., 2003; Ratcliff et al., 2007b). These extremely poor clinical responses are markedly different from those apparent from the same locations in the 1990s, suggesting an evolving trait of increasingly drug-tolerant parasites.

Evidence for declining chloroquine efficacy against *P. vivax*, albeit less pronounced as that observed in New Guinea, has also been reported from other parts of the Indonesia archipelago (Sutanto et al., 2009, 2010), South Korea (Lee et al., 2009), Myanmar (Guthmann et al., 2008), Thailand (Phyo et al., 2011; Rijken et al., 2011), Ethiopia (Yohannes et al., 2011; Ketema et al., 2011), Madagascar (Barnadas et al., 2008), India (Srivastava et al., 2008) and Brazil (de Santana Filho et al., 2007). Indeed there is now evidence that the efficacy of chloroquine against *P. vivax* is declining in most locations where vivax malaria is endemic (Price et al., 2009; Douglas et al., 2010).

In endemic regions where populations have limited access to health care, the inability to reliably eradicate the liver stages of the parasite and emerging chloroquine resistance by the blood stages of parasite, can result in *P. vivax* becoming a chronic relapsing disease exacerbated by a complex array of co-morbidities: nutritional, genetic and immunological disorders. In this context, high-grade chloroquine resistance is likely to make a significant contribution to reports of severe and fatal vivax malaria, which was previously assumed to be a benign infection (Price et al., 2009). The effects of declining drug efficacy are manifestly worse in the most vulnerable populations, such as young children and pregnant women (Poespoprodjo et al., 2008, 2009).

6.1.2. Non-chloroquine treatment regimens of P. vivax

Almost all antimalarial drugs with activity against *P. falciparum* demonstrate intrinsic activity against the asexual stages of *P. vivax*. The main exception to this being the antifolate drugs which exhibit low potency (Darlow et al., 1982) and vulnerability to the rapid development of drug resistance (Pukrittayakamee et al., 2000; Tjitra et al., 2002). Clinical trials of mefloquine (Maguire et al., 2006a), atovaquone + proguanil (Lacy et al., 2002), halofantrine (Baird et al., 1995a), piperaquine (Ratcliff et al., 2007a; Karunajeewa et al., 2008b), artesunate (Ratcliff et al., 2007a; Karunajeewa et al., 2008b) and pyronaridine (Poravuth et al., 2011) show these antimalarial agents to exhibit good efficacy against chloroquine-resistant *P. vivax*. These clinical studies are supported by *in vitro* studies conducted in an area of high-grade CQR (Price et al., 2010; Russel et al., 2008). However, a degree of *in vitro* and *in vivo* cross-resistance has been observed between chloroquine and amodiaquine (Hasugian et al., 2009), and this may account for the high rates of recurrence in clinical studies conducted in Papua, Indonesia (Hasugian et al., 2007) and Papua New Guinea (Karunajeewa et al., 2008b).

6.1.3. Artemisinin combination therapy for P. vivax *infection*

Over the past decade, global antimalarial policy has made a major shift to artemisinin-based combination therapies (ACTs) as the treatment of choice for uncomplicated falciparum malaria. By 2009, ACT had been adopted as policy in 81 malarious countries as first-line therapy although certainly with varying degrees of implementation success (World Health Organization, 2009b). During this period, most countries have maintained a dual policy for uncomplicated malaria with chloroquine plus primaquine remaining the mainstay of treatment for vivax malaria. In areas where chloroquine efficacy against *P. vivax* is declining, the World Health Organization (WHO) guidelines recommend ACT plus primaquine as an appropriate alternative (World Health Organization, 2010), however, to date, this has only been adopted by The Solomon Islands, Vanuatu, Papua New Guinea (PNG) and Indonesia (World Health Organization, 2009b). The reluctance to embrace ACT for vivax malaria reflects a number of important perceptions. Some policy makers argue that ACT is an unnecessary and expensive choice for a disease which can be readily treated in most areas with chloroquine (Whitty et al., 2008). Although the combination of chloroquine plus primaquine results in adequate efficacy against low-level CQR isolates (Baird et al., 1995a), the underlying level and extent of CQR has almost certainly been underestimated (Baird, 2009; Price et al., 2009; Douglas et al., 2010). If high-grade resistance to CQR continues to rise, then this default option will no longer be tenable.

As has been explained, the activity of primaquine against relapse when partnered with newer ACTs in radical cure is unknown. There is a potential that the introduction of these treatments for vivax malaria could result in diminished efficacy against relapse (Baird, 2011). A recent trial in Indonesian soldiers repatriated from Papua to a non-endemic area of Java, however, demonstrated superb efficacy of primaquine against relapse when partnered with dihydroartemisinin–piperaquine for radical cure (Sutanto, in preparation). Despite these caveats, there is a growing opinion advocating the benefits of a unified ACT-based therapy for both *P. falciparum* and *P. vivax* in co-endemic regions. The same rationale of combination therapy to reduce the emergence of drug-resistant *P. falciparum*, applies to *P. vivax* (White, 1999). Misidentification of species and under-diagnosis of mixed parasitaemia is rife, hence continued recommendation of segregated therapeutic strategies for this parasite species will inevitably result in inadvertent use of chloroquine for *P. falciparum* infections, with potentially catastrophic consequences in individual patients. A unified treatment strategy would streamline antimalarial procurement and distribution systems and provide greater impetus for drug companies to reduce ACT manufacturing costs. As the cost of ACTs falls and their availability increases, the advantages of a pragmatic, unified ACT-based treatment protocol for both *P. falciparum* and *P. vivax* in all co-endemic regions becomes compelling (Douglas et al., 2010).

The efficacy of ACT for treating *P. vivax* monoinfection has been assessed in 13 published clinical trials. Including seven studies of dihydroartemisinin + piperaquine (DHA + PIP) (Ratcliff et al., 2007a; Karunajeewa et al., 2004, 2008a, 2008b; Hasugian et al., 2007; Hung et al., 2004; Awab et al., 2010), five studies of artemether + lumefantrine (AM + LUM) (Yohannes et al., 2011; Ratcliff et al., 2007a; Karunajeewa et al., 2008b; Li et al., 1999; Krudsood et al., 2007), three studies of artesunate + sulfadoxine + pyrimethamine (AS + SP) (Tjitra et al., 2002; Karunajeewa et al., 2008b; Kolaczinski et al., 2007), one study each of artesunate + amodiaquine (AS + AQ) (Hasugian et al., 2007), and pyronaridine with either dihydroartemisinin (LeYuan and QingXia, 2001) or artesunate (Poravuth et al., 2011). Most of these were conducted in Asia in areas known to be under threat from CQR *P. vivax*.

A hallmark of the artemisinin derivatives is their high potency and ability to affect a rapid reduction of initial parasite biomass, a property which is apparent in both *P. falciparum* and *P. vivax* (Pukrittayakamee et al., 2000). All of the published ACT clinical trials confirmed the rapid clearance of *P. vivax* asexual stages (Sinclair et al., 2011). The main difference between treatment regimens was high variability in the rates of recurrent parasitaemia between 28 and 63 days of follow-up. The risk of recurrent parasitemia was greatest in patients treated with artesunate + sulfadoxine + pyrimethamine (67%) (Karunajeewa et al., 2008b), artesunate + amodiaquine (48%) (Hasugian et al., 2007) and artemether + lumefantrine (57–70%) (Ratcliff et al., 2007a; Karunajeewa et al., 2008b). Unlike its utility in assessing *P. falciparum* treatment outcomes, genotyping is of limited benefit in discriminating the cause of recurrent *P. vivax* parasitaemia since these can originate from recrudescence, reinfections or relapse (Chen et al., 2007). Evidence of poor efficacy of monotherapy SP and amodiaquine suggest that in the case of AS + SP and AS + AQ, high recurrence rates by day 28 are likely to reflect recrudescence or relapse of insensitive parasites although one cannot prove this with current genotyping methods. The high rates of treatment failure with AM + LUM are remarkable with 57–70% of Papuan patients having recurrent infection if primaquine was delayed until day 28 (Ratcliff et al., 2007a; Karunajeewa et al., 2008b). These figures compare only 2.6% of Thai patients having recurrent parasitaemia when AM + LUM was administered concurrently with primaquine (Krudsood et al., 2007). These data suggest that the high rates of recurrence with AM + LUM monotherapy are due to relapsing rather than recrudescent infections.

6.1.4. The role of post-treatment prophylaxis

PTP refers to the ability of a treatment regimen to maintain antimalarial concentrations in the blood above the MIC of the parasite for a period of time during which recurrent infections are prevented or delayed. In falciparum malaria therapeutics, the benefits of PTP have been observed particularly in high-transmission regions from Africa; the greater the risk of recurrence, the more apparent this prophylactic effect is likely to become (Arinaitwe et al., 2009). In vivax-endemic regions, the rates of relapse following treatment of blood stage *P. vivax* infection vary considerably but can reach almost 90% within 63 days in equatorial regions (Douglas et al., 2011; Karunajeewa et al., 2008b; Hasugian et al., 2007). These rates of recurrences are comparable to *P. falciparum* reinfections in high transmission settings in Africa. In this context, the pharmacokinetics and dynamics of the initial schizontocidal treatment have significant effects on the subsequent clinical and parasitological response following vivax treatment (White, 2008).

None of the current ACTs have any efficacy against the hypnozoite stages of *P. vivax*, but those containing slowly eliminated partner drugs, such as piperaquine (half-life ~28 days) and mefloquine (~12 days), provide a clinically significant period of PTP against blood-stage parasites, which results in a delay in relapse and reinfections and a longer period free from symptomatic malaria. The risk of relapse and benefits of PTP are also apparent in patients presenting with falciparum malaria. In regions co-endemic for both species, *P. vivax* parasitaemia can occur in over half of patients treated with a rapidly-eliminated antimalarial drug for acute falciparum malaria within 9 weeks of follow-up (Douglas et al., 2011). The use of slowly eliminated antimalarials such as mefloquine or piperaquine reduces markedly the risk of these early heterologous relapses (Douglas et al., 2011; Karunajeewa et al., 2008b; Mayxay et al., 2006; Smithuis et al., 2006, 2010).

The short-term benefits of PTP can be seen in a reduction in the risk of anaemia associated with the greater length of time for haematological recovery (Ratcliff et al., 2007a; Hasugian et al., 2007). Gametocyte carriage, which is almost always associated with recurrent asexual parasitaemia, is also lower, potentially reducing the transmission potential to the mosquito vector (Phyo et al., 2011). The long-term benefits of prolonged post-exposure prophylaxis are less clear. A single inoculation of *P. vivax* sporozoites can seed the liver with multiple hypnozoites. Although slowly eliminated antimalarials will suppress the first, and possibly the second relapse in equatorial regions, it remains to be shown whether this will reduce the total number of relapses or simply delay the occurrence of the next one. These benefits of PTP, unfortunately, reflect our limited ability to provide definitive cure of the dormant liver stages of *P. vivax*. If hypnozoitocidal treatment can be combined with ACTs in a reliable, safe, and effective way, then the superior efficacy against *P. vivax* afforded by the longer-acting combinations would be limited to a reduction in the rate of post-exposure reinfection that, in most vivax endemic regions, is reasonably low.

6.2. Treatment of *P. vivax* Exoerythrocytic Stages

The ability of *P. vivax* to lie dormant in asymptomatic individuals presents perhaps the greatest challenge for the clinical management of infection. This difficulty impacts the success of control as well (Douglas et al., 2010; Baird and Hoffman, 2004). Hypnozoites are insensitive to most blood schizontocidal drugs, including those that show activity against liver-stage schizonts, such as atovaquone + proguanil. The 8-aminoquinoline antimalarial drugs exert hypnozoitocidal activity at nanomolar concentrations; only primaquine is licensed (Galappaththy et al., 2007), tafenoquine is in clinical development and bulaquine has been abandonded (Elmes et al., 2008; Krudsood et al., 2006). Apart from these agents, the pipeline for new hypnozoitocidal candidates is remarkably dry (Wells et al., 2010).

The World Health Organization recommends a dose of 0.25–0.5 mg/kg primaquine daily for 14 days for radical treatment of vivax malaria in those who are not G6PDd (World Health Organization, 2010). However, the evidence base from which these recommendations are derived is extremely limited. The experimental and clinical evidence behind the total dose effect has been detailed above and elsewhere (Alving et al., 1960; Clyde and McCarthy, 1977). Trials have demonstrated that the standard low-dose regimen of primaquine (total dose of 3.5 mg/kg) fails to prevent relapses in many different endemic locations (White, 2011). For this reason, the most recent WHO antimalarial guidelines recommended a high-dose regimen of 7 mg/kg (equivalent to an adult dose of 30 mg per day for 14 days). In practice, most endemic countries continue to recommend the lower total dose out of fear of harm to unscreened G6PDd patients.

Antimalarial policy is prioritised towards eliminating the initial threat to the patient and avoiding adverse outcomes, rather than preventing future recurrences, which in most endemic settings are often regarded as inevitable. For this reason, the most important factor driving primaquine dosing policy are concerns over its safety and tolerability rather than its treatment efficacy. Primaquine has a well-described association with upper gastrointestinal discomfort. At a dose of more than 200 mg, all individuals complain of immediate and severe cramping, but this proportion is reduced to 30% of adults receiving 30 mg and 10% of those receiving 15 mg (Clayman et al., 1952). Administering primaquine with food improves tolerance of even the highest doses of the drug. In double blind randomized studies, 30 mg primaquine daily for as long as 50 weeks was as well tolerated as chloroquine or a placebo (Baird et al., 2001; Weiss et al., 1995).

An important constraint on the global deployment of primaquine is its potential to cause haemolysis in patients with G6PDd, which typically occurs in 2–15% (and up to 40%) of patients in endemic zones (Nkhoma et al., 2009). Individuals who have <10% of normal enzyme activity may be at risk of threatening haemolysis (Pannacciulli et al., 1965), but serious haemolysis has also occurred at higher levels of residual enzyme activity (Ziai et al., 1967). Relative susceptibility to primaquine may not be characterized on the basis of enzyme activity alone. Primaquine sensitivity phenotypes have been characterized in only three of the many dozens of clinically significant variants (A-, Mediterranean, and Mahidol variants). Much more work is required to establish the relationship between the residual enzyme activity and primaquine sensitivity phenotypes. This important issue is considered in greater depth in Chapter 4 of Volume B of this series.

Treatment of soldiers returning from the Korean War with vivax malaria revealed that African-Americans were more prone to a relatively mild and self-limiting primaquine-induced haemolysis compared to Caucasian soldiers. The 15 mg base daily dose (×14 days), while not efficacious against the experimental challenges with the Chesson strain, was highly efficacious against those Korean infections (Baird and Hoffman, 2004). The 15 mg regimen in African-American soldiers caused relatively mild and self-limiting haemolysis (Alving et al., 1960) and was thus adopted as the treatment of choice in patients not screened for G6PDd. The US Navy and Army did not begin routine screening of sailors and soldiers for G6PDd until 1981 and 2004, respectively. This regimen became the global standard of treatment, although a 5-day regimen (15 mg daily) found favour in some nations on the basis of a single poorly controlled trial (Cedillos et al., 1978). Only relatively recent well-controlled trials demonstrated the 5-day regimen to be almost completely inefficacious against relapse (Gogtay et al., 1999;Yadav and Ghosh, 2002). Moreover, the inadequacy of the 15 mg/14 days regimen against strains like Chesson has been a long-acknowledged problem. Studies consistently show high risk of relapse after such therapy in patients infected in Southeast Asia or, especially, the island of New Guinea (Elliott et al., 2004; Jelinek et al., 1995). Authorities now recommend the 30 mg regimen, with G6PDd screening, as first-line treatment regardless of geographic origin.

The radical cure of *P. vivax* in patients known to be G6PDd is difficult. Current WHO guidelines recommend for patients with >10% residual G6PD enzyme activity (WHO Class III or IV G6PDd variant), a weekly dose of 0.75 mg/kg for 8 weeks to reduce the risk of haemolysis whilst retaining efficacy (Alving et al., 1960; Leslie et al., 2008). The weekly dosing schedule was derived from studies in the USA in a small number of adults with the mildly primaquine-sensitive African A-G6PDd variant (Alving et al., 1960). Since host vulnerability to haemolysis varies between over 100 different G6PDd variants (Clyde, 1981), the available evidence is inadequate to ensure the universal safety of a 0.75 mg/kg dose either as a single dose, as advocated for reducing the transmission of falciparum malaria, or a weekly dose for the radical cure of vivax malaria (Baird and Surjadjaja, 2011).

There are as yet no reliable bedside tests for G6PDd. Hence, many health care providers elect to administer primaquine without G6PD testing or, probably more often, simply adopt a blanket policy to omit the prescription of anti-relapse therapy altogether. Clinicians may initiate low-dose treatment with primaquine with instructions to stop medication should the patient notice signs of haematuria.

The most recent WHO malaria treatment guidelines suggest that primaquine should be avoided in children <4 years of age and in pregnant women (World Health Organization, 2010). The rationale behind the 4-year-old cut off is unclear and unreferenced, but may reflect caution in the face of inadequate data rather than overt toxicity. However, the consequences of denying young children anti-relapse therapy also need to be factored into such decisions that profoundly shape policy and practice since the adverse consequences of recurrent vivax malaria in early life may be very considerable (Poespoprodjo et al., 2009). In practice, national malaria control programs often advocate deployment of 14 days primaquine in children as young as 1-year old, but studies are urgently needed to examine the risk vs.

benefit of primaquine in children, particularly infants and those with mild and moderate G6PDd.

Advocating long treatment courses of primaquine represents the greatest pragmatic challenge to the successful deployment of radically curative therapy. Adherence within the context of a clinical trial has been reported as acceptable (Leslie et al., 2004), but outside of study settings is almost certainly poor (Khantikul et al., 2009). One strategy to overcome this is use of shorter courses of a higher daily dose of primaquine to improve adherence. In Thailand, directly observed primaquine (1 mg/kg/day) administered over 7 days was well-tolerated and reduced relapses by day 28 to 4% (vs. >60% norm) (Krudsood et al., 2008). This is encouraging but since many relapses following therapy of the rapid relapsing Chesson strain with 8-aminoquinolines typically occur well beyond 1 month (Fig. 4.3), longer follow-up is needed to distinguish whether relapse was prevented or deferred. If the efficacy, tolerability and safety of short-course, high-dose primaquine regimens can be assured across the range of endemic

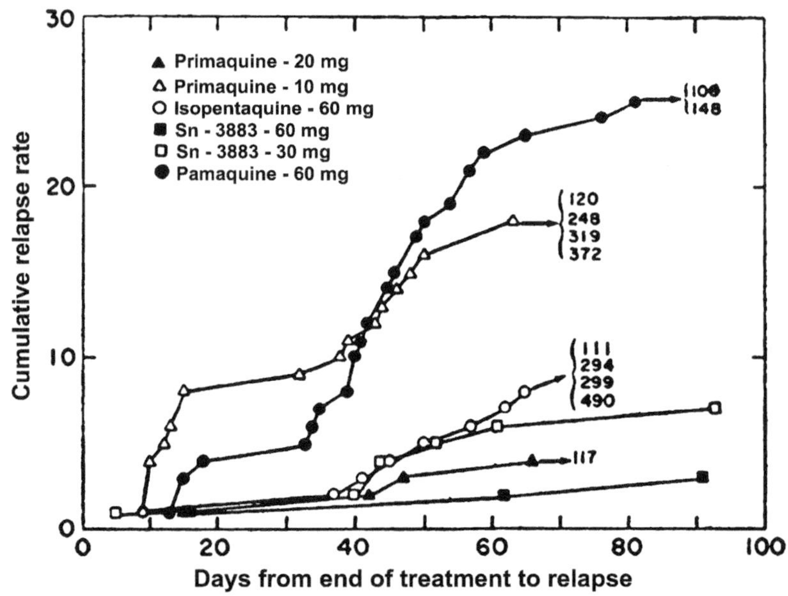

Figure 4.3 Data illustrating failure of various 8-aminoquinolines to achieve cure against relapse of Chesson strain *Plasmodium vivax* in experimentally challenged American prisoner volunteers (Cooper et al., 1953). Although approximately 80% of subjects thus challenged will relapse within 28 days without 8-aminoquinoline therapy, this graphic illustrates delay of relapse as the rule with inadequate anti-relapse therapy. *Reprinted with permission of the American Journal of Tropical Medicine & Hygiene.*

settings, then this new primaquine regimen would be a major advance in malaria treatment improving adherence to and thus the effectiveness of anti-relapse therapy. Alternatively, patient education campaigns and even routine direct observation of therapy may also be beneficial and warrant further development.

The development of a reliable and safe radical cure of *P. vivax* that can be pragmatically deployed in endemic country settings is one of the top priorities of vivax therapeutics. Once achieved, novel deployment strategies must then be studied to target the transmission cycle and reduce the overall burden of disease. Individuals infected with hypnozoites and/or undetectable or asymptomatic asexual parasitemia, constitute a major reservoir of infection. This is highlighted by observations of high rates of *P. vivax* recurrence following treatment for supposedly pure *P. falciparum* infections, presumably from occult infections (Douglas et al., 2011). This argues for the provision of effective anti-hypnozoite treatment to *all* patients with malaria in co-endemic regions rather than just those with confirmed *P. vivax* infections (Douglas et al., 2011).

7. TREATMENT OF SEVERE AND COMPLICATED VIVAX MALARIA

Severe vivax malaria, in the absence of mixed *P. falciparum* infection, has been reported with increasing frequency from India, Brazil, Venezuela, Afghanistan, Pakistan, India, Indonesia, and Papua New Guinea (Poespoprodjo et al., 2009; Islam and Qamruddin, 1995; Munteis et al., 1997; Torres et al., 1997; Perren et al., 1998; Pukrittayakamee et al., 1998; Carlini et al., 1999; Curlin et al., 1999; Tanios et al., 2001; Lawn et al., 2003; Habib and Singh, 2004; Kochar et al., 2005; Kotwal et al., 2005; Lomar et al., 2005; Saleri et al., 2006; Agarwal et al., 2007; Kumar et al., 2007; Maguire et al., 2007; Price et al., 2007a; Barcus et al., 2007; Genton et al., 2008; Manning et al., 2011; Tjitra et al., 2008; Nurleila et al., 2012; Shaikh et al., 2012). The most frequent manifestations are severe anaemia and acute respiratory distress. Less commonly, cerebral malaria, kidney injury, shock, and coagulopathy have also been reported. Although there have been no prospective clinical trials, severe vivax malaria has been successfully managed with various regimens including oral or intravenous quinine, intravenous artesunate, chloroquine, quinidine, quinine plus doxycycline, mefloquine, and artemether plus chloroquine (Genton et al., 2008). Until controlled clinical trials are conducted, severe non-falciparum malaria should be managed similarly to severe falciparum malaria in intensive care settings with intravenous artesunate or quinine. This is particularly important in co-endemic regions where mixed infections may often be missed by the microscopist.

8. EVALUATING DRUG RESISTANCE IN VIVAX MALARIA

Following an initial period of drug development in the 1940s and 1950s, the emphasis on drug evaluation of *P. vivax* declined and between 1966 and 2002, only 11% (47/435) of published antimalarial drug trials assessed antimalarial efficacy in patients with vivax malaria (Myint et al., 2004). Studies of antimalarial efficacy focused primarily on *P. falciparum* through the second half of the twentieth century. WHO protocols for the evaluation of *P. falciparum* have undergone extensive revision over that period; in general, these methodologies have been extended to include the study of *P. vivax* rather than specifically being adapted to the idiosyncracies of this parasite. The *in vivo* and *ex vivo* assessment of *P. vivax* drug susceptibility are far more challenging than *P. falciparum* (Price et al., 2009; Baird, 2004), reflecting fundamental differences in the biology of these two species that need to be specifically addressed. Indeed when the *in vivo* and *in vitro* methods for evaluating *P. falciparum* drug resistance are applied to *P. vivax*, they yield misleading or ambiguous findings.

8.1. *In vivo* Assessment of Schizontocidal Therapy

Assessment of the acute response to drug treatment of *P. vivax* is the same as that for *P. falciparum*. However, the interpretation of late parasitological outcome is confounded by the occurrence of relapses (as well as reinfection). Recurrent *P. vivax* parasitemia following initial treatment can arise from reinfection, relapse, or recrudescence. Molecular analysis, which has been so valuable in quantifying true recrudescence in clinical trials of *P. falciparum*, is not helpful in *P. vivax* studies since it is currently not possible to distinguish between recrudescence and relapse of the same parasite strain. *Plasmodium vivax* genotyping methodologies have been developed (Koepfli et al., 2009) but their application to clinical trials is not straightforward. A recurrent infection with a parasite with an identical *P. vivax* genotype can represent either a recrudescence or relapse from a brood of hypnozoites originating from the same inoculation with sporozoites (Chen et al., 2007; Imwong et al., 2007). Likewise, a mismatched genotype may be either a reinfection or a relapse. This problem can significantly confound clinical efficacy studies of hypnozoitocidal drugs since recurrence may appear from either (1) recrudescence despite therapy with blood schizontocide(s) (genetic match); (2) reinfection (genetic mismatch) or (3) relapse despite hypnozoitocidal therapy (genetic match or mismatch).

Considerable geographical differences exist between strains of *P. vivax* with regard to the absolute risk of relapse and the timing at which these occur. These range from 50% to 80% relapse within 3–6 weeks of an initial infection with Chesson strain to <5–20% risk occurring at 12 months in temperate strains (White, 2011). These relapse rates are clearly observed in the early clinical trials in which a curative but short-acting quinine regimen was used as a positive control for therapeutic activity of primaquine (Wiselogle, 1946). The clinical response with a long-acting antimalarial drug such as chloroquine is very different. A standard dose of chloroquine (total dose of 25 mg/kg) should eliminate all of the initial parasite biomass in blood, and its sustained effective blood concentrations prevent early reinfection or relapse. Clinical studies demonstrate that following a standard dose against sensitive parasites, the first recurrences are not observed until approximately 35 days following treatment; at this time, the mean whole blood concentrations of chloroquine have fallen to ~100 ng/ml (Baird et al., 1997). This figure concurs with previous studies to determine the MIC of CQS parasite which estimated a whole blood value between 90 and 120 ng/ml (Berliner et al., 1948a; Rombo et al., 1985).

The current 28-day *in vivo* test of chloroquine works on the basic principle that if the treatment regimen was well absorbed, no parasite recurrence should be observed within the duration of follow-up, regardless of its origin. Pharmacokinetic analysis, for instance, a measurement at day 7, can help to confirm adequate drug absorption (White et al., 2008) and establish whether recurrent infections occur in the face of blood concentrations known to exceed the MIC of the parasite (Baird et al., 1997). In practice, pharmacokinetic analyses are often omitted because of logistical constraints or cost. Anti-relapse therapy with primaquine is often administered concomitantly with a blood schizontocidal study drug. Since primaquine has both hypnozoitocidal and blood schizontocidal activity, its early administration can augment initial asexual parasite clearance and thus mask early indicators of declining blood schizontocidal drug efficacy. Primaquine should therefore be delayed until after the 28-day follow-up.

Although such *in vivo* tests have a useful positive predictive value in identifying populations where declining drug efficacy is evolving, it has some notable shortcomings. Early clinical resistance to long-acting antimalarial drugs can often manifest up to 63 days after treatment, and such a response would not be detected if follow-up is limited to 28 days. Furthermore, treatment failure reflects the ability of the recurrent parasite to grow in blood concentrations above a sensitive MIC, but this does not necessarily equate to the initial parasite biomass being CQR. Hence, caution is needed when attempting to extrapolate the initial biomarkers of resistance with subsequent treatment failure. In other words, the initial parasitemia may have been completely sensitive to treatment but was followed by relapse of a drug-resistant strain. The genotype and phenotype of parasites appearing in the primary parasitaemia remains ambiguous, except in the instance of therapeutic success and a classification as drug sensitive. The resistant phenotype and genotype may only be applied to the parasite populations later emerging in the presence of ordinarily effective concentrations of chloroquine.

Clinical trials of other blood schizontocidal agents add another layer of complexity to the interpretation. Most recent studies have adapted a pragmatic approach comparing the overall risk of recurrent *P. vivax* infection within a defined period of time. Such *in vivo* assessments have a public health merit, reflecting both blood schizontocidal and PTP. However, they do not permit the true efficacy to be gauged. Instead, the unadjusted risk of recurrence needs to be interpreted in light of the concomitant use of primaquine, the timing of recurrence, available molecular data, blood schizontocidal drug concentrations, entomological inoculation rates, and the epidemiology of relapse in the study area.

8.2. *Ex vivo* Assessment of Schizontocidal Therapy

8.2.1. Ex vivo *drug susceptibility testing*

Ex vivo drug susceptibility assays provide an alternative means of quantifying drug resistance of *Plasmodium* sp. free from the confounders of adherence to therapy and its absorption or metabolism, host immunity, relapse, and reinfection. In *P. falciparum*, a multitude of methods has been used to quantify drug activity, and significant difficulties persist in comparing results between laboratories and methodologies (Bacon et al., 2007). The analysis of *P. vivax* is complicated by its preferential invasion of young red blood cells, which limits its reproductive capacity and the ability to culture the parasite *ex vivo* (Udomsangpetch et al., 2007). To date, *P. vivax* cannot be sustained in *ex vivo* culture or grown from cryopreserved isolates for more than 1–2 cycles of asexual growth. To overcome this, *ex vivo* studies of *P. vivax* have focused on techniques evaluating the inhibitory effect of blood schizontocides on fresh parasites direct from the human host after short-term cultivation (Tasanor et al., 2006; Russell et al., 2003; Chotivanich et al., 2004; Druilhe et al., 2007). A variety of methods have been used to quantify parasite growth, including a modified schizont maturation test (Russel et al., 2008), enzymatic assays (Druilhe et al., 2007) and RTPCR (Russell et al., 2003).

A major confounding factor in the interpretation of these results is a marked stage specificity of action, which varies between drugs and *Plasmodium* sp. This property has been demonstrated for chloroquine activity in *P. vivax*, with parasites at trophozoite stage showing almost complete tolerance of chloroquine (Russel et al., 2008; Sharrock et al., 2008). IC_{50} estimates derived from parasite isolates with a high proportion of trophozoites are likely to be far greater than the same isolate tested at predominantly ring stage. Unlike *P. falciparum* infections, the parasitaemia of *P. vivax* presents with most stages of development detectable in peripheral blood smears, the degree of asynchronicity varying with inter-related host factors (such as immunity and age of the patient) and local transmission intensity. The *in vitro* drug response, therefore, needs to be interpreted according to the initial stage composition of the parasite population and the duration of the assay (Russel et al., 2008). The differences in quantifying *in vitro* drug susceptibility may explain some of the discrepancies between results of different laboratories, in which derived thresholds for chloroquine resistance have varied from 22 to 100 nM (Druilhe et al., 2007; Congpuong et al., 2002; Suwanarusk et al., 2007).

The stage specificity of action of chloroquine although apparent in *P. falciparum* is far less pronounced than that observed in *P. vivax*. This may reveal fundamental differences in the mechanism of action of the drug between these two species. Chloroquine activity in *P. falciparum* is thought to act via inhibiting haemozoin polymerisation, a central process by which young ring forms detoxifying digested haemoglobin in the parasite food vacuole. However, in *P. vivax*, haemozoin does not appear until more mature trophozoite and schizont stages, the very stages which appear tolerant of chloroquine. The findings of Nomura et al. (2001) emphasize fundamental differences in the mechanism of resistance to chloroquine by these two species and the need for deeper exploration of the cellular and molecular processes underlying therapeutic failure by resistance.

Development of a standardized *ex vivo* assay for assessing the therapeutic susceptibility of *P. vivax* needs to be a research priority. Recent funding opportunities have stimulated renewed efforts aimed at generating laboratory methods capable of sustaining *P. vivax* in continuous *ex vivo* culture.

The success of these will have a huge impact in transforming the current *ex vivo* assays into a more robust *in vitro* test that will permit its application repeatedly to the same parasite isolate. However, despite its limitations, the current schizont maturation assay has shown utility in discriminating parasite populations with different degrees of chloroquine resistance (Hasugian et al., 2009; Suwanarusk et al., 2007; Chotivanich et al., 2009), characterizing drug susceptibility profiles (Chotivanich et al., 2009; Suwanarusk et al., 2008) and the screening of susceptibility to therapeutic agents (Price et al., 2010; Lek-Uthai et al., 2008; Marfurt et al., 2011a, 2011b). This work permits definitive drug resistance phenotyping, and thus, the vitally important search for its genetic determinants.

8.2.2. Genetic basis of P. vivax *drug resistance*

Preliminary studies on drug resistance in *P. vivax* have focused on the orthologues of the transporter genes *pfcrt* and *pfmdr1*, known to be key determinants in *P. falciparum* resistance. The role of *pvcrt-o* in chloroquine-resistant *P. vivax* is not immediately apparent. Nomura et al. (2001) compared *pfcrt* homologues from a number of species and although synteny was observed between *P. falciparum* and *P. vivax*, they found no relationship between mutations in *pvcrt-o* and the clinical. Using an experimental system, in which *pvcrt-o* was transfected into *P. falciparum*, Sa et al. (2006) demonstrated a potential role for *pvcrt-o* in chloroquine transport across membranes, a threonine mutation at position 76 in their *Dictyostelium* system being associated with impaired accumulation of chloroquine. However, the relationship between *pvcrt-o* sequence and either clinical or *in vitro* phenotype was not apparent in studies of more than 330 field isolates from Thailand, Indonesia and Madagascar (Barnadas et al., 2008; Suwanarusk et al., 2007). An interesting study of returning travellers with severe vivax malaria contracted from the Brazilian Amazonia, found twofold increased *pvmdr1* copy number and 21-fold increased copy number of *pvcrt-0* compared to a patient with non-severe vivax malaria (Fernandez-Becerra et al., 2009). No studies have evaluated the role of *pvcrt-o* amplification as a marker of CQR in this parasite.

More progress has been made correlating CQR *P. vivax* with single-nucleotide polymorphisms of *pvmdr1*. Field studies of a large number of clinical isolates demonstrated that a Y976F *pvmdr1* polymorphism was present in nearly all patients presenting to a clinic in Papua Indonesia, where high-grade CQR is known to be present (Suwanarusk et al., 2007). The same study showed that the same mutation was only present in 25% of isolates in Thai isolates from an area where chloroquine retains high efficacy, and this was manifest by lower *in vitro* susceptibility in parasites with the mutation. These observations are supported by work from Papua New Guinea where the Y976F mutation was correlated with the *in vivo* response (Marfurt et al., 2008).

However, since chloroquine resistance can occur in isolates with wild-type *pvmdr1* (Suwanarusk et al., 2007; Sa et al., 2005; Kim et al., 2011b), and cure can be achieved in the presence of the 976 mutation (Barnadas et al., 2008), *pvmdr1* mutations are likely to be at best only minor determinants of resistance. Whereas the *pvmdr1* 976 mutation was found at greater prevalence in Papuan isolates, *pvmdr1* amplification had a greater prevalence in Thailand (21%) compared to Indonesian (0%), and was associated with a twofold reduction in *in vitro* susceptibility to mefloquine (Suwanarusk et al., 2008). Imwong et al. (2008) found amplification of *pvmdr1* in *P. vivax* isolates, and proposed that this represented parasite selection associated with greater use of mefloquine in Thailand and Myanmar.

The roles of *DHFR* and *DHPS* have been well characterized in *P. falciparum* and the relationship with clinical and *in vitro* drug susceptibility established. Sequential acquisition of mutations in *pvdhfr* and *pvdhps* has also observed where *P. vivax* has been exposed to intense antifolate pressure (Tjitra et al., 2002; Karunajeewa et al., 2008b; Fernandez-Becerra et al., 2009; Rungsihirunrat et al., 2008; Zakeri et al., 2009). However, in view of the lack of antifolate efficacy against most *P. vivax* and logistical difficulties in anti-folate *in vitro* testing, no studies to date have addressed phenotypic correlation with prevalent genotypes. Studies with limited numbers of patients with *P. vivax* and erroneously treated with sulfadoxine plus pyrimethamine, however, suggested uniform sensitivity of strains having wild-type DHFR and DHPS (Hastings et al., 2004; Hastings et al., 2005).

8.2.3. Non-human primates

8.2.3.1. Human parasites

A means of assessing the therapeutic response to chloroquine and other drugs by human strains of *P. vivax* in owl monkeys (*Aotus trivirgatus*) was detailed by Rossan et al. (1975) and Schmidt (1978). Several other species of owl monkey were proven receptive to hosting *P. vivax*: *Aotus lemurinus griseimembra*, *A. vociferans*, *A. nancymai* and *A. azarae boliviensis*. One species of squirrel money, *Saimiri boliviensis*, also hosts blood-stage *P. vivax*. A variety of other New and Old World monkeys (tamrin, spider monkey, gibbon, and white-faced monkey) have proven more or less receptive to *P. vivax* infection (Coatney, 1971), but the owl and squirrel monkeys have proven the most durable with respect to reliability of the infection and suitability to humane captivity. Chimpanzees (*Pan troglodytes*), unlike monkey hosts, experience both the acute attack and relapses (Krotoski et al., 1986), but high cost of humane captivity and large size, apart from the ethical challenges of experimentation on sentient animals, makes chimpanzees unsuitable for this purpose.

With the lone exception of chimpanzees, adapting *P. vivax* to these animals is laborious and often fails. Use of splenectomized animals greatly improves the likelihood of success. However, in experiments involving *Aotus lemurinus griseimembra*, investigators found no difference in susceptibility between spleen intact and splenectomized monkeys (Jordan-Villegas et al., 2005). Moreover, they found no differences in infection success between the adapted Salvador I strain of *P. vivax* and two isolates from Colombia. The unnatural host–parasite relationship, the absence of a spleen, and possible confounding or bias in selection for monkey-adapted strains of human parasites need all to be considered in interpreting experimentation with chemotherapeutics.

Tables 4.8 and 4.9 list the therapeutic responses of several strains of *P. vivax* established long before onset of clinical resistance to chloroquine (Rossan et al., 1975; Schmidt, 1978). These may thus be considered standard responses of chloroquine-sensitive parasites to chloroquine therapy in these animals. Investigators using owl or squirrel monkeys tend to report total dose of drug rather than mg/kg (see rhesus monkey data in earlier tables), because these adult monkeys typically weigh at or near 1 kg. The effectiveness of an approximately 10 mg dose ranged from none to complete by strain, whereas a 25 mg dose of chloroquine proved uniformly effective (also see references Collins et al., 2005, 1998; Sullivan et al., 2001). Thus, Collins et al. (2000) considered failure of a 30 mg dose definitive evidence of resistance to chloroquine by *P. vivax* in *Aotus* or *Saimiri* monkeys.

Table 4.8 Cure Rates of Various Doses of Chloroquine against Chloroquine-Sensitive Strains of *Plasmodium vivax* in Blood stage-Challenged *Aotus trivirgatus*

Strain	Total Chloroquine Dose (mg)	Monkeys	Cure rate ($)
New Guinea Chesson	9	2	0
	18	4	100
	36	2	100
Vietnam Palo Alto	9	4	0
	18	5	80
	36	4	100

Adapted from Schmidt (1978)

Table 4.9 Cure Rates of 10 mg or 25 mg Doses of Chloroquine against Chloroquine-Sensitive Achiote Strain of *Plasmodium vivax* in Blood stage-Challenged *Aotus trivirgatus* or *Saimiri sciureus* ()

Animal	Number	Dose (mg)	Cure rate (%)
Aotus trivirgatus	6	25	100
Saimiri sciureus	10	25	100
	4	10	100

Adapted from Rossan et al. (1975)

A sufficient knowledge base informs the usual range of effective doses for chloroquine-sensitive *P. vivax* in the *Aotus* or *Saimiri* models. Clinically phenotyped isolates from travellers infected in Papua New Guinea (AMRU-1) and Sumatra, Indonesia (Indonesia I/CDC) often recrudesced with 30 mg or 15 mg, respectively (Rieckmann et al., 1989; Collins et al., 1992; Sullivan et al., 1999; Obaldia et al., 1997). Later, the Indonesia XIX/CDC strain from patients in Indonesian Papua (clinically phenotyped as chloroquine-resistant) was inoculated into 11 monkeys and either failed to clear parasitemia (1 animal) or recrudesced (8 animals) following 30 mg treatments of chloroquine (Collins et al., 2000). Tafenoquine has been evaluated against blood stages of chloroquine-resistant AMRU-1 strain in splenectomized *Aotus* monkeys (Obaldia et al., 1997; Cooper et al., 1994). Tafenoquine, though apparently slow-acting on the AMRU-1 strain blood stages, was nonetheless more rapidly acting that chloroquine against the chloroquine-sensitive Chesson strain in *Aotus* monkeys (Cooper et al., 1994).

8.2.3.2. *Plasmodium cynomolgi*

Although not directly useful in assessing clinical resistance of human parasite isolates, the therapeutic responses of *P. cynomolgi* to various blood schizontocides provides a frame of reference for relative potency of the drugs against a parasite very close to *P. vivax* but in its natural host. Schmidt et al. (1982) reported hundreds of experimental infections of three distinct strains of *P. cynomolgi* (M, B, and R_0) in rhesus macaques treated by a wide variety of blood schizontocides. The investigators evaluated several important variables, including route of challenge, size of inoculum, anopheline species, source and preparation of inoculum, route of drug administration and duration of dosing. The report represents an appeal for and technical foundation supporting standardization in the evaluation of chemotherapies aimed at *P. vivax* in humans. The drugs evaluated include quinine, atabrine, amodiaquine, chloroguanide and pyrimethamine blood schizontocides, along with pamaquine, pentaquine, isopentaquine, and primaquine hypnozoitocides. The investigators close their report by writing, '...*with respect to the major features of disease and responses to standard chemotherapeutic agents, infections with sporozoites and trophozoites of P. cynomolgi in rhesus monkeys are remarkably similar to infections with New Guinea Chesson strain of P. vivax in human volunteers. Infections with these strains are not identical in every respect. They are sufficiently alike, however, to give strong support to the use of the P. cynomolgi-rhesus monkey model in the search for improved prophylactic, radical curative, and suppressive drugs*'.

This superb model, however, has not been widely applied in evaluating blood schizontocides for potential use in humans. Instead, what typically occurs is the development and licensure of drugs for acute falciparum malaria being evaluated against *P. vivax* in patients. This has been the case with mefloquine, atovaquone–proguanil, and each of the ACTs as they matured in development. That approach, pragmatic as it may be, will miss partnering with primaquine or some other hypnozoitocide for demonstration of the safety and efficacy of the radical cure required against vivax malaria in the real world. As the extensive independent work of Schmidt and Rossan demonstrated, this animal model may offer a practical means of evaluating blood schizontocide–hypnozoitocide partners prior to use in relatively more costly and risky clinical trials.

8.3. Primaquine

8.3.1. In vivo

High rates of *P. vivax* recurrence have been observed following primaquine therapy, but there is no standardized means of affirming resistance in such patients. The existence of truly primaquine-resistant hypnozoites has never been unequivocally demonstrated (Galappaththy et al., 2007; Goller et al., 2007). Measuring the therapeutic response to primaquine is by no means impossible, and in infected travellers returned to non-endemic zones, relatively straightforward. The difference between great difficulty and relative ease is confounding by risk of reinfection. Reinfection represents the primary obstacle in assessing the therapeutic response to primaquine. Relapse and reinfection may each occur during the necessary months of post-treatment follow up. Unlike *P. falciparum*, where reinfection and recrudescence represent the relatively easily distinguished possible sources of new parasitaemia, the hypnozoite of *P. vivax* imposes complexity and ambiguity. The limitations with molecular genotyping as a tool of discernment have already been explained in this chapter.

Another difficulty in assessing the therapeutic response to primaquine is the highly variable natural rate of relapse among strains of *P. vivax* (see Chapter 2). If local strains relapse in only 10% of infections (as in India), for example, a 10% relapse rate among treated patients is not 90% efficacy, but no efficacy at all. In contrast, a 10% relapse rate in patients infected in Southeast Asia or Oceania, where most strains relapse at least once, may well approach 90% real efficacy. The natural relapse rate must be known to estimate therapeutic efficacy.

The conduct of studies in non-endemic zones eliminates confounding by reinfection, but nonetheless requires knowledge of relapse risk (without primaquine) in order to estimate the efficacy of primaquine in treated patients. Studies in non-endemic zones typically involve a series of patients infected at various locations abroad. Unless all of the patients were infected at the same location, efforts to estimate primaquine efficacy in infected travellers may be considered untenable. The value of such study populations may be limited to compelling case demonstrations of resistance to primaquine in individual patients. The approach to assessing primaquine resistance thus differs greatly depending upon endemic vs. non-endemic settings. In brief, endemic settings may yield estimates of primaquine efficacy in populations but cannot demonstrate a primaquine-resistant strain in any given patient; and the converse is true in non-endemic settings.

The two sections that follow each expresses analytical approaches that could perhaps be applied in assessing therapeutic responses to primaquine. Neither has been assessed in practice, much less validated. These should not be construed as recommended for implementation. Instead, the exercise of considering the problem provides useful insights, principally regarding the obstacles, pitfalls, and gaps in understanding precisely what makes such assessments so challenging.

8.3.1.1. Primaquine resistance in endemic zones

Assessing the therapeutic response to primaquine in endemic zones, where reinfection likely occurs, requires two essential features:

(1) Data that estimate the natural rate of relapse in the study population.

(2) Data that estimate the natural rate of reinfection in the study population, or have confidence that reinfection is either highly unlikely, or accept reinfection as equally distributed among treatment groups.

Meeting both of these requirements permits assessment of the therapeutic response in populations of patients. In other words, patients require randomization to at least two or three treatment arms in order to assess the therapeutic response to a single radical cure regimen.

In the instance of a trial including a reinfection control group, the groups would resemble these in design and intent:

A) Blood schizontocide only for therapy of acute vivax malaria (relapse and reinfection control).

B) Blood schizontocide plus high-dose primaquine (reinfection control).

C) Blood schizontocide plus standard primaquine (evaluation arm).

This approach assumes that the blood schizontocide(s) has complete efficacy in all arms, and that the high-dose primaquine has complete efficacy against relapse in the reinfection control. Few therapeutic options today come with such certainty. Nonetheless, if such treatments were applied, new parasitaemias among these groups may be classified as follows:

A = parasitaemias arising from relapse and reinfection;

B = parasitaemias arising from reinfection;

C = parasitaemias arising from reinfection and failure of primaquine;

where A − B = risk of relapse without primaquine, and C − B = risk of primaquine treatment failure, then [(A − B) − (C − B)]/(A − B)) = proportion of parasitaemias from relapse prevented by the primaquine in the evaluation arm, i.e. the therapeutic response to standard primaquine therapy. This approach may only apply to first relapse and cannot manage multiple relapses in any given subject. The model also does not account for possible confounding by long- vs. short-elimination half-life blood schizontocidal treatments. Alternatively, a design without reinfection control would look like this:

X = parasitaemias arising from reinfection and relapse (no primaquine);

Y = parasitaemias arising from reinfection and failure of primaquine;

Where X − Y = relapses from failure of primaquine; X − Y/X = proportion of parasitaemias prevented by primaquine therapy. A low rate of reinfection, e.g. <5% of recurrences, would have minimal impact on the estimate of efficacy. However, as risk of reinfection rises, the estimate of efficacy becomes increasingly confounded in the negative direction.

Whether using reinfection control or not, in any given subject of study in endemic settings, a single infection cannot be unambiguously classified as a relapse. Phenotyping resistance to primaquine for any isolate is not currently possible in this setting.

8.3.1.2. Primaquine resistance in returned travellers

Removing reinfection as a confounder by evaluating primaquine therapeutic responses in travellers repatriated to non-endemic zones eases the difficulty of demonstrating resistance. However, efficacy cannot be estimated in such populations unless they come in relatively large numbers from the same locality, thereby providing a means to estimate the natural relapse rate among them. In individual patients coming from scattered locations, risk of relapse linked to each patient may be unknowable. Thus, demonstrating resistance to primaquine in them becomes a task of proving that a recurrent parasitaemia occurred despite fully compliant and naturally metabolized standard primaquine therapy, i.e. knowing the patient was infected by a primaquine-resistant strain of *P. vivax*. Investigators should resist the temptation to put denominators under these treatment failures. The absence of relapse following primaquine therapy in such patients may be a consequence of primaquine-sensitive hypnozoites, or simply the absence of hypnozoites. Treatment efficacy in the absence of certainty regarding natural risk of relapse is unknowable even if individual cases of primaquine-resistance are demonstrable.

Patients free of risk of reinfection and experiencing a recurrent parasitaemia within 2 weeks–3 years of starting standard primaquine therapy for a primary attack may be considered possibly infected by a primaquine-resistant strain of *P. vivax*. Standard therapy should be a total dose of 3 or 6 mg/kg (Korea, India, Americas vs tropical Asia or Oceania) delivered in daily doses over 2 weeks. Some means of assuring adherence should be implemented. Plasma or blood levels of primaquine are not helpful in demonstrating resistance beyond simply confirming adherence: blood levels of 8-aminoquinolines do not correlate with therapeutic effect. There may be human genetic mutations to CYP genes that profoundly impact metabolism of primaquine, and as these become better characterized and understood, they should certainly be considered as a possible basis of therapeutic failure.

8.3.2. Ex vivo

Methods to assess therapeutic responses of *P. vivax* isolates in *in vitro* hepatic cell lines have been described (Chattopadhyay et al., 2010; Sattabongkot et al., 2006; Millet et al., 1989; Mazier et al., 1983). No procedure or hepatocyte line has yet been standardized or wholly validated. The requirement for viable sporozoites of *P. vivax* poses a daunting challenge to such systems. Not only must these be obtained from patients (or chimpanzees) via anopheline mosquitoes fed on their blood but also each such isolation will bring a distinct strain of parasite to bear (except in chimpanzees challenged with a known strain). This makes standardizing therapeutic responses observed in hepatocyte lines very difficult. A possible and partial solution would be the efficient harvest of as many sporozoites as possible derived from a single patient at a single time, thus maximizing the number of assays that could be evaluated without confounding by strain variability (Hoffman). This approach requires an almost industrial approach to the problem.

Understanding and segregating tissue schizontocidal vs. hypnozoitocidal activity is crucial in *ex vivo* assessment of hypnozoitocidal activity. Preventing the development of hypnozoites (tissue schizontocidal) fundamentally differs from killing extant hypnozoites (hypnozoitocidal). Causal prophylaxis with primaquine, for example, likely works by killing active early liver stages destined to be either primary tissue schizonts or hypnozoites. This is tissue schizontocidal activity and is achieved at much lower doses relative to standard radical cure of patients with acute vivax malaria (Section 4). They already harbour hypnozoites and primaquine therapy must destroy them. *Ex vivo* systems that demonstrate inhibition of hypnozoite formation by drug exposure during or soon after introduction of sporozoites do not offer insights relevant to radical cure. Systems capturing hypnozoitocidal activity must demonstrate inhibition of hypnozoite activation by drugs introduced to the system following their maturation to dormant hypnozoites. This represents a much steeper technical challenge than demonstration of inhibition of hypnozoite formation.

8.3.2.1. Non-human primates

Plasmodium vivax in non-human primates does not form viable hypnozoites, with the exception of chimpanzees. This renders their evaluation for susceptibility to primaquine virtually impossible by this means. Although technically feasible with chimpanzees, this approach would be extremely costly, and socially and ethically challenging. Infecting a chimpanzee with an isolate or strain would likely be a very rare event compelled by extreme scientific interest, need, and potential benefit. Federal legislation being considered by the US Congress in 2012, however, would prohibit such experimentation.

9. RESEARCH AND DEVELOPMENT URGENCIES

This chapter presents the complexities and challenges of the current state of the diagnosis and treatment of vivax malaria. Failing chloroquine and the challenge of deploying primaquine reliably in a safe and effective manner, lie at the heart of vivax chemotherapeutics. However, progress has been made, particularly in the past 5 years with a growing awareness among malariologists of the importance of *P. vivax*. Long withheld fiscal resources have begun to be committed, and scientists and clinicians may now unleash those and their collective intellect upon this remarkably complex problem.

Fielding safe and effective radical cure represents the greatest urgency across a very broad front of necessary work. In the short and expedient term, a course justified by both the stakes and the prospects for success, existing therapies must be optimized for safety and efficacy in patients, especially among the G6PDd, pregnant and small children. As given in Section 3, this is by no means a trivial task. It requires the kind of focus, determination and commitment exhibited by Alving, his colleagues and sponsors, but without a global war spurring progress. It also requires grasp of very complex issues underpinning rational and efficient experimental and clinical approaches. Above all, the community of malariology should accept the heavy burden of bringing therapies to bear against the problem of vivax malaria as it occurs in endemic zones; that is, where specific diagnosis of infection and screening against dangerous contraindications is very often absent.

9.1. Proof of Efficacy

The loss of chloroquine–primaquine for radical cure, real or looming, demands proof of efficacy for new partners to primaquine (and tafenoquine) in this essential chemotherapeutic tool. The apparent reliance of primaquine upon its co-drug for efficacy raises the possibility that not all new blood schizontocides will be suitable partner drugs in radical cure. These new therapies against the acute attack must be evaluated with primaquine and its efficacy and safety against relapse. The recent trial in Indonesian soldiers already described (Sutanto, in preparation) provides hopeful evidence that other drugs do indeed permit good primaquine efficacy against relapse despite 60 years of continuous and widespread improper use. The effective co-drug in that instance, from dihydroartemisinin–piperaquine treatment, was almost certainly the very long-lived piperaquine component. Primaquine was not administered to those subjects until day 28, the approximate elimination half-life of piperaquine. This single trial firmly points to dihydroartemisinin–piperaquine combined with primaquine as a highly efficacious radical cure, and in so doing addresses the most conspicuous gap in vivax malaria chemotherapeutics. These promising results should embolden further such efforts.

9.2. Proof of Safety

Standard primaquine dosing (210–420 mg over 14 days, or shorter) will not be proven safe in unscreened G6PDd patients. That is an issue for strategies aimed at minimizing harm, like less threatening dosing or G6PDd diagnostics (see Minimizing Harm below). Proof of safety with standard primaquine dosing in G6PD normal patients is required for new partners with primaquine in radical cure. As the harmful drug–drug interaction between atabrine and pamaquine clearly demonstrated in 1943, similar poor outcomes combining these families of compounds is possible. Lack of PK/PD data on DHA-PP and primaquine impelled the day 28 dosing of primaquine in the Indonesian study discussed. Concurrent administration would be the preferred approach in order to maximize both possible synergies and convenience of dosing. Blood schizontocides intended for partnering with primaquine require PK/PD studies in healthy volunteers to guide and affirm safe dosing strategies.

9.3. Proof of Effectiveness

As occurred with pamaquine nearly a century ago, malaria chemotherapists today strive to balance risk of harm against risk of poor adherence with primaquine. A high-dose/short-duration regimen that generates fear in patients and health care workers would likely prove as ineffective as low-dose/long-duration strategies. In the absence of non-threatening doses of primaquine or a practical means of excluding the vulnerable (see Section 9.4), this problem will persist. Malariologists should strive to develop the evidence required to buttress strategies aimed at levels of effectiveness that successfully challenge the hypnozoite reservoir in endemic zones. Mitigating fear, for example, may be accomplished with practical early detection of haemolysis and cessation of dosing. Likewise, demonstrations of the impact of successful primaquine strategy in communities may provide the motivation for providers and patients alike to strive for fuller adherence. Confidence in the success of primaquine therapy, as in almost all human endeavours, would inspire fuller commitment to it.

9.4. Minimizing Risk of Harm

Two 8-aminoquinoline drugs, primaquine and tafenoquine, represent the only likely agents of anti-relapse therapy for the coming decade. The threat that these drugs pose to unscreened patients with G6PDd severely limits their effectiveness in combating endemic malaria. Shorter duration, higher dose regimens of primaquine, like pamaquine before it, were not considered because of the threat such dosing posed to unscreened G6PDd patients. Tafenoquine progressed through 30 years of development almost solely on the promise of shorter duration dosing and, presumably, assured exclusion of G6PDd patients from perhaps catastrophic exposure to drug.

Malaria drug developers may solve the problem of 8-aminoquinoline toxicity in G6PDd patients by simply mining the phenomenon of co-drug synergy to discover, perhaps, a co-drug (whether it happens to be a blood schizontocide or not) that provides complete efficacy against hypnozoites at doses below the threshold of harm to the most sensitive G6PDd variants. This phenomenon, recognized in 1955 as the hypnozoitocide development program was drawing to a close, was not further leveraged to improve the therapeutic index of primaquine. Schmidt began an exploration of co-drug synergy in the 1980s for that expressed purpose, but the Army abandoned tafenoquine for radical cure in favour of a chemoprophylaxis development strategy. A full exploration of co-drug synergy with primaquine and tafenoquine should at last be carried out.

Alving's unpublished 1949 report (Alving et al., 1949) to his sponsors describing the complete abrogation of severe haemolytic sensitivity to an 8-aminoquinoline (isopentaquine) by the addition of methylene blue to the treatment suggests yet another approach. Rather than dose reduction, detoxifying the 8-aminoquinoline at standard dosing appears possible. Ironically, the logical starting point for that exploration is precisely that which launched modern antimalarial drugs – the aniline dye methylene blue. This compound is a licensed therapy for a number of indications, including reversal of the profound methaemoglobinemia of nitrite poisoning. It is at least plausible that this therapeutic effect could uncouple the oxidative events that lead to 8-aminoquinoline-induced haemolysis. Confirming and optimizing this effect at high doses of primaquine, along with co-formulation, would perhaps provide universally safe and effective therapy against relapse.

Another approach to minimizing harm is to effectively exclude those at risk. Among patients and subjects considered good candidates for the treatment, the 8-aminoquinolines are remarkably safe and well tolerated when administered with food, even in relatively massive doses (e.g. a 200 mg single dose (Clayman et al., 1952) or 30 mg daily for 50 weeks (Fryauff et al., 1995)). A point-of-care G6PD diagnostic device would open up high-dose/short-duration dosing for most patients. Such development is at last underway and shows promise (Kim et al., 2011a).

9.5. New Drug Discovery

Unless the new blood schizontocides being discovered for falciparum malaria fail against vivax malaria, there is no practical need for such discovery solely for *P. vivax*. That highly sophisticated pipeline has produced good therapies for the acute attack and is likely to continue doing so. Necessary drug discovery for vivax malaria is effectively limited to hypnozoitocides, and so far, only the 8-aminoquinolines (and their attendant toxicity issues) have shown broad activity as a class. However, that search has had important limitations, and still does.

Pamaquine was not identified on the basis of screening for anti-relapse activity. Likewise, primaquine was selected for superior activity against relapse in humans from a field of just a few dozen candidate 8-aminoquinolines chosen solely on the basis of toxicity data irrelevant to the G6PDd problem. Activity against hypnozoites and toxicity in G6PDd erythrocytes never entered the development algorithm. An effort by the US Army during the 1970s and 1980s, however, was informed by screening diverse chemicals for anti-relapse activity against *P. cynomolgi* in rhesus monkeys. The broad failure of chemicals not belonging to the 8-aminoquinolines, and the broad success of those that did, brought resignation to an enduring problem of haemolytic toxicity in radical cure of vivax malaria.

The US Army faced this problem by striving to find a means of estimating relative haemolytic toxicity in a pre-clinical *in vitro* system (Baird et al., 1984, 1986a, 1986b; Baird, 1984). The mechanism of primaquine-induced haemolysis, at the time (and in some quarters, still) believed to be generation of an intracellular oxidizing stress generated by primaquine metabolites, could not be reconciled with those *in vitro* data. Oxidative injury did not correlate with the oxidative activity of 8-aminoquinolines, very similar to methylene blue (Baird, 1984). An unknown mechanism of haemolysis was at work in G6PDd red blood cells. The failure to elucidate that process effectively crippled pre-clinical screening of 8-aminoquinoline candidates for haemolytic toxicity. Tafenoquine advanced through development with almost no understanding of its potential haemolytic toxicity relative to primaquine. In 2012, more than 30 years after the entry of tafenoquine into clinical studies that critical assessment awaits the completion of dose challenge studies in healthy G6PDd volunteers.

Drug discovery of hypnozoitocides, in the absence of a stroke of extraordinary luck in the very limited screening capacity made available with *P. cynomolgi* in rhesus macaques, will remain limited to 8-aminoquinolines. Discovery of a means of pre-clinical screening against haemolytic toxicity will be an essential element of that complex task. That will likely require elucidation of the molecular basis of 8-aminoquinoline haemolytic toxicity and rational leveraging of that understanding in pre-clinical screening.

10. SUMMARY

No other infectious disease so seriously impacting so many people as vivax malaria requires reliance upon such poor chemotherapeutic tools and strategies unimproved in over 60 years. Over the past 20 years, *P. vivax* has developed resistance to chloroquine, which has spread over most vivax-endemic regions and in some places, has already achieved dominance over sensitive strains. The radical cure of hypnozoite stages threatens dangerous toxicity as prescribed, and poor evidence of contemporary therapeutic efficacy further inhibits access to this essential therapy. The research community has ignored safe and effective anti-relapse therapy for those at greatest risk. The inadequate chemotherapeutics for vivax malaria in residents of endemic zones exacerbates the morbidity and associated mortality of this infection and deeply undermines efforts to eliminate it. This chapter outlines the rationale and nature of further investigations aimed at mitigating the serious deficiencies in the diagnosis and treatment of *P. vivax*.

REFERENCES

Agarwal, R., Nath, A., Gupta, D., 2007. Noninvasive ventilation in *Plasmodium vivax* related ALI/ARDS. Intern. Med. 46, 2007–2011.

Alving, A.S., Craige Jr., B., Pullman, T.N., Whorton, C.M., Jones, R.J., et al., 1948. Procedures used at Stateville penitentiary for the testing of potential antimalarial agents. J. Clin. Invest. 27, 2–5.

Alving, A.S., Eichelberger, L., Arnold, J., Edgcomb, J., 1949. The clinical testing of antimalarial drugs at Stateville Penitentiary: semi annual report 1948-49. NIH Malaria Report No 87 USPHS Antimalarial Grant No 198.

Alving, A.S., Arnold, J., Hockwald, R.S., Clayman, C.B., Dern, R.J., et al., 1955. Potentiation of the curative action of primaquine in vivax malaria by quinine and chloroquine. J. Lab. Clin. Med. 46, 301–306.

Alving, A.S., Rucker, K., Flanagan, C.L., et al., 1959. Observations of primaquine in the prophylaxis and cure of vivax malaria. In: Proceedings of the 6th International Congress on Tropical Medicine and Malaria 7: 203–209.

Alving, A.S., Johnson, C.F., Tarlov, A.R., Brewer, G.J., Kellermeyer, R.W., et al., 1960. Mitigation of the haemolytic effect of primaquine and enhancement of its action against exoerythrocytic forms of the Chesson strain of *Plasmodium vivax* by intermittent regimens of drug administration: a preliminary report. Bull. World Health Organ. 22, 621–631.

Arinaitwe, E., Sandison, T.G., Wanzira, H., Kakuru, A., Homsy, J., et al., 2009. Artemether-lumefantrine versus dihydroartemisinin-piperaquine for falciparum malaria: a longitudinal, randomized trial in young Ugandan children. Clin. Infect. Dis. 49, 1629–1637.

Arnold, J., Alving, A.S., Hockwald, R.S., Clayman, C.B., Dern, R.J., et al., 1954. The effect of continuous and intermittent primaquine therapy on the relapse rate of Chesson strain vivax malaria. J. Lab. Clin. Med. 44, 429–438.

Arnold, J., Alving, A.S., Hockwald, R.S., Clayman, C.B., Dern, R.J., et al., 1955. The antimalarial action of primaquine against the blood and tissue stages of falciparum malaria (Panama, P-F-6 strain). J. Lab. Clin. Med. 46, 391–397.

Awab, G.R., Pukrittayakamee, S., Imwong, M., Dondorp, A.M., Woodrow, C.J., et al., 2010. Dihydroartemisinin-piperaquine versus chloroquine to treat vivax malaria in Afghanistan: an open randomized, non-inferiority, trial. Malar. J. 9, 105.

Bacon, D.J., Jambou, R., Fandeur, T., Le Bras, J., Wongsrichanalai, C., et al., 2007. World Antimalarial Resistance Network (WARN) II: in vitro antimalarial drug susceptibility. J. Infect. Dis. 6, 120.

Baird, J.K., Hoffman, S.L., 2004. Primaquine therapy for malaria. Clin. Infect. Dis. 39, 1336–1345.

Baird, J.K., Surjadjaja, C., 2011. Consideration of ethics in primaquine therapy against malaria transmission. Trends Parasitol. 27, 11–16.

Baird, J.K., Decker-Jackson, J.E., Davidson, D.E.J., 1984. An in vitro micro-volume procedure for rapid measurement of erythrocytic hexose monophosphate shunt activity. Int. J. Biochem. 16, 1049–1052.

Baird, J.K., Davidson, D.E.J., Decker-Jackson, J.E., 1986a. Oxidative activity of hydroxylated primaquine analogs. Non-toxicity to glucose-6-phosphate dehydrogenase-deficient human red blood cells in vitro. Biochem. Pharmacol. 35, 1091–1098.

Baird, J.K., McCormick, G.J., Canfield, C.J., 1986b. Effects of nine synthetic putative metabolites of primaquine on activity of the hexose monophosphate shunt in intact human red blood cells in vitro. Biochem. Pharmacol. 35, 1099–1106.

Baird, J.K., Basri, H., Bangs, M.J., Subianto, B., Patchen, L.C., et al., 1991. Resistance to chloroquine by *Plasmodium vivax* in Irian Jaya, Indonesia. Am. J. Trop. Med. Hyg. 44, 547–552.

Baird, J.K., Basri, H., Subianto, B., Fryauff, D.J., McElroy, P.D., et al., 1995a. Treatment of chloroquine resistance *Plasmodium vivax* with chloroquine and primaquine or halofantrine. J. Infect. Dis. 171, 1678–1682.

Baird, J.K., Fryauff, D.J., Basri, H., Bangs, M.J., Subianto, B., et al., 1995b. Primaquine for prophylaxis against malaria among nonimmune transmigrants in Irian Jaya, Indonesia. Am. J. Trop. Med. Hyg. 52, 479–484.

Baird, J.K., Leksana, B., Masbar, S., Fryauff, D.J., Sutanihardja, M.A., et al., 1997. Diagnosis of resistance to chloroquine by *Plasmodium vivax*: timing of recurrence and whole blood chloroquine levels. Am. J. Trop. Med. Hyg. 56, 621–626.

Baird, J.K., Lacy, M.D., Basri, H., Barcus, M.J., Maguire, J.D., et al., 2001. Randomized, parallel placebo-controlled trial of primaquine for malaria prophylaxis in Papua, Indonesia. Clin. Infect. Dis. 33, 1990–1997.

Baird, J.K., Fryauff, D.J., Hoffman, S.L., 2003. Primaquine for prevention of malaria in travelers. Clin. Infect. Dis. 37, 1659–1667.

Baird, J.K., 1984. Methylene blue-mediated hexose monophosphate shunt stimulation in human red blood cells in vitro: independence from intracellular oxidative injury. Int. J. Biochem. 16, 1053–1058.

Baird, J.K., 2004. Chloroquine resistance in *Plasmodium vivax*. Antimicrob. Agents Chemother. 48, 4075–4083.

Baird, J.K., 2007. Neglect of *Plasmodium vivax* malaria. Trends Parasitol. 23, 533–539.

Baird, J.K., 2008. Real-world therapies and the problem of vivax malaria. N. Engl. J. Med. 359, 2601–2603.

Baird, J.K., 2009. Resistance to therapies for infection by *Plasmodium vivax*. Clin. Microbiol. Rev. 22, 508–534.

Baird, J.K., 2010. Eliminating malaria – all of them. Lancet 376, 1883–1885.

Baird, J.K., 2011. Resistance to chloroquine unhinges vivax malaria therapeutics. Antimicrob. Agents Chemother. 55, 1827–1830.

Baird, J.K., 2012a. Elimination therapies for the endemic malarias. Curr. Infect. Dis. Rep.10.1007/s11908-012-0250-z.

Baird, J.K., 2012b. Tropical health sustainability. In: Meyers, R.A. (Ed.), Encyclopedia of Sustainability Science and Technology, Springer Verlag, New York, pp. 11134–11163.

Barcus, M.J., Basri, H., Picarima, H., Manyakori, C., Sekartuti, et al., 2007. Demographic risk factors for severe and fatal vivax and falciparum malaria among hospital admissions in northeastern Indonesian Papua. Am. J. Trop. Med. Hyg. 77, 984–991.
Barnadas, C., Ratsimbasoa, A., Tichit, M., Bouchier, C., Jahevitra, M., et al., 2008. *Plasmodium vivax* resistance to chloroquine in Madagascar: clinical efficacy and polymorphisms in pvmdr1 and pvcrt-o genes. Antimicrob. Agents Chemother. 52, 4233–4240.
Berliner, R.W., Earle, D.P., Taggart, J.V., Zubrod, C.G., Welch, W.J., et al., 1948a. Studies on the chemotherapy of the human malarias. VI. The physiological disposition, antimalarial activity, and toxicity of several derivatives of 4-aminoquinoline. J. Clin. Invest. 27, 98–107.
Berliner, R.W., Earle, D.P.J., Taggart, J.V., Atchley, A., Shannon, J.A., 1948b. Studies on the chemotherapy of the human malarias. VII. The antimalarial activity of pamaquine. J. Clin. Invest. 27, 108–113.
Beutler, E., 2008. Glucose-6-phosphate dehydrogenase deficiency: a historical perspective. Blood 111, 16–24.
Brosius, O.T., 1927. Plasmochin in malaria. Section II p. 26–81. In: Medical Department, Sixteenth Annual Report United Fruit Company, Boston.
Carlini, M.E., White Jr., A.C., Atmar, R.L., 1999. Vivax malaria complicated by adult respiratory distress syndrome. Clin. Infect. Dis. 28, 1182–1183.
Carmona-Fonseca, J., Maestre, A., 2009. Prevention of *Plasmodium vivax* malaria recurrence: efficacy of the standard total dose of primaquine administered over 3 days. Acta Trop. 112, 188–192.
Cedillos, R.A., Warren, M., Jeffery, G.M., 1978. Field evaluation of primaquine in the control of *Plasmodium vivax*. Am. J. Trop. Med. Hyg. 27, 466–472.
Chandramohan, D., Jaffar, S., Greenwood, B., 2002. Use of clinical algorithms for diagnosing malaria. Trop. Med. Int. Health 7, 45–52.
Chattopadhyay, R., Velmurugan, S., Chakiath, C., Andrews Donkor, L., Milhous, W., et al., 2010. Establishment of an in vitro assay for assessing the effects of drugs on the liver stages of *Plasmodium vivax* malaria. PLoS ONE 5, e14275.
Chen, N., Auliff, A., Rieckmann, K., Gatton, M., Cheng, Q., 2007. Relapses of *Plasmodium vivax* infection result from clonal hypnozoites activated at predetermined intervals. J. Infect. Dis. 195, 934–941.
Chen, J.H., Lu, F., Lim, C.S., Kim, J.Y., Ahn, H.J., et al., 2010. Detection of *Plasmodium vivax* infection in the Republic of Korea by loop-mediated isothermal amplification (LAMP). Acta Trop. 113, 61–65.
Chotivanich, K., Udomsangpetch, R., Chierakul, W., Newton, P.N., Ruangveerayuth, R., et al., 2004. In vitro efficacy of antimalarial drugs against *Plasmodium vivax* on the western border of Thailand. Am. J. Trop. Med. Hyg. 70, 395–397.
Chotivanich, K., Sattabongkot, J., Choi, Y.K., Park, J.S., Sritabal, J., et al., 2009. Antimalarial drug susceptibility of *Plasmodium vivax* in the Republic of Korea. Am. J. Trop. Med. Hyg. 80, 902–904.
Clark, W.M., 1946. History of the co-operative wartime program. A Survey of Antimalarial Drugs 1941–1945, vol. IJW Edwards, Ann Harbor, Michigan, pp. 2–57.
Clayman, C.B., Arnold, J., Hockwald, R.S., Yount, E.H.J., Edgcomb, J.H., et al., 1952. Toxicity of primaquine in Caucasians. J. Am. Med. Assoc. 149, 1563–1568.
Clyde, D.F., McCarthy, V.C., 1977. Radical cure of Chesson strain vivax malaria in man by 7, not 14, days of treatment with primaquine. Am. J. Trop. Med. Hyg. 26, 562–563.
Clyde, D.F., 1981. Clinical problems associated with the use of primaquine as a tissue schizontocidal and gametocytocidal drug. Bull. World Health Organ. 59, 391–395.
Coatney, G.R., Cooper, W.C., Ruhe, D.S., 1948. Studies in human malaria. VI. The organization of a program for testing potential antimalarial drugs in prisoner volunteers. Am. J. Epidemiol. 47, 112–119.

Coatney, G.R., 1963. Pitfalls in a discovery: the chronicle of chloroquine. Am. J. Trop. Med. Hyg. 12, 121–128.

Coatney, G.R., 1971. Primate Malarias. US Government Printing Office, Washington, DC.

Collins, W.E., Schwartz, I.K., Skinner, J.C., Morris, C., Filipski, V.K., 1992. The susceptibility of the Indonesian I/CDC strain of *Plasmodium vivax* to chloroquine. J. Parasitol. 78, 344–349.

Collins, W.E., Richardson, B.B., Morris, C.L., Sullivan, J.S., Galland, G.G., 1998. Salvador II strain of *Plasmodium vivax* in Aotus monkeys and mosquitoes for transmission-blocking vaccine trials. Am. J. Trop. Med. Hyg. 59, 29–34.

Collins, W.E., Sullivan, J.S., Fryauff, D.J., Kendall, J., Jennings, V., et al., 2000. Adaptation of a chloroquine-resistant strain of *Plasmodium vivax* from Indonesia to New World monkeys. Am. J. Trop. Med. Hyg. 62, 491–495.

Collins, W.E., Sullivan, J.S., Galland, G.G., Williams, A., Nace, D., et al., 2005. Observations on the Vietnam Palo Alto strain of *Plasmodium vivax* in two species of Aotus monkeys. J. Parasitol. 91, 461–463.

Comfort, N., 2009. The prisoner as model organism: malaria research at Stateville Penitentiary. Stud. Hist. Philos. Biol. Biomed. Sci. 40, 190–203.

Congpuong, K., Na-Bangchang, K., Thimasarn, K., Tasanor, U., Wernsdorfer, W.H., 2002. Sensitivity of *Plasmodium vivax* to chloroquine in Sa Kaeo Province, Thailand. Acta Trop. 83, 117–121.

Cooper, W.C., Myatt, A.V., Hernandez, T., Jeffery, G.M., Coatney, G.R., 1953. Studies in human malaria. XXXI. Comparison of primaquine, isopentaquine, SN-3883, and pamaquine as curative agents against Chesson strain vivax malaria. Am. J. Trop. Med. Hyg. 2, 949–957.

Cooper, R.D., Milhous, W.K., Rieckmann, K.H., 1994. The efficacy of WR238605 against the blood stages of a chloroquine resistant strain of *Plasmodium vivax*. Trans. R. Soc. Trop. Med. Hyg. 88, 691–692.

Craige Jr., B., Alving, A.S., et al., 1947a. The Chesson strain of *Plasmodium vivax* malaria; relationship between prepatent period, latent period and relapse rate. J. Infect. Dis. 80, 228–236.

Craige, B.J., Jones, R.J., Whorton, C.M., Pullman, T.N., Alving, A.S., et al., 1947b. Clinical standardization of pamaquine (plasmochin) in mosquito-induced vivax malaria, Chesson strain. Am. J. Trop. Med. Hyg. 27, 309–315.

Curlin, M.E., Barat, L.M., Walsh, D.K., Granger, D.L., 1999. Noncardiogenic pulmonary edema during vivax malaria. Clin. Infect. Dis. 28, 1166–1167.

Darlow, B., Vrbova, H., Gibney, S., Jolley, D., Stace, J., et al., 1982. Sulfadoxine-pyrimethamine for the treatment of acute malaria in children of Papua New Guinea. II. *Plasmodium vivax*. Am. J. Trop. Med. Hyg. 31, 10–13.

Davidson Jr., D.E., Ager, A.L., Brown, J.L., Chapple, F.E., Whitmire, R.E., et al., 1981. New tissue schizontocidal antimalarial drugs. Bull. World Health Organ. 59, 463–479.

de Santana Filho, F.S., Arcanjo, A.R., Chehuan, Y.M., Costa, M.R., Martinez-Espinosa, F.E., et al., 2007. Chloroquine-resistant *Plasmodium vivax*, Brazilian Amazon. Emerg. Infect. Dis. 13, 1125–1126.

Di Lorenzo, A., Marx, R.L., Alving, A.S., Jones Jr., R., 1953. Korean vivax malaria. IV. Curative effect of 15 milligrams of primaquine daily for 7 days. Am. J. Trop. Med. Hyg. 2, 983–984.

Douglas, N.M., Anstey, N.M., Angus, B.J., Nosten, F., Price, R.N., 2010. Artemisinin combination therapy for vivax malaria. Lancet Infect. Dis. 10, 405–416.

Douglas, N.M., Nosten, F., Ashley, E.A., Phaiphun, L., van Vugt, M., et al., 2011. *Plasmodium vivax* recurrence following falciparum and mixed species malaria: risk factors and effect of antimalarial kinetics. Clin. Infect. Dis. 52, 612–620.

Dow, G.S., Gettayacamin, M., Hansukjariya, P., Imerbsin, R., Komcharoen, S., et al., 2011. Radical curative efficacy of tafenoquine combination regimens in *Plasmodium cynomolgi*-infected Rhesus monkeys (Macaca mulatta). Malar. J. 10, 212.

Downs, W.G., Harper, P.A., Lisansky, E.T., 1947. Malaria and other insect-borne diseases in the South Pacific campaign. II. Epidemiology of insect borne diseases in Army troops. Am. J. Trop. Med. Hyg. 27, 69–89.

Druilhe, P., Brasseur, P., Blanc, C., Makler, M., 2007. Improved assessment of *Plasmodium vivax* response to antimalarial drugs by a colorimetric double-site plasmodium lactate dehydrogenase antigen capture enzyme-linked immunosorbent assay. Antimicrob. Agents Chemother. 51, 2112–2116.

Ehrman, F.C., Ellis, J.M., Young, M.D., 1945. *Plasmodium vivax* Chesson Strain. Science 101, 377.

Elliott, J.H., O'Brien, D., Leder, K., Kitchener, S., Schwartz, E., et al., 2004. Imported *Plasmodium vivax* malaria: demographic and clinical features in nonimmune travelers. J. Travel. Med. 11, 213–217.

Elmes, N.J., Nasveld, P.E., Kitchener, S.J., Kocisko, D.A., Edstein, M.D., 2008. The efficacy and tolerability of three different regimens of tafenoquine versus primaquine for post-exposure prophylaxis of *Plasmodium vivax* malaria in the Southwest Pacific. Trans. R. Soc. Trop. Med. Hyg. 102, 1095–1101.

Feachem, R.G., Phillips, A.A., Targett, G.A., Snow, R.W., 2010. Call to action: priorities for malaria elimination. Lancet 376, 1517–1521.

Fernandez-Becerra, C., Pinazo, M.J., Gonzalez, A., Alonso, P.L., del Portillo, H.A., et al., 2009. Increased expression levels of the pvcrt-o and pvmdr1 genes in a patient with severe *Plasmodium vivax* malaria. Malar. J. 8, 55.

Field, J.W., 1938. Notes on the Chemotherapy of Malaria. Bulletins from the Institute for Medical Research Federated Malay States Printing, Kuala Lumpur 180 pp.

Freedman, D.O., Weld, L.H., Kozarsky, P.E., Fisk, T., Robins, R., et al., 2006. Spectrum of disease and relation to place of exposure among ill returned travelers. N. Engl. J. Med. 354, 119–130.

Fryauff, D.J., Baird, J.K., Basri, H., Sumawinata, I., Purnomo, et al., 1995. Randomised placebo-controlled trial of primaquine for prophylaxis of falciparum and vivax malaria. Lancet 346, 1190–1193.

Galappaththy, G.N., Omari, A.A., Tharyan, P., 2007. Primaquine for preventing relapses in people with *Plasmodium vivax* malaria. Cochrane Database Syst. Rev., CD004389.

Gbotosho, G.O., Happi, C.T., Ganiyu, A., Ogundahunsi, O.A., Sowumni, A., et al., 2009. Potential contribution of prescription practices to the emergence and spread of chloroquine resistance in south-west Nigeria: caution in the use of artemisinin combination therapy. Malar. J. 8, 313.

Genton, B., D'Acremont, V., Rare, L., Bae, K., Reeder, J.C., et al., 2008. *Plasmodium vivax* and mixed infections are associated with severe malaria in children: a prospective cohort study from Papua New Guinea. PLoS Med. 5, 6.

Gogtay, N.J., Desai, S., Kamtekar, K.D., Kadam, V.S., Dalvi, S.S., et al., 1999. Efficacies of 5- and 14-day primaquine regimens in the prevention of relapses in *Plasmodium vivax* infections. Ann. Trop. Med. Parasitol. 93, 809–812.

Goller, J.L., Jolley, D., Ringwald, P., Biggs, B.A., 2007. Regional differences in the response of *Plasmodium vivax* malaria to primaquine as anti-relapse therapy. Am. J. Trop. Med. Hyg. 76, 203–207.

Grassi, B., Feletti, R., 1890. Parasites malariques ches oiseaux. Arch Ital de Biologie 13, 297–300.

Greenwood, D., 1995. Conflicts of interest: the genesis of synthetic antimalarial agents in peace and war. J. Antimicrob. Chemother. 36, 857–872.

Guthmann, J.P., Pittet, A., Lesage, A., Imwong, M., Lindegardh, N., et al., 2008. *Plasmodium vivax* resistance to chloroquine in Dawei, southern Myanmar. Trop. Med. Int. Health 13, 91–98.

Habib, A.G., Singh, K.S., 2004. Respiratory distress in nonimmune adults with imported malaria. Infection 32, 356–359.

Hastings, M.D., Porter, K.M., Maguire, J.D., Susanti, I., Kania, W., et al., 2004. Dihydrofolate reductase mutations in *Plasmodium vivax* from Indonesia and therapeutic response to sulfadoxine plus pyrimethamine. J. Infect. Dis. 189, 744–750.

Hastings, M.D., Maguire, J.D., Bangs, M.J., Zimmerman, P.A., Reeder, J.C., et al., 2005. Novel *Plasmodium vivax* dhfr alleles from the Indonesian Archipelago and Papua New Guinea: association with pyrimethamine resistance determined by a *Saccharomyces cerevisiae* expression system. Antimicrob. Agents Chemother. 49, 733–740.

Hasugian, A.R., Purba, H.L., Kenangalem, E., Wuwung, R.M., Ebsworth, E.P., et al., 2007. Dihydroartemisinin-piperaquine versus artesunate-amodiaquine: superior efficacy and posttreatment prophylaxis against multidrug-resistant *Plasmodium falciparum* and *Plasmodium vivax* malaria. Clin. Infect. Dis. 44, 1067–1074.

Hasugian, A.R., Tjitra, E., Ratcliff, A., Siswantoro, H., Kenangalem, E., et al., 2009. In vivo and in vitro efficacy of amodiaquine monotherapy for treatment of infection by chloroquine-resistant *Plasmodium vivax*. Antimicrob. Agents Chemother. 53, 1094–1099.

Hill, D.R., Baird, J.K., Parise, M.E., Lewis, L.S., Ryan, E.T., et al., 2006. Primaquine: Report from CDC expert meeting on malaria chemoprophylaxis I. Am. J. Trop. Med. Hyg. 75, 402–415.

Hozhabri, S., Luby, S.P., Rahbar, M.H., Akhtar, S., 2002. Clinical diagnosis of *Plasmodium falciparum* among children with history of fever, Sindh, Pakistan. Int. J. Infect. Dis. 6, 233–235.

Hung, T.Y., Davis, T.M., Ilett, K.F., Karunajeewa, H., Hewitt, S., et al., 2004. Population pharmacokinetics of piperaquine in adults and children with uncomplicated falciparum or vivax malaria. Br. J. Clin. Pharmacol. 57, 253–262.

Imwong, M., Snounou, G., Pukrittayakamee, S., Tanomsing, N., Kim, J.R., et al., 2007. Relapses of *Plasmodium vivax* infection usually result from activation of heterologous hypnozoites. J. Infect. Dis. 195, 927–933.

Imwong, M., Pukrittayakamee, S., Pongtavornpinyo, W., Nakeesathit, S., Nair, S., et al., 2008. Gene amplification of the multidrug resistance 1 gene of *Plasmodium vivax* isolates from Thailand, Laos, and Myanmar. Antimicrob. Agents Chemother. 52, 2657–2659.

Islam, N., Qamruddin, K., 1995. Unusual complications in benign tertian malaria. Trop. Geogr. Med. 47, 141–143.

Jeffery, G.M., 1958. Infectivity to mosquitoes of *Plasmodium vivax* following treatment with chloroquine and other antimalarials. Am. J. Trop. Med. Hyg. 7, 207–211.

Jelinek, T., Nothdurft, H.D., Von Sonnenburg, F., Loscher, T., 1995. Long-term efficacy of primaquine in the treatment of vivax malaria in nonimmune travelers. Am. J. Trop. Med. Hyg. 52, 322–324.

Jensen, M., Mehlhorn, H., 2009. 75 years of Resochin in the Fight Against Malaria. Dusseldorf University Press, German 102 pp..

Jones Jr., R., Jackson, L.S., Di Lorenzo, A., Marx, R.L., Levy, B.L., et al., 1953. Korean vivax malaria. III. Curative effect and toxicity of primaquine in doses from 10 to 30 mg. daily. Am. J. Trop. Med. Hyg. 2, 977–982.

Jordan-Villegas, A., Zapata, J.C., Perdomo, A.B., Quintero, G.E., Solarte, Y., et al., 2005. *Aotus lemurinus* griseimembra monkeys: a suitable model for *Plasmodium vivax* sporozoite infection. Am. J. Trop. Med. Hyg. 73, 10–15.

Karunajeewa, H., Lim, C., Hung, T.Y., Ilett, K.F., Denis, M.B., et al., 2004. Safety evaluation of fixed combination piperaquine plus dihydroartemisinin (Artekin) in Cambodian children and adults with malaria. Br. J. Clin. Pharmacol. 57, 93–99.

Karunajeewa, H.A., Ilett, K.F., Mueller, I., Siba, P., Law, I., et al., 2008a. Pharmacokinetics and efficacy of piperaquine and chloroquine in Melanesian children with uncomplicated malaria. Antimicrob. Agents Chemother. 52, 237–243.

Karunajeewa, H.A., Mueller, I., Senn, M., Lin, E., Law, I., et al., 2008b. A trial of combination antimalarial therapies in children from Papua New Guinea. N. Engl. J. Med. 359, 2545–2557.

Ketema, T., Getahun, K., Bacha, K., 2011. Therapeutic efficacy of chloroquine for treatment of *Plasmodium vivax* malaria cases in Halaba district, South Ethiopia. Parasit. Vectors 4, 46.

Khantikul, N., Butraporn, P., Kim, H.S., Leemingsawat, S., Tempongko, M.A., et al., 2009. Adherence to antimalarial drug therapy among vivax malaria patients in northern Thailand. J. Health Popul. Nutr. 27, 4–13.

Kim, S., Nguon, C., Guillard, B., Duong, S., Chy, S., et al., 2011a. Performance of the CareStart^{TM} G6PD deficiency screening test, a point-of-care diagnostic for primaquine theraphy screening. PLoS ONE 6, 12.

Kim, Y.K., Kim, C., Park, I., Kim, H.Y., Choi, J.Y., et al., 2011b. Therapeutic efficacy of chloroquine in *Plasmodium vivax* and the pvmdr1 polymorphisms in the Republic of Korea under mass chemoprophylaxis. Am. J. Trop. Med. Hyg. 84, 532–534.

Kimura, M., Miyake, H., Kim, H.S., Tanabe, M., Arai, M., et al., 1995. Species-specific PCR detection of malaria parasites by microtiter plate hybridization: clinical study with malaria patients. J. Clin. Microbiol. 33, 2342–2346.

Kochar, D.K., Saxena, V., Singh, N., Kochar, S.K., Kumar, S.V., et al., 2005. *Plasmodium vivax* malaria. Emerg. Infect. Dis. 11, 132–134.

Koepfli, C., Mueller, I., Marfurt, J., Goroti, M., Sie, A., et al., 2009. Evaluation of *Plasmodium vivax* genotyping markers for molecular monitoring in clinical trials. J. Infect. Dis. 199, 1074–1080.

Kolaczinski, K., Durrani, N., Rahim, S., Rowland, M., 2007. Sulfadoxine-pyrimethamine plus artesunate compared with chloroquine for the treatment of vivax malaria in areas co-endemic for *Plasmodium falciparum* and *P. vivax*: a randomised non-inferiority trial in eastern Afghanistan. Trans. R. Soc. Trop. Med. Hyg. 101, 1081–1087.

Kotwal, R.S., Wenzel, R.B., Sterling, R.A., Porter, W.D., Jordan, N.N., et al., 2005. An outbreak of malaria in US Army Rangers returning from Afghanistan. J. Am. Med. Assoc. 293, 212–216.

Krotoski, W.A., Garnham, P.C., Cogswell, F.B., Collins, W.E., Bray, R.S., et al., 1986. Observations on early and late post-sporozoite tissue stages in primate malaria. IV. Pre-erythrocytic schizonts and/or hypnozoites of Chesson and North Korean strains of *Plasmodium vivax* in the chimpanzee. Am. J. Trop. Med. Hyg. 35, 263–274.

Krudsood, S., Wilairatana, P., Tangpukdee, N., Chalermrut, K., Srivilairit, S., et al., 2006. Safety and tolerability of elubaquine (bulaquine, CDRI 80/53) for treatment of *Plasmodium vivax* malaria in Thailand. Korean J. Parasitol. 44, 221–228.

Krudsood, S., Tangpukdee, N., Muangnoicharoen, S., Thanachartwet, V., Luplertlop, N., et al., 2007. Clinical efficacy of chloroquine versus artemether-lumefantrine for *Plasmodium vivax* treatment in Thailand. Korean J. Parasitol. 45, 111–114.

Krudsood, S., Tangpukdee, N., Wilairatana, P., Phophak, N., Baird, J.K., et al., 2008. High-dose primaquine regimens against relapse of *Plasmodium vivax* malaria. Am. J. Trop. Med. Hyg. 78, 736–740.

Kumar, S., Melzer, M., Dodds, P., Watson, J., Ord, R., 2007. *P. vivax* malaria complicated by shock and ARDS. Scand. J. Infect. Dis. 39, 255–256.

Lacy, M.D., Maguire, J.D., Barcus, M.J., Ling, J., Bangs, M.J., et al., 2002. Atovaquone/proguanil therapy for *Plasmodium falciparum* and *Plasmodium vivax* malaria in Indonesians who lack clinical immunity. Clin. Infect. Dis. 35, e92–95.

Lawn, S.D., Krishna, S., Jarvis, J.N., Joet, T., Macallan, D.C., 2003. Case reports: pernicious complications of benign tertian malaria. Trans. R. Soc. Trop. Med. Hyg. 97, 551–553.

Lee, S.W., Lee, M., Lee, D.D., Kim, C., Kim, Y.J., et al., 2009. Biological resistance of hydroxychloroquine for *Plasmodium vivax* malaria in the Republic of Korea. Am. J. Trop. Med. Hyg. 81, 600–604.

Lek-Uthai, U., Suwanarusk, R., Ruengweerayut, R., Skinner-Adams, T.S., Nosten, F., et al., 2008. Stronger activity of human immunodeficiency virus type 1 protease inhibitors against clinical isolates of *Plasmodium vivax* than against those of *P. falciparum*. Antimicrob. Agents Chemother. 52, 2435–2441.

Leslie, T., Rab, M.A., Ahmadzai, H., Durrani, N., Fayaz, M., et al., 2004. Compliance with 14-day primaquine therapy for radical cure of vivax malaria–a randomized placebo-controlled trial comparing unsupervised with supervised treatment. Trans. R. Soc. Trop. Med. Hyg. 98, 168–173.

Leslie, T., Mayan, I., Mohammed, N., Erasmus, P., Kolaczinski, J., et al., 2008. A randomised trial of an eight-week, once weekly primaquine regimen to prevent relapse of *Plasmodium vivax* in Northwest Frontier Province, Pakistan. PLoS ONE 3, 8.

LeYuan, S., QingXia, H., 2001. Observation on efficacy of pyronaridine phosphate and combined with dihydroartemisinin for treatment of malaria in Eritrea, Africa. Chin. J. Parasitic. Dis. Control 14, R382. 331.

Li, X., Li, C., Che, L., Liu, X., Li, Z., et al., 1999. Observation on efficacy of artemether compound against vivax malaria. Zhongguo Ji Sheng Chong Xue Yu Ji Sheng Chong Bing Za Zhi 17, 175–177.

Lomar, A.V., Vidal, J.E., Lomar, F.P., Barbas, C.V., de Matos, G.J., et al., 2005. Acute respiratory distress syndrome due to vivax malaria: case report and literature review. Braz. J. Infect. Dis. 9, 425–430.

Lu, F., Gao, Q., Zhou, H., Cao, J., Wang, W., et al., 2011. Molecular test for vivax malaria with loop-mediated isothermal amplification method in central China. Parasitol. Res. 110, 2439–2444.

Lysenko, A.Y., 1960. Use of quinocide in treatment and prophylaxis of vivax malaria. Bull. World Health Organ. 22, 641–662.

Maguire, J.D., Krisin, Marwoto, H., Richie, T.L., Fryauff, D.J., et al., 2006a. Mefloquine is highly efficacious against chloroquine-resistant *Plasmodium vivax* malaria and *Plasmodium falciparum* malaria in Papua, Indonesia. Clin. Infect. Dis. 42, 1067–1072.

Maguire, J.D., Lederman, E.R., Barcus, M.J., O'Meara, W.A., Jordon, R.G., et al., 2006b. Production and validation of durable, high quality standardized malaria microscopy slides for teaching, testing and quality assurance during an era of declining diagnostic proficiency. Malar. J. 5, 92.

Maguire, J.D., Fenton, M.E., Susanti, A.I., Walker, J.B., 2007. *Plasmodium vivax*-associated acute respiratory distress syndrome after extended travel in Afghanistan. Travel Med. Infect. Dis. 5, 301–305.

Manning, L., Laman, M., Law, I., Bona, C., Aipit, S., et al., 2011. Features and prognosis of severe malaria caused by *Plasmodium falciparum, Plasmodium vivax* and mixed *Plasmodium* species in Papua New Guinean children. PLoS ONE 6, e29203.

Marfurt, J., de Monbrison, F., Brega, S., Barbollat, L., Muller, I., et al., 2008. Molecular markers of in vivo *Plasmodium vivax* resistance to amodiaquine plus sulfadoxine-pyrimethamine: mutations in pvdhfr and pvmdr1. J. Infect. Dis. 198, 409–417.

Marfurt, J., Chalfein, F., Prayoga, P., Wabiser, F., Kenangalem, E., et al., 2011a. Ex vivo activity of histone deacetylase inhibitors against multidrug-resistant clinical isolates of *Plasmodium falciparum* and *P. vivax*. Antimicrob. Agents Chemother. 55, 961–966.

Marfurt, J., Chalfein, F., Prayoga, P., Wabiser, F., Kenangalem, E., et al., 2011b. Ex vivo drug susceptibility of ferroquine against chloroquine-resistant isolates of *Plasmodium falciparum* and *P. vivax*. Antimicrob. Agents Chemother. 55, 4461–4464.

Marsh, K., 2010. Research priorities for malaria elimination. Lancet 376, 1626–1627.

Mayxay, M., Thongpraseuth, V., Khanthavong, M., Lindegardh, N., Barends, M., et al., 2006. An open, randomized comparison of artesunate plus mefloquine vs. dihydroartemisinin-piperaquine for the treatment of uncomplicated *Plasmodium falciparum* malaria in the Lao People's Democratic Republic (Laos). Trop. Med. Int. Health 11, 1157–1165.

Mazier, D., Landau, I., Miltgen, F., Druilhe, P., Guguen-Guillouzo, C., et al., 1983. In vitro infestation of human hepatocytes by sporozoites of *Plasmodium vivax*: schizogony and liberation of merozoites capable of infesting human erythrocytes. Ann. Parasitol. Hum. Comp. 58, 405–406.

Mendis, K., Sina, B.J., Marchesini, P., Carter, R., 2001. The neglected burden of Plasmodium vivax malaria. Am. J. Trop. Med. Hyg. 64, 97–106.

Millet, P., Collins, W.E., Herman, L., Cochrane, A.H., 1989. *Plasmodium vivax*: in vitro development of exoerythrocytic stages in squirrel monkey hepatocytes and inhibition by an anti-*P. cynomolgi* monoclonal antibody. Exp. Parasitol. 69, 91–93.

Milne, L.M., Kyi, M.S., Chiodini, P.L., Warhurst, D.C., 1994. Accuracy of routine laboratory diagnosis of malaria in the United Kingdom. J. Clin. Pathol. 47, 740–742.

Most, H., Kane, C.A., et al., 1946a. Combined quinine-plasmochin treatment of vivax malaria; effect of relapse rate. Am. J. Med. Sci. 212, 550–560.

Most, H., London, I.M., et al., 1946b. Chloroquine for treatment of acute attacks of vivax malaria. J. Am. Med. Assoc. 131, 963–967.

Munteis, E., Mellibovsky, L., Marquez, M.A., Minguez, S., Vazquez, E., et al., 1997. Pulmonary involvement in a case of *Plasmodium vivax* malaria. Chest 111, 834–835.

Mwangi, T.W., Mohammed, M., Dayo, H., Snow, R.W., Marsh, K., 2005. Clinical algorithms for malaria diagnosis lack utility among people of different age groups. Trop. Med. Int. Health 10, 530–536.

Myint, H.Y., Tipmanee, P., Nosten, F., Day, N.P., Pukrittayakamee, S., et al., 2004. A systematic overview of published antimalarial drug trials. Trans. R. Soc. Trop. Med. Hyg. 98, 73–81.

Nkhoma, E.T., Poole, C., Vannappagari, V., Hall, S.A., Beutler, E., 2009. The global prevalence of glucose-6-phosphate dehydrogenase deficiency: a systematic review and meta-analysis. Blood Cells Mol. Dis. 42, 267–278.

Nomura, T., Carlton, J.M., Baird, J.K., del Portillo, H.A., Fryauff, D.J., et al., 2001. Evidence for different mechanisms of chloroquine resistance in 2 *Plasmodium* species that cause human malaria. J. Infect. Dis. 183, 1653–1661.

Nurleila, S., Syafruddin, D., Elyazar, I.R.F., Baird, J.K., 2012. Serious and fatal illness associated with falciparum and vivax malaria among patients admitted to hospital at West Sumba in Eastern Indonesia. Am. J. Trop. Med. Hyg. 87, 41–49.

Obaldia 3rd, N., Rossan, R.N., Cooper, R.D., Kyle, D.E., Nuzum, E.O., et al., 1997. WR 238605, chloroquine, and their combinations as blood schizonticides against a chloroquine-resistant strain of *Plasmodium vivax* in Aotus monkeys. Am. J. Trop. Med. Hyg. 56, 508–510.

Office of the Surgeon General, 1943. The drug treatment of malaria, suppressive and clinical. Circular Letter No. 153 J. Am. Med. Assoc. 123, 205–208.

O'Meara, W.P., Barcus, M., Wongsrichanalai, C., Muth, S., Maguire, J.D., et al., 2006. Reader technique as a source of variability in determining malaria parasite density by microscopy. Malar. J. 5, 118.

Pannacciulli, I., Tizianello, A., Ajmar, F., Salvidio, E., 1965. The course of experimentally induced hemolytic anemia in a primaquine sensitive Caucasian. Blood 25, 92–95.

Path, 2011. Staying the course? Malaria Research and Development in a Time of Economic Uncertainty. Seattle PATH, 87 pp.

Perisse, A.R., Strickland, G.T., 2008. Usefulness of clinical algorithm as screening process to detected malaria in low-to-moderate transmission areas of scarce health related resources. Acta Trop. 107, 224–229.

Perren, A., Beretta, F., Schubarth, P., 1998. ARDS in *Plasmodium vivax* malaria. Schweiz. Med. Wochenschr. 128, 1020–1023.

Phyo, A.P., Lwin, K.M., Price, R.N., Ashley, E.A., Russell, B., et al., 2011. Dihydroartemisinin-piperaquine versus chloroquine in the treatment of *Plasmodium vivax* malaria in Thailand: a randomized controlled trial. Clin. Infect. Dis. 53, 977–984.

Poespoprodjo, J.R., Fobia, W., Kenangalem, E., Lampah, D.A., Warikar, N., et al., 2008. Adverse pregnancy outcomes in an area where multidrug-resistant *Plasmodium vivax* and *Plasmodium falciparum* infections are endemic. J. Infect. Dis. 46, 1374–1381.

Poespoprodjo, J.R., Fobia, W., Kenangalem, E., Lampah, D.A., Hasanuddin, A., et al., 2009. Vivax malaria: a major cause of morbidity in early infancy. Clin. Infect. Dis. 48, 1704–1712.

Poravuth, Y., Socheat, D., Rueangweerayut, R., Uthaisin, C., Pyae Phyo, A., et al., 2011. Pyronaridine-artesunate versus chloroquine in patients with acute *Plasmodium vivax* malaria: a randomized, double-blind, non-inferiority trial. PLoS ONE 6, e14501.

Powell, R.D., Brewer, G.J., 1967. The effects of pyrimethamine, chloroquine, and primaquine against exoerythrocytic forms of a strain chloroquine-resistant *Plasmodium falciparum* from Thailand. Am. J. Trop. Med. Hyg. 16, 693–698.

Price, R.N., Douglas, N.M., 2010. Maximizing the public health benefit of antimalarials. Lancet Infect. Dis. 10, 654–655.

Price, R., Nosten, F., Simpson, J.A., Luxemburger, C., Phaipun, L., et al., 1999. Risk factors for gametocyte carriage in uncomplicated falciparum malaria. Am. J. Trop. Med. Hyg. 60, 1019–1023.

Price, L., Planche, T., Rayner, C., Krishna, S., 2007a. Acute respiratory distress syndrome in *Plasmodium vivax* malaria: case report and review of the literature. Trans. R. Soc. Trop. Med. Hyg. 101, 655–659.

Price, R.N., Tjitra, E., Guerra, C.A., Yeung, S., White, N.J., et al., 2007b. Vivax malaria: neglected and not benign. Am. J. Trop. Med. Hyg. 77, 79–87.

Price, R.N., Douglas, N.M., Anstey, N.M., 2009. New developments in *Plasmodium vivax* malaria: severe disease and the rise of chloroquine resistance. Curr. Opin. Infect. Dis. 22, 430–435.

Price, R.N., Marfurt, J., Chalfein, F., Kenangalem, E., Piera, K.A., et al., 2010. In vitro activity of pyronaridine against multidrug-resistant *Plasmodium falciparum* and *Plasmodium vivax*. Antimicrob. Agents Chemother. 54, 5146–5150.

Pukrittayakamee, S., Chantra, A., Vanijanonta, S., White, N.J., 1998. Pulmonary oedema in vivax malaria. Trans. R. Soc. Trop. Med. Hyg. 92, 421–422.

Pukrittayakamee, S., Chantra, A., Simpson, J.A., Vanijanonta, S., Clemens, R., et al., 2000. Therapeutic responses to different antimalarial drugs in vivax malaria. Antimicrob. Agents Chemother. 44, 1680–1685.

Ratcliff, A., Siswantoro, H., Kenangalem, E., Maristela, R., Wuwung, R.M., et al., 2007a. Two fixed-dose artemisinin combinations for drug-resistant falciparum and vivax malaria in Papua, Indonesia: an open-label randomised comparison. Lancet 369, 757–765.

Ratcliff, A., Siswantoro, H., Kenangalem, E., Wuwung, M., Brockman, A., et al., 2007b. Therapeutic response of multidrug-resistant *Plasmodium falciparum* and *P. vivax* to chloroquine and sulfadoxine-pyrimethamine in southern Papua, Indonesia. Trans. R. Soc. Trop. Med. Hyg. 101, 351–359.

Rieckmann, K.H., Davis, D.R., Hutton, D.C., 1989. *Plasmodium vivax* resistance to chloroquine?. Lancet 2, 1183–1184.

Rijken, M.J., Boel, M.E., Russell, B., Imwong, M., Leimanis, M.L., et al., 2011. Chloroquine resistant vivax malaria in a pregnant woman on the western border of Thailand. Malar. J. 10, 113.

Rocco, F., 2004. Quinine: Malaria and the Quest for a Cure that Changed the World. Harper Perennial, New York 384 pp.

Rombo, L., Ericsson, O., Alván, G., Lindström, B., Gustafsson, L.L., et al., 1985. Chloroquine and desethylchloroquine in plasma, serum, and whole blood: problems in assay and handling of samples. Ther. Drug. Monit. 7, 211–215.

Rossan, R.N., Young, M.D., Baerg, D.C., 1975. Chemotherapy of *Plasmodium vivax* in Saimiri and Aotus models. Am. J. Trop. Med. Hyg. 24, 168–173.

Ruhe, D.S., Cooper, R.D., Coatney, G.R., Josephson, E.S., 1949. Studies in human malaria. XV. The therapeutic action of pamaquine (Plasmochin) against St. Elizabeth strain vivax malaria. Am. J. Hyg. 49, 367–373.
Rungsihirunrat, K., Sibley, C.H., Mungthin, M., Na-Bangchang, K., 2008. Geographical distribution of amino acid mutations in *Plasmodium vivax* DHFR and DHPS from malaria endemic areas of Thailand. Am. J. Trop. Med. Hyg. 78, 462–467.
Russel, B., Chalfein, G., Prasetyorini, B., Kenangalem, E., Piera, K., et al., 2008. Determinants of in vitro drug susceptibility testing of *Plasmodium vivax*. Antimicrob. Agents Chemother. 52, 1040–1045.
Russell, B.M., Udomsangpetch, R., Rieckmann, K.H., Kotecka, B.M., Coleman, R.E., et al., 2003. Simple in vitro assay for determining the sensitivity of *Plasmodium vivax* isolates from fresh human blood to antimalarials in areas where *P. vivax* is endemic. Antimicrob. Agents Chemother. 47, 170–173.
Sa, J.M., Nomura, T., Neves, J., Baird, J.K., Wellems, T.E., et al., 2005. *Plasmodium vivax*: allele variants of the mdr1 gene do not associate with chloroquine resistance among isolates from Brazil, Papua, and monkey-adapted strains. Exp. Parasitol. 109, 256–259.
Sa, J.M., Yamamoto, M.M., Fernandez-Becerra, C., de Azevedo, M.F., Papakrivos, J., et al., 2006. Expression and function of pvcrt-o, a *Plasmodium vivax* ortholog of pfcrt, in *Plasmodium falciparum* and *Dictyostelium discoideum*. Mol. Biochem. Parasitol. 150, 219–228.
Saleri, N., Gulletta, M., Matteelli, A., Caligaris, S., Tomasoni, L.R., et al., 2006. Acute respiratory distress syndrome in *Plasmodium vivax* malaria in traveler returning from Venezuela. J. Travel. Med. 13, 112–113.
Sattabongkot, J., Yimamnuaychoke, N., Leelaudomlipi, S., Rasameesoraj, M., Jenwithisuk, R., et al., 2006. Establishment of a human hepatocyte line that supports in vitro development of the exo-erythrocytic stages of the malaria parasites *Plasmodium falciparum* and *P. vivax*. Am. J. Trop. Med. Hyg. 74, 708–715.
Schlitzer, M., 2007. Malaria chemotherapeutics part I: history of antimalarial drug development, currently used therapeutics, and drugs in clinical development. Chem. Med. Chem. 2, 944–986.
Schmidt, L.H., Fradkin, R., Vaughan, D., Rasco, J., 1977. Radical cure of infections with *Plasmodium cynomolgi*: a function of total 8-aminoquinoline dose. Am. J. Trop. Med. Hyg. 26, 1116–1128.
Schmidt, L.H., Fradkin, R., Genther, C., Rossan, R.N., Squires, W., 1982. Responses of sporozoite-induced and trophozoite-induced infections to standard antimalarials. Am. J. Trop. Med. Hyg. 31, 646–665.
Schmidt, L.H., 1978. *Plasmodium falciparum* and *Plasmodium vivax* infections in the owl monkey (*Aotus trivirgatus*). II. Responses to chloroquine, quinine, and pyrimethamine. Am. J. Trop. Med. Hyg. 27, 703–717.
Schmidt, L.H., 1981. Comparative efficacies of quinine and chloroquine as companions to primaquine in a curative drug regimen. Am. J. Trop. Med. Hyg. 30, 20–25.
Schmidt, L.H., 1983. Appraisals of compounds of diverse chemical classes for capacities to cure infections with sporozoites of *Plasmodium cynomolgi*. Am. J. Trop. Med. Hyg. 32, 231–257.
Schmidt, L.H., 1985. Enhancement of the curative activity of primaquine by concomitant administration of mirincamycin. Antimicrob. Agents Chemother. 27, 151–157.
Shaikh, S., Memon, H., Shaikh, A., Ahmed, I., Iohano, B., et al., 2012. Severe disease in children hospitalized with a diagnosis of *Plasmodium vivax* in south-eastern Pakistan. Malar. J. 11, 144.
Shannon, J.A., 1946. Chemotherapy in malaria. Bull. N.Y. Acad. Med. 22, 345–357.
Sharrock, W.W., Suwanarusk, R., Lek-Uthai, U., Edstein, M.D., Kosaisavee, V., et al., 2008. *Plasmodium vivax* trophozoites insensitive to chloroquine. Malar. J. 7, 94.

Sinclair, D., Gogtay, N., Brand, F., Olliaro, P., 2011. Artemisinin-based combination therapy for treating uncomplicated *Plasmodium vivax* malaria. Cochrane Database Syst. Rev. CD008492.
Sinton, J.A., Bird, W., 1928. Studies in malaria with special reference to treatment. IX. Plasmoquine in the treatment of malaria. Indian J. Med. Res. 16, 159–177.
Sinton, J.A., Smith, S., Pottinger, D., 1930. Studies in malaria with special reference to treatment. XII. Further researches into the treatment of chronic benign tertian malaria with plasmoquine and quinine. Indian J. Med. Res. 17, 793–814.
Smithuis, F., Kyaw, M.K., Phe, O., Aye, K.Z., Htet, L., et al., 2006. Efficacy and effectiveness of dihydroartemisinin-piperaquine versus artesunate-mefloquine in falciparum malaria: an open-label randomised comparison. Lancet 367, 2075–2085.
Smithuis, F., Kyaw, M.K., Phe, O., Win, T., Aung, P.P., et al., 2010. Effectiveness of five artemisinin combination regimens with or without primaquine in uncomplicated falciparum malaria: an open-label randomised trial. Lancet Infect. Dis. 10, 673–681.
Srivastava, H.C., Yadav, R.S., Joshi, H., Valecha, N., Mallick, P.K., et al., 2008. Therapeutic responses of *Plasmodium vivax* and *P. falciparum* to chloroquine, in an area of western India where *P. vivax* predominates. Ann. Trop. Med. Parasitol. 102, 471–480.
Sullivan, J.S., Morris, C.L., Richardson, B.B., Galland, G.G., Jennings, V.M., et al., 1999. Adaptation of the AMRU-1 strain of *Plasmodium vivax* to Aotus and Saimiri monkeys and to four species of anopheline mosquitoes. J. Parasitol. 85, 672–677.
Sullivan, J.S., Strobert, E., Yang, C., Morris, C.L., Galland, G.G., et al., 2001. Adaptation of a strain of *Plasmodium vivax* from India to New World monkeys, chimpanzees, and anopheline mosquitoes. J. Parasitol. 87, 1398–1403.
Sumawinata, I.W., Bernadeta, Leksana, B., Sutamihardja, A., Purnomo, et al., 2003. Very high risk of therapeutic failure with chloroquine for uncomplicated *Plasmodium falciparum* and *P. vivax* malaria in Indonesian Papua. Am. J. Trop. Med. Hyg. 68, 416–420.
Sutanto, I., Suprijanto, S., Nurhayati, Manoempil, P., Baird, J.K., 2009. Resistance to chloroquine by *Plasmodium vivax* at Alor in the Lesser Sundas Archipelago in eastern Indonesia. Am. J. Trop. Med. Hyg. 81, 338–342.
Sutanto, I., Endawati, D., Ling, L.H., Laihad, F., Setiabudy, R., et al., 2010. Evaluation of chloroquine therapy for vivax and falciparum malaria in southern Sumatra, western Indonesia. Malar. J. 9 (52), 52.
Sutanto, I., Efficacy of primaquine when combined with quinine or dihydroartemisinin-piperaquine against relapse, in preparation.
Suwanarusk, R., Russell, B., Chavchich, M., Chalfein, F., Kenangalem, E., et al., 2007. Chloroquine resistant *Plasmodium vivax*: in vitro characterisation and association with molecular polymorphisms. PLoS ONE 2, e1089.
Suwanarusk, R., Chavchich, M., Russell, B., Jaidee, A., Chalfein, F., et al., 2008. Amplification of pvmdr1 associated with multidrug-resistant *Plasmodium vivax*. J. Infect. Dis. 198, 1558–1564.
Swartz, E., Parise, M., Kozarsky, P., Cetron, M., 2003. Delayed onset malaria - implications for chemoprophylaxis in travelers. N. Engl. J. Med. 349, 1510–1516.
Tanios, M.A., Kogelman, L., McGovern, B., Hassoun, P.M., 2001. Acute respiratory distress syndrome complicating Plasmodium vivax malaria. Crit. Care. Med. 29, 665–667.
Tao, Z.Y., Zhou, H.Y., Xia, H., Xu, S., Zhu, H.W., et al., 2011. Adaptation of a visualized loop-mediated isothermal amplification technique for field detection of *Plasmodium vivax* infection. Parasit. Vectors 4, 115.
Tasanor, O., Ruengweerayut, R., Sirichaisinthop, J., Congpuong, K., Wernsdorfer, W.H., et al., 2006. Clinical-parasitological response and in-vitro sensitivity of *Plasmodium vivax* to chloroquine and quinine on the western border of Thailand. Trans. R. Soc. Trop. Med. Hyg. 100, 410–418.
Taylor, N., 1945. Chinchona in Java: The Story of Quinine. Greenberg, New York, pp. 87.

Tjitra, E., Baker, J., Suprianto, S., Cheng, Q., Anstey, N.M., 2002. Therapeutic efficacies of artesunate-sulfadoxine-pyrimethamine and chloroquine-sulfadoxine-pyrimethamine in vivax malaria pilot studies: relationship to *Plasmodium vivax* dhfr mutations. Antimicrob. Agents Chemother. 46, 3947–3953.

Tjitra, E., Anstey, N.M., Sugiarto, P., Warikar, W., Kenangalem, E., et al., 2008. Multidrug-resistant *Plasmodium vivax* associated with severe and fatal malaria: a prospective study in Papua, Indonesia. PLoS Med. 5, 6.

Torres, J.R., Perez, H., Postigo, M.M., Silva, J.R., 1997. Acute non-cardiogenic lung injury in benign tertian malaria. Lancet 350, 31–32.

Udomsangpetch, R., Somsri, S., Panichakul, T., Chotivanich, K., Sirichaisinthop, J., et al., 2007. Short-term in vitro culture of field isolates of *Plasmodium vivax* using umbilical cord blood. Parasitol. Int. 56, 65–69.

Weiss, W.R., Oloo, A.J., Johnson, A., Koech, D., Hoffman, S.L., 1995. Daily primaquine is effective for prophylaxis against falciparum malaria in Kenya: comparison with mefloquine, doxycycline, and chloroquine plus proguanil. J. Infect. Dis. 171, 1569–1575.

Wells, T.N., Burrows, J.N., Baird, J.K., 2010. Targeting the hypnozoite reservoir of *Plasmodium vivax*: the hidden obstacle to malaria elimination. Trends Parasitol. 26, 145–151.

Whitby, M., Wood, G., Veenendaal, J.R., Rieckmann, K., 1989. Chloroquine-resistant *Plasmodium vivax*. Lancet 2, 1395.

White, N.J., Olliaro, P.L., 1996. Strategies for the prevention of antimalarial drug resistance: rationale for combination chemotherapy for malaria. Parasitol. Today 12, 399–401.

White, N.J., Stepniewska, K., Barnes, K., Price, R.N., Simpson, J., 2008. Simplified antimalarial therapeutic monitoring: using the day-7 drug level?. Trends Parasitol. 24, 159–163.

White, N., 1999. Antimalarial drug resistance and combination chemotherapy. Philos. Trans. R. Soc. Lond. B Biol. Sci. 354, 739–749.

White, N.J., 2008. How antimalarial drug resistance affects post-treatment prophylaxis. Malar. J. 7, 9.

White, N.J., 2011. Determinants of relapse periodicity in *Plasmodium vivax* malaria. Malar. J. 10, 297.

Whitty, C.J., Chandler, C., Ansah, E., Leslie, T., Staedke, S.G., 2008. Deployment of ACT antimalarials for treatment of malaria: challenges and opportunities. Malar. J. 7 (Suppl. 1), S7.

WHO Global Malaria Programme, 2009. Malaria Rapid Diagnostic Test performance. Summary results of WHO product testing of malaria RDTs: Round 1 and 2 (2008-2009). Available: http://www2wprowhoint/NR/rdonlyres/E4823BBF-69FC-45DE-B924-1A9FCDD7406C/0/RoundIII_SummaryFINALpdf.

Wilson, M.E., Weld, L.H., Boggild, A., Keystone, J.S., Kain, K.C., et al., 2007. Fever in returned travelers: results from the GeoSentinel Surveillance Network. Clin Infect Dis 44, 1560–1568.

Wiselogle, F.Y., 1946. A Survey of Antimalarial Drug, 1941–1945. Michigan, Ann Arbor.

World Health Organization, 2009a. Malaria microscopy quality assurance manual. Version 1. 125 p.

World Health Organization, 2009b. World Malaria Report 2009. 1–66.

World Health Organization, 2010. Guidelines for the Treatment of Malaria, Second ed. WC 770: 191 p.

World Health Organization, 2012. WHO Global Malaria Programme: Information Note on Recommended Selection Criteria for Procurement of Malaria Rapid Diagnostic Tests (RDTs). WHO 12 April 2012. Available at: http://wwwwhoint/malaria/diagnosis_treatment/diagnosis/RDT_selection_criteriapdf.

Yadav, R.S., Ghosh, S.K., 2002. Radical curative efficacy of five-day regimen of primaquine for treatment of *Plasmodium vivax* malaria in India. J. Parasitol. 88, 1042–1044.

Yohannes, A.M., Teklehaimanot, A., Bergqvist, Y., Ringwald, P., 2011. Confirmed vivax resistance to chloroquine and effectiveness of artemether-lumefantrine for the treatment of vivax malaria in Ethiopia. Am. J. Trop. Med. Hyg. 84, 137–140.
Zakeri, S., Motmaen, S.R., Afsharpad, M., Djadid, N.D., 2009. Molecular characterization of antifolates resistance-associated genes, (dhfr and dhps) in *Plasmodium vivax* isolates from the Middle East. Malar. J. 8, 20.
Ziai, M., Amirhakimi, G.H., Reinhold, J.G., Tabatabee, M., Gettner, M.E., et al., 1967. Malaria prophylaxis and treatment in G-6-PD deficiency. An observation on the toxicity of primaquine and chloroquine. Clin. Pediatr. 6, 242–243.

CHAPTER FIVE

Chemotherapeutic Strategies for Reducing Transmission of *Plasmodium vivax* Malaria

Nicholas M. Douglas*,†,‡,1, George K. John*, Lorenz von Seidlein†, Nicholas M. Anstey†,‡, Ric N. Price*,†,‡

*Centre for Tropical Medicine, Nuffield Department of Clinical Medicine, University of Oxford, Oxford, UK

†Global Health Division, Menzies School of Health Research and Charles Darwin University, Darwin, Australia

‡Department of Infectious Diseases, Division of Medicine, Royal Darwin Hospital, Darwin, Australia

[1]Corresponding author: Email: nicholas.douglas@gmail.com

Contents

Abstract

Effective use of anti-malarial drugs is key to reducing the transmission potential of *Plasmodium vivax*. In patients presenting with symptomatic disease, treatment with potent and relatively slowly eliminated blood schizontocidal regimens administered concurrently with a supervised course of 7 mg/kg primaquine over 7–14 days has potential to exert the greatest transmission-blocking benefit. Given the spread of chloroquine-resistant *P. vivax* strains, the artemisinin combination therapies dihydroartemisinin + piperaquine and artesunate + mefloquine are currently the most assured means of preventing *P. vivax* recrudescence. Preliminary evidence suggests that, like chloroquine, these combinations potentiate the hypnozoitocidal effect of primaquine, but further supportive evidence is required. In view of the high rate of *P. vivax* relapse following falciparum infections in co-endemic regions, there is a strong argument for broadening current radical cure policy to include the administration of hypnozoitocidal doses of primaquine to patients with *Plasmodium falciparum* malaria. The most important reservoir for *P. vivax* transmission is likely to be very low-density, asymptomatic

Advances in Parasitology, Volume 80
ISSN 0065-308X, http://dx.doi.org/10.1016/B978-0-12-397900-1.00005-0

infections, the majority of which will arise from liver-stage relapses. Therefore, judicious mass administration of hypnozoitocidal therapy will reduce transmission of *P. vivax* to a greater extent than strategies focused on treatment of symptomatic patients. An efficacious hypnozoitocidal agent with a short curative treatment course would be particularly useful in mass drug administration campaigns.

1. INTRODUCTION

Plasmodium vivax threatens 2.85 billion people worldwide and infects between 130 and 435 million people per year (Guerra et al., 2010; Hay et al., 2004; Price et al., 2007; Baird, 2007). By far, the greatest clinical and numerical burden of disease is in Central and Southeast Asia. *Plasmodium vivax* has unique characteristics that make it relatively refractory to existing malaria control measures. The most important of these is its ability to relapse from dormant liver forms (hypnozoites). As elimination efforts 'preferentially' reduce the burden of falciparum malaria, *P. vivax* is becoming a proportionately more common cause of malaria (Feachem et al., 2010; Nosten et al., 2000; Rodriguez et al., 2011; Gama et al., 2011).

Anti-malarial drugs have an important role in reducing transmission of malaria. However, drug discovery and development of novel agents has focussed almost entirely on *Plasmodium falciparum*. Little priority has been given to development of compounds especially for *P. vivax*. There are several reasons for this. First, blood schizontocidal drugs that are active against *P. falciparum* are also active against the blood stages of *P. vivax*. Second, our inability to sustain *P. vivax* in continuous *ex vivo* culture significantly restricts high-throughput *in vitro* drug-susceptibility testing (Mueller et al., 2009). Third, there is no standardised approach to assay anti-malarial activity against hypnozoites. *Plasmodium vivax* has long been regarded as a benign infection and, therefore, has attracted less attention than its more virulent cousin *P. falciparum*. However, recent studies have challenged this paradigm, highlighting *P. vivax* as a significant cause of morbidity and mortality (reviewed in Chapter 3) and socioeconomic disruption. It is clear that the long-standing neglect of this species is no longer tenable (Price et al., 2007; Baird, 2007; Tjitra et al., 2008; Genton et al., 2008; Beg et al., 2008; Barcus et al., 2007; Lampah et al., 2011; Kochar et al., 2005, 2009; Alexandre et al., 2010; Flower et al., 2011; Tan et al., 2008; Bassat and Alonso, 2011; Anstey et al., 2009).

During the 1920s, *P. vivax* was used extensively for malariatherapy of patients with neurosyphilis (Austin et al., 1992). Many of these patients, as well as 'volunteers' from penitentiaries, were enrolled in trials of early anti-malarial compounds (see Chapter 4 for a comprehensive historical account of drug development and use in *P. vivax* malaria). Chloroquine was first synthesised in Germany in 1934 and subsequently tested by Sioli in four neurosyphilis patients treated with vivax malaria. Due to perceived toxicity, it did not become widely available until the end of the Second World War. By the 1950s, chloroquine was being used increasingly as a reliable prophylactic and blood schizontocidal treatment for both *P. vivax* and *P. falciparum* malaria. It remains the mainstay of *P. vivax* blood-stage treatment in most parts of the vivax-endemic world (Douglas et al., 2010).

The anti-relapse properties of primaquine were recognised in the 1950s following observations of its ability to prevent late recurrences of vivax malaria in troops returning from the Korean War (Alving et al., 1953). Remarkably, anti-relapse treatment is still reliant on primaquine, a drug that is contraindicated in infants and pregnant women and can cause life-threatening haemolysis in those with glucose-6-phosphate dehydrogenase (G6PD) deficiency. The pipeline for new, safer hypnozoitocidal alternatives is dismally bare (Medicines for Malaria Venture, 2011; Anthony et al., 2012). This situation is compounded by a lack of point-of-care tests for G6PD deficiency, as well as the poor sensitivity of rapid diagnostic tests for *P. vivax* (Arevalo-Herrera et al., 2010; World Health Organization et al., 2009). Currently *P. vivax* attracts less than 5% of the US National Institutes of Health malaria research budget (Carlton et al., 2011).

It is time to re-focus research efforts on *P. vivax* and, perhaps most urgently, on the optimal means of reducing transmission of this highly evolved parasite. Decreasing the interaction between humans and mosquito vectors will play a crucial role in this endeavour. Anti-malarial drugs will also be critically important components of the elimination arsenal. In this chapter we focus on how best to use currently available anti-malarials to reduce transmission of *P. vivax* including in the management of symptomatic malaria due to both *P. vivax* and *P. falciparum* and in mass drug administration (MDA) campaigns.

2. TRANSMISSION OF *P. VIVAX*

Mosquito transmission of *P. vivax* begins when a female anopheline ingests mature gametocytes from an infected human. Once inside the insect's midgut, the male *Plasmodium* gamete fertilises the female gamete initiating the process of sporogony. Assuming a mean ambient temperature of 25 °C, sporozoites concentrate in the mosquito's salivary glands within 9 days (the shortest sporogonic period of any of the *Plasmodium* species that infect humans (Beier, 1998)). If the mosquito subsequently takes another blood meal, these infectious sporozoites will be inoculated into the subcutaneous tissue, thus initiating a new human *P. vivax* infection.

In *P. vivax*, *Plasmodium ovale* and *Plasmodium malariae* infections, a proportion of erythrocytic merozoites will begin replicating sexually to produce male and female offspring immediately after release from the liver. In clinical trials of *P. vivax* therapy, the proportion of patients who are gametocytaemic on enrolment typically varies from 60 to 95% (Huh et al., 2011; Ratcliff et al., 2007; Hasugian et al., 2007). By contrast, gametocytogenesis in *P. falciparum* infections is delayed by 1 or 2 weeks with respect to the initiation of blood-stage asexual parasite replication (Stepniewska et al., 2008). Consequently, *P. vivax*, but not *P. falciparum* infections, are usually transmissible before symptom onset, and thus before anti-malarial treatment is commenced.

The defining feature of *P. vivax* malaria, and one that plays an important part in its refractoriness to current malaria control strategies, is its ability to relapse from dormant liver forms. After arrival in the human liver, an unknown, and probably variable, proportion of sporozoites differentiate to become dormant hypnozoites. These inert forms reawaken at intervals following the initial bout of malaria giving rise to recurrent blood-stage infections. The frequency and pattern of these relapses vary widely by geographic region, and even within the same locale, ranging from three weekly relapses in many equatorial regions to relapse intervals of 6 months or greater in many temperate regions (Hill and Amatuzio, 1949; Coatney et al., 1950; Hankey et al., 1953). Biphasic relapse patterns characterised by long periods of latency followed by a burst of frequent relapses have also been described and some strains do not cause a primary blood-stage infection at all (Hankey et al., 1953). The determinants of the periodicity of these relapses are unclear but one leading hypothesis is that febrile malaria episodes 'reawaken' dormant hypnozoites in a self-fulfilling cycle (see Chapter 2) (White, 2011; Douglas et al., 2011; Looareesuwan et al., 1987). The duration over which relapses from a single inoculation occur is not entirely clear but induced infections with the Chesson strain, prevalent in the island of New Guinea, have been shown to relapse for up to 5 years following initial inoculation (provided the development of homologous immunity is limited by prompt blood schizontocidal treatment of each relapse) (Hill and Amatuzio, 1949). In the absence of anti-malarial treatment, non-immune adults will develop clinical tolerance to persistent *P. vivax* parasitaemia in the space of 3–8 weeks (Boyd, 1938). Homologous immunity strong enough to suppress peripheral parasitaemia below patent levels requires an extended period of persistent parasitaemia, in many cases longer than 200 days (Coatney et al., 1950). Sterilising immunity does eventually develop (Van den Eede et al., 2011) but probably follows suppressive immunity by a substantial period of time. Since even sub-patent *P. vivax* infections can be transmitted relatively efficiently, asymptomatic individuals are likely to be a very important source of *P. vivax* transmission in endemic settings (Bousema and Drakeley, 2011; Jeffery, 1952; Bharti et al., 2006).

Unlike *P. falciparum* gametocytes, *P. vivax* gametocytes are short-lived. In the absence of anti-malarial treatment, vivax gametocytes are estimated to circulate in the peripheral bloodstream for a maximum of 3 days (Bousema and Drakeley, 2011). Therefore, *P. vivax* gametocytaemia does not persist for any significant length of time following clearance of asexual parasitaemia by anti-malarial drugs. The mathematical relationship between asexual parasite density and gametocyte density does not appear to differ between primary and recurrent infections (Douglas et al., in submission). The factors that determine the presence of gametocytaemia and, therefore, the overall likelihood of transmission of a *P. vivax* inoculum, are those that determine the presence of asexual parasitaemia. The most important of these determinants are the relapse pattern of the particular *P. vivax* strain, the strength of homologous immunity and the anti-malarial drug regimen received. Since *P. vivax* gametocytaemia mirrors asexual parasitaemia, one would assume that the impact of anti-malarial treatment on the total area under the asexual parasitaemia–time curve for a given inoculation could be used as a primary indicator of a regimen's potential to reduce transmission of *P. vivax*.

All patients with patent *P. vivax* monoinfections or mixed species infections involving *P. vivax* will have hypnozoites and are, therefore, at risk of multiple future relapses. Even in low-transmission areas, a significant proportion of those without patent *P. vivax* parasitaemia will also harbour *P. vivax* hypnozoites. Acute infection with *P. falciparum* is a strong risk factor for activation of hypnozoites in these individuals (Looareesuwan et al., 1987). Table 5.1 shows the risk of *P. vivax* parasitaemia following treatment of *P. falciparum* infection with artemether + lumefantrine in several co-endemic regions (Ratcliff et al., 2007; Hasugian et al., 2007; Douglas et al., 2011; Smithuis et al., 2010; Karunajeewa et al., 2008; McGready et al., 2008; Hutagalung et al., 2005; Lefevre et al., 2001; van Vugt et al., 1998, 1999; Mayxay et al., 2004; Stohrer et al., 2004; Denis et al., 2006; Thriemer et al., 2010; van den Broek et al., 2005; Haque et al., 2007; Thapa et al., 2007; Assefa et al., 2010; Hwang et al., 2011). This particular artemisinin

Table 5.1 Recurrence of *P. vivax* Parasitaemia Following Treatment of *P. falciparum* Malaria with Artemether + Lumefantrine

Reference	Duration of Follow-up (days)	Number of Patients Completing Follow-up	Number (%) with Recurrent *P. vivax**	Number (%) with Recurrent *P. falciparum* (PCR Unadjusted)	Percentage of Recurrences due to *P. vivax*
Southeast Asia and Oceania					
Thailand					
McGready et al. (2008)	42	124	37[§] (29.8)	43 (34.7)	46.3%
Hutagalung et al. (2005)	42	225	90 (40.0)	27 (12.0)	76.9%
Lefevre et al. (2001)	28	155	6 (3.9)	7 (4.5)	46.2%
van Vugt et al. (1999)[#]	28	90	11 (12.2)	13 (14.4)	45.8%
van Vugt et al. (1999)[#]	28–63[‖]	142	43	8	
van Vugt et al. (1998)	63	248	83 (35.5)	61 (24.6)	57.6%
Papua New Guinea					
Karunajeewa et al. (2008)	42	102	66 (64.9)	5 (4.9)	93.0%
Myanmar					
Smithuis et al. (2010)	63	137	62 (45.3)	15 (11.0)	80.5%
Laos					
Mayxay et al. (2004)	42	107	5 (4.7)	14 (13.1)	26.3%
Stohrer et al. (2004)	42	47	2 (4.1)	13 (26.5)	13.3%
Indonesia					
Ratcliff et al. (2007)	42	219	67 (30.6)	29 (13.2)	69.8%

Cambodia					
Denis et al. (2006)[##]	28	49	2 (4.1)	15 (30.6)	11.8%
Denis et al. (2006)[##]	28	77	1 (1.3)	12 (15.6)	7.7%
Indian Subcontinent					
Bangladesh					
Thriemer et al. (2010)	42	73	4 (5.5)	6 (8.2)	40%
van den Broek et al. (2005)	42	119	25 (21.0)	20 (16.8)	55.6%
Haque et al. (2007)	42	67	7 (10.4)	10 (14.9)	41.2%
Nepal					
Thapa et al. (2007)[†]	28	64	9[¶] (13.6)	0 (0.0)	100.0%
Africa					
Ethiopia					
Assefa et al. (2010)	28	85	5 (5.9)	3 (3.5)	62.5%
Hwang et al. (2011)	42	112	8 (7.1)	1 (0.9)	88.8%

Abbreviations: AM + LUM; artemether + lumefantrine, PCR; polymerase chain reaction.
[*]When patients were excluded due to recurrent *P. vivax* infection or *P. falciparum* reinfection during the follow-up period, these numbers were added to the eligible participants.
[#]Evaluated three regimens of artemether + lumefantrine at two locations with variable follow-up.
[##]Contained details of two separate studies.
[§]Before delivery or day 42 if later.
[‖]Variable follow-up periods used for different participants.
[¶]*P. vivax* infection detected by PCR.
[†]Includes five patients with mixed infection by PCR at enrolment.

combination therapy was chosen for demonstration purposes because it is highly effective for eliminating the blood stages of both *P. vivax* and *P. falciparum* (thus minimising the risk of recrudescence) and is cleared relatively rapidly from the bloodstream ($t_{1/2}$ for artesunate is approximately 30 min and for lumefantrine is approximately 3.2 days). The latter property ensures that concentrations of the drugs will be negligible by day 16, well before the first *P. vivax* relapse is likely to appear (approximately day 21 for tropical strains). In Myanmar, 45% of patients treated for *P. falciparum* monoinfections with AM + LUM had a recurrence of *P. vivax* by day 63 (Smithuis et al., 2010). On the Thai-Myanmar border, the corresponding proportion at day 63 following AM + LUM was 53.8% (Douglas et al., 2011) and in Papua, New Guinea, at day 42 was 64.7% (Karunajeewa et al., 2008). These proportions far exceeded basal *P. vivax* infection rates in the respective study locations and given the potency of both artesunate and lumefantrine against *P. vivax*, the recurrences were very unlikely to have been recrudescences from sub-patent *P. vivax* blood-stage infections. In many of these studies, the risk of *P. vivax* following *P. falciparum* malaria was greater than the rate of *P. falciparum* failure and, therefore, the most commonly transmitted parasite following falciparum malaria, paradoxically, was *P. vivax* (median percentage of recurrences due to *P. vivax* = 51%, range 7.7–100%).

The most likely explanation for heterologous *P. vivax* relapse is that the physiological derangement associated with falciparum malaria awakens dormant hypnozoites acquired during previous *P. vivax* infections (White, 2011; Douglas et al., 2011; Looareesuwan et al., 1987). Alternatively, heterologous *P. vivax* relapses may arise from simultaneous *P. vivax* inoculations that never reach patency (Douglas et al., 2011; Looareesuwan et al., 1987). Whatever the cause, *P. vivax* infection following falciparum malaria carries significant morbidity and is likely to have an important role in sustaining *P. vivax* transmission in co-endemic regions (Douglas et al., 2011). Modifying the probability that *P. vivax* relapse occurs should, therefore, be an important consideration in *P. falciparum* as well as *P. vivax* malaria therapy.

3. REDUCING *P. VIVAX* TRANSMISSION BY TREATING SYMPTOMATIC DISEASE

A primary opportunity to reduce total parasitaemia associated with a *P. vivax* inoculation is the point at which a patient presents with symptomatic infection. There are three aspects to acute treatment of vivax malaria which have the potential to reduce total parasitaemia. The first is eradication of blood-stage parasites, the second is post-treatment prophylaxis against recurrent blood-stage infection and the third is prevention of future relapses using hypnozoitocidal drugs. We consider these separately.

3.1. Eradicating Blood-Stage Parasites

Interrupting acute *P. vivax* infections is possible using a range of different anti-malarial drugs. All drugs that are commonly used for falciparum malaria are active against both the asexual and sexual stages of *P. vivax* but there is significant variability in their potency (Pukrittayakamee et al., 2004, 2008). In the absence of acquired resistance, the artemisinin derivatives have the highest potency followed by chloroquine, the other quinolines (including mefloquine, quinine and primaquine), the anti-bacterial anti-malarials clindamycin, tetracycline, doxycycline and azithromycin followed by sulfadoxine + pyrimethamine (SP) (Pukrittayakamee et al., 2004; Darlow et al., 1982). The potencies of the ACT partner drugs, lumefantrine and piperaquine, have not been compared head-to-head with the aforementioned anti-malarials *in vivo* but *in vitro* experiments suggest that they fall below chloroquine but well above the antibiotic anti-malarials (Lux et al., 2003, 2011; Chotivanich et al., 2009). *In vitro* studies in Papua, Indonesia, an area endemic for chloroquine-resistant *P. vivax*, demonstrate that the potency of amodiaquine is similar to lumefantrine and piperaquine (Russell et al., 2008). *In vitro* and *in vivo* evidence suggests that development of chloroquine resistance confers cross-resistance to amodiaquine (Hasugian et al., 2007, 2009; Russell et al., 2008).

Drugs with greater potency clear asexual parasitaemia, and therefore, gametocytaemia, faster than drugs with lower potency. However, the relative transmission-blocking benefits of more rapid clearance alone will be minimal since gametocytes capable of transmission will almost certainly have been present in the peripheral circulation for a number of days prior to presentation (Huh et al., 2011; Ratcliff et al., 2007; Hasugian et al., 2007). Furthermore, assuming complete parasitological cure, the majority of parasitaemia associated with a single inoculation is likely to be related to relapses rather than the primary episode (Hill and Amatuzio, 1949; Hankey et al., 1953). Apart from primaquine, none of the blood schizontocides used for the treatment of vivax malaria has activity against hypnozoites and, therefore, cannot eliminate the potential for future relapses. A 1-week course of 60 mg primaquine daily will clear blood-stage *P. vivax* parasitaemia

(Pukrittayakamee et al., 2010) but its use as monotherapy would be neither optimal nor advisable.

Prevention of recrudescence by total eradication of asexual blood-stage parasitaemia is critically important for reducing *P. vivax* transmission. For a drug to completely sterilise the blood, it needs to be at or above the minimum inhibitory concentration for as many asexual cycles as it takes to remove the very last parasite, or at least to reduce parasitaemia to a level at which the human's immune system can remove the residuum (White, 1999). Parasite multiplication rate, drug potency and drug elimination half-life govern the total duration of treatment required, which in turn is a key determinant of drug adherence. Optimal patient adherence can theoretically be achieved by using a single-dose drug regimen.

Chloroquine has high potency against sensitive *P. vivax* strains and is slowly eliminated allowing a practical 3-day treatment regimen. It is cheap and well tolerated and, therefore, has been the drug of choice for vivax malaria since the mid-twentieth century. Resistance of *P. vivax* to this drug is spreading from a presumed epicentre in New Guinea (Douglas et al., 2010; Baird, 2009). In the worst-affected areas such as in Papua New Guinea, Eastern Indonesia, the western parts of India and Western Thailand, a significant proportion of *P. vivax* infections will persist, and thus recrudesce, in the face of high chloroquine concentrations (Karunajeewa et al., 2008; Baird et al., 1995, 1997; Sumawinata et al., 2003; Sutanto et al., 2009; Rijken et al., 2011; Singh, 2000). In Papua New Guinea and Eastern Indonesia, over half of the *P. vivax* infections recrudesce within 28 days of treatment (Karunajeewa et al., 2008; Baird et al., 1995, 1997; Sumawinata et al., 2003; Sutanto et al., 2009). Although the degree of chloroquine resistance is less severe in other vivax-endemic areas, there is clear evidence that the efficacy of chloroquine is declining in many of these places (Douglas et al., 2010).

Currently, the most effective schizontocidal treatment of *P. vivax* is a 3-day course of an artemisinin derivative in combination with a highly active, but more slowly eliminated, partner drug (Ratcliff et al., 2007; Hasugian et al., 2007; Karunajeewa et al., 2008; Awab et al., 2010; Poravuth et al., 2011; Phyo et al., 2011). The rapidly eliminated artemisinin component quickly clears the bulk of the parasite biomass and the more slowly eliminated partner drug persists long enough to remove any remaining parasites. Artemisinin-combination therapies currently recommended by the World Health Organization for falciparum malaria include artemether plus lumefantrine (AM + LUM), artesunate plus amodiaquine (AS + AQ), artesunate plus mefloquine (AS + MQ), artesunate plus sulfadoxine–pyrimethamine

(AS + SP) and dihydroartemisinin plus piperaquine (DHA + PIP). *Plasmodium vivax* resistance to the artemisinins has not been described but cross-resistance of chloroquine-resistant *P. vivax* strains to amodiaquine results in high rates of recurrent infections following AS + AQ in regions such as Eastern Indonesia and Papua New Guinea (Hasugian et al., 2007; Karunajeewa et al., 2008). Although SP has low inherent efficacy against *P. vivax* (Darlow et al., 1982), AS + SP is efficacious in the absence of SP resistance (Tjitra et al., 2002); however, this combination is not recommended for *P. vivax* due to the widespread presence of antifolate resistance in this species (Karunajeewa et al., 2008; Tjitra et al., 2002; Imwong et al., 2003). Although the initial therapeutic response of *P. vivax* following treatment with AM + LUM, AS + MQ or DHA + PIP is generally excellent, the rates of *P. vivax* recurrence within 42 days vary considerably. This reflects the variability in the duration of post-treatment prophylaxis against recurrent infection provided by the partner drug.

3.2. Post-treatment Prophylaxis

Slowly eliminated blood schizontocidal drugs that remain above minimum inhibitory concentrations for a long time after administration have the potential to reduce markedly recurrent *P. vivax* parasitaemia following acute infection (Douglas et al., 2010, 2011; Ratcliff et al., 2007; Karunajeewa et al., 2008). On the island of New Guinea, where relapses tend to occur every 3 weeks, the slowly eliminated combination dihydroartemisinin + piperaquine (half-life of piperaquine ~28 days (Hung et al., 2004; Tarning et al., 2008)) is associated with a two- to fourfold lower risk of recurrence than artemether + lumefantrine (half-life of lumefantrine ~3.2 days (Ezzet et al., 2000)) within 42 days of treatment (Ratcliff et al., 2007; Karunajeewa et al., 2008). Where relapses occur less frequently, the period of post-treatment prophylaxis is less likely to cover the first relapse and, therefore, the benefit of a long period of time above minimum inhibitory concentration will be limited to a brief period of refractoriness to reinfection.

Increasing anti-malarial drug resistance permits parasites to grow in greater concentrations of drug. In low-grade resistance, concentrations of drug may be sufficient to eradicate most of the initial parasite biomass, but the period of time for which the drug is above minimum inhibitory concentration decreases. The first manifestation of such resistance is a reduction in the duration of post-treatment prophylaxis and late recrudescent infections. The proportionate relationship between asexual stages and gametocytes does not appear to differ between primary and recurrent

P. vivax infections and, therefore, this potentially translates to a shorter period free from gametocytaemia. Early chloroquine resistance might explain the greater parasitological cure rate observed with DHA + PIP relative to CQ in Afghanistan despite the longer terminal elimination half-life and greater potency of chloroquine against sensitive *P. vivax* strains when compared with piperaquine (Lu et al., 2011; Awab et al., 2010; Krishna and White, 1996). Assessments of the effectiveness of artesunate + amodiaquine (half-life 1–3 weeks (Krishna and White, 1996)) in regions where chloroquine resistance has emerged demonstrate that the post-treatment prophylactic effect of this combination has also been truncated in some regions (Ratcliff et al., 2007; Karunajeewa et al., 2008), presumably because of cross-resistance to amodiaquine.

The risk of relapse and potential benefits of post-treatment prophylaxis are not restricted to patients with acute *P. vivax* monoinfection. Use of slowly eliminated anti-malarials, such as mefloquine or piperaquine, markedly reduces the risk of early heterologous *P. vivax* relapses in patients presenting initially with pure *P. falciparum* infection (Douglas et al., 2011). The benefits of post-treatment prophylaxis are generally short-lived, with almost all anti-malarial drugs reduced to sub-therapeutic concentrations within 63 days. Although reducing early recurrent *P. vivax* infections may impact upon early gametocyte carriage and transmission potential (Phyo et al., 2011), the long-term benefits of prolonged post-exposure prophylaxis are unknown. It is not clear whether suppressing the first relapse reduces the total number of relapses, and hence total parasitaemia, associated with a single inoculation, or simply delays the occurrence of the first one. In regions where reinoculation of *P. vivax* occurs frequently, this is an academic question as the liver will continue to be seeded with new hypnozoites throughout life. Chemotherapeutic suppression of any relapses in these regions should reduce the life-time burden of parasitaemia and, therefore, should be associated with reduced transmission of *P. vivax*. This proposition remains speculative. Proof would best be achieved by conducting long-term, cluster-randomised studies of repeated treatment with either a rapidly or a slowly eliminated artemisinin combination therapy.

3.3. Eradicating Hypnozoites

Primaquine, an 8-aminoquinoline, is the only licensed drug with activity against hypnozoites and, therefore, the only means of preventing all future relapses from previously acquired hypnozoites. Given the dearth of new candidates in the hypnozoitocidal drug discovery pipeline, primaquine is likely to remain the sole option for at least the next few years (Medicines for Malaria Venture, 2011; Anthony et al., 2012). Although primaquine was first approved over 50 years ago, there are many areas of uncertainty surrounding its toxicity profile, optimal dosing and mechanism of action. Its efficacy is hard to gauge since geographical differences in recurrence rates may reflect acquired resistance, inherent differences in strain susceptibility or simply differences in relapse patterns.

Early work by Alving et al. showed that concurrent administration of primaquine with either chloroquine or quinine was associated with greater efficacy than serial administration of quinine followed by primaquine (Alving et al., 1955). It is currently not known whether other blood schizontocides, such as the artemisinin combination therapies, also have this potentiating effect. Several trials have shown good efficacy of artemisinin derivatives either in combination with or closely followed by primaquine for preventing *P. vivax* recurrences up to day 28 (Krudsood et al., 2007, 2008; Hamedi et al., 2004; Dao et al., 2007; Silachamroon et al., 2003; Wilairatana et al., 1999). The short duration of follow-up in these trials precludes an assessment of the risk of late relapse.

The efficacy of primaquine is determined by the total dose of drug received as opposed to the concentrations achieved in the blood (Leslie et al., 2008; Clyde and McCarthy, 1977; Alving et al., 1960). A weekly primaquine dosing regimen given over 8 weeks after effective blood schizontocidal treatment is as efficacious as the same total dose of primaquine given over 2 weeks (Leslie et al., 2008). The optimal total dose required to prevent future relapses is uncertain but appears to differ by geographic location (Goller et al., 2007). A dose of 210 mg over 14 days (~3.5 mg/kg) has been associated with close to 100% protection against future relapse in patients with *P. vivax* infection acquired in parts of South Korea, South and Central America and India (Alving et al., 1953; Coatney et al., 1953; Gogtay et al., 1999; Miller et al., 1974). Twice that total dose (~7 mg/kg) may be required to reliably prevent relapses in those with the frequently relapsing Chesson strain, known to be prevalent in Papua New Guinea and Papua, Indonesia (Baird, 2009; Clyde and McCarthy, 1977; Goller et al., 2007; Coatney and Getz, 1962; Collins and Jeffery, 1996). Based on the limited available evidence, the expert committees of the World Health Organization recommend 0.25 mg/kg primaquine daily over 14 days except in Oceania and Southeast Asia, where a dose of 0.5 mg/kg/day is recommended (World Health Organization, 2010). Further work to clarify relapse patterns and the efficacy of different doses of primaquine in vivax-endemic regions is essential to provide evidence-based advice from which to optimise current policies and practices. Given the wide variation in relapse patterns both within the same geographic location and between different locations, follow-up of patients in all clinical studies of anti-relapse therapies must be prolonged (John et al., 2012).

Several practical issues limit the effectiveness of primaquine for preventing *P. vivax* relapses. First, adherence to the standard 2-week primaquine regimen is poor (Baird and Rieckmann, 2003). A small number of clinical trials have shown that direct supervision of primaquine treatment is associated with a significant benefit over unsupervised therapy (Grietens et al., 2010; Maneeboonyang et al., 2011; Takeuchi et al., 2010; Leslie et al., 2004). These studies are likely to underestimate the benefit of supervision since even the unsupervised patients in these trials were more likely to have been compliant than those in the community, purely by virtue of their study participation (the Hawthorne effect). Compressing a 14-day course of primaquine into 7 days (still much longer than typical blood schizontocide courses) may help to improve patient adherence without compromising efficacy (Krudsood et al., 2008; Clyde and McCarthy, 1977). The same total dose over 3 days has not been shown to be as efficacious and if the drug is not taken with food can cause excessive gastrointestinal side effects (Carmona-Fonseca and Maestre, 2009). Routine direct observation of long courses of primaquine is unlikely to be practicable in most malaria-endemic settings. Alternative approaches to improving adherence such as education and poster campaigns warrant further exploration.

A second practical concern is that primaquine can cause significant haemolysis in patients with G6PD deficiency (Nkhoma et al., 2009). At least 140 mutations in the gene coding for G6PD have been described, each conferring varying degrees of enzyme inactivation (Beutler and Vulliamy, 2002). Patients with severe deficiency are at risk of life-threatening haemolysis after one dose of primaquine, whereas those with mild deficiency will only experience a minor, self-limiting drop in haemoglobin (Baird and Hoffman, 2004; Reeve et al., 1992). It is not known what threshold level of G6PD deficiency confers an unacceptable risk of haemolysis. Furthermore, there is currently no rapid test for reliably identifying and grading patients with G6PD deficiency. Safety studies in those with G6PD deficiency and development of rapid, affordable and accurate tests for G6PD activity are arguably the highest priorities in the field of vivax-malaria research today.

The 2010 World Health Organization guidelines recommend that children under the age of 4 years should not receive primaquine (World

Health Organization, 2010). The exact rationale for this age cutoff is not clear. In most endemic settings, young patients are at greatest risk of recurrent *P. vivax* infections and have the highest rates of morbidity and mortality, particularly with regard to anaemia (Douglas et al., 2011, 2012; Phimpraphi et al., 2008; Poespoprodjo et al., 2009). A recent review of the literature revealed 11 clinical studies of primaquine therapy that recruited patients under 4 years of age. Although age-stratified tolerability was not reported, the overall safety profile of primaquine in G6PD normal individuals was excellent and no worse than for other anti-malarial drugs (John et al., 2012). Some policymakers, recognising the importance of providing anti-relapse therapy, have opted to recommend prescription of primaquine in children as young as 1 year of age. As far as we are aware, no national malaria control guidelines advocate primaquine for those <1-year old. The lack of provision of radical cure for infants represents not only a failure to reduce morbidity in the most vulnerable population but also a lost opportunity to impact on *P. vivax* transmission.

In view of the high rates of *P. vivax* relapse following treatment of falciparum malaria in co-endemic regions, there is a rationale for administering terminal prophylaxis in patients presenting without peripheral *P. vivax* parasitaemia. The proposition of giving hypnozoitocidal doses of primaquine to those with falciparum malaria raises several ethical issues (Baird, 2011; Baird and Surjadjaja, 2010; Price and Douglas, 2010). There is currently no test to determine the presence of *P. vivax* hypnozoites and, therefore, a proportion of patients with *P. falciparum* malaria would receive primaquine without any individual benefit. Unlike other anti-malarials in use today, primaquine has activity against all stages of *P. falciparum* gametocytes (White, 2008). A single dose has been associated with a significant reduction in gametocyte carriage when compared with artemisinin combination therapy alone and the same will be true of a full curative course provided for the purpose of eradicating liver stages of *P. vivax* (Smithuis et al., 2010; Price and Douglas, 2010). Whether the chance of individual benefit from preventing *P. vivax* relapses and the population benefit of reduced transmissibility of both *P. vivax* and *P. falciparum* outweigh the individual risks of primaquine in those with falciparum malaria requires further study.

The only potential hypnozoitocidal alternatives to primaquine on the horizon are tafenoquine, bulaquine and NPC1161B. Tafenoquine is a synthetic analogue of primaquine with a half-life of approximately 14 days (compared to 4–6 h for primaquine) (Brueckner et al., 1998). A total dose of 1200 mg over 3 days has been shown to be as efficacious at preventing

P. vivax relapses as a 14-day course of primaquine at a dose of 22.5 mg/day in combination with doxycycline (Elmes et al., 2008). The shortened treatment duration may improve patient adherence, however tafenoquine has the same propensity to cause haemolysis in those with G6PD deficiency as primaquine and is also contraindicated in pregnant women (Crockett and Kain, 2007). Given its long elimination half-life, reversal of haemolysis, should it occur, is likely to be more difficult with tafenoquine than with primaquine. Bulaquine is a prodrug of primaquine which is marketed in India as a 5-day regimen of 25 mg/day for prevention of *P. vivax* relapses (Krudsood et al., 2006). The small number of clinical trials of this drug to date have had methodological weaknesses and there appears to be little interest in developing this agent further (Krudsood et al., 2006; Valecha et al., 2001). NPC1161B, another 8-aminoquinoline, is in Phase I development. It has shown good tissue schizontocidal activity against *P. cynomolgi* (a close relative of *P. vivax*) in rhesus monkeys but no published data exist on its use in humans (LaMontagne et al., 1982).

4. REDUCING *P. VIVAX* TRANSMISSION USING MASS DRUG ADMINISTRATION

Evidence from co-endemic regions shows that on average, individuals gain immunity to *P. vivax* malaria more rapidly than to *P. falciparum* malaria and consequently, a higher proportion of *P. vivax* infections is asymptomatic (Tjitra et al., 2008; Genton et al., 2008; Luxemburger et al., 1996; Michon et al., 2007; Karyana et al., 2008). Studies from Southern Papua, The Solomon Islands and Vanuatu suggest that between 71% and 97% of all microscopically patent *P. vivax* infections in these regions are asymptomatic (Karyana et al., 2008; Harris et al., 2010; Kinzer et al., 2010). The proportion of sub-patent *P. vivax* infections that are asymptomatic is likely to be even higher. Passive detection and treatment of individuals with symptomatic vivax malaria will clearly have limited impact on *P. vivax* transmission. Alternative chemotherapeutic strategies for reducing transmission of asymptomatic infections include mass screening and treatment (MSAT) and MDA.

Certain biological features of *P. vivax* infections present unique challenges for the MSAT strategy in particular. Sub-patent *P. vivax* blood-stage infections can be transmitted relatively efficiently (Bousema and Drakeley, 2011; Jeffery, 1952; Bharti et al., 2006). Cross-sectional studies employing PCR-based diagnostic methodology show that microscopy underestimates the true prevalence of *P. vivax* parasitaemia and, therefore, suggest that subpatent infections are an important source of *P. vivax* transmission (Harris et al., 2010; Gunasekera et al., 2011; Poschl et al., 2010; Steenkeste et al., 2010). Active case detection based on light microscopy alone will miss this important group of potentially infectious individuals, and thus have limited efficacy. In Yanomami people living in the Amazon rainforest, a microscopy-based active case detection and treatment campaign during the late 1960s resulted in a 46% reduction in the incidence of malaria over a 3-year period. However, during this time, the ratio of *P. falciparum* to *P. vivax* infections fell from 42% to 24% indicating that the campaign had been much less effective against *P. vivax* than *P. falciparum* (Macauley, 2005).

Newer, more sensitive, means of detecting low-density *Plasmodium* infections may increase the yield of active detection and treatment efforts. Rapid diagnostic tests are simple and relatively cheap. Some have been shown to be highly sensitive for low-density *P. falciparum* infections but their performance at detecting low-density *P. vivax* infections remains poor (World Health Organization et al., 2009). Loop-mediated isothermal amplification (LAMP) has greater sensitivity for *P. vivax* infections than microscopy and is cheaper, simpler and faster than PCR-based methods (Poschl et al., 2010; Chen et al., 2010). LAMP may have a role both in clinics and as a tool for detection of asymptomatic infections in endemic regions (Poschl et al., 2010; Chen et al., 2010; Sirichaisinthop et al., 2011). Nevertheless, LAMP is not yet widely available and although cheaper than PCR, is likely to remain prohibitively expensive and too slow for widespread use, at least in the near future. For highly resourced regions, PCR remains the most sensitive means of detecting *P. vivax* malaria and, therefore, hypothetically, the best means of maximising the yield of active case detection and treatment campaigns. Again, given logistical and financial constraints, PCR is unlikely to be used in clinical practice.

Our inability to determine whether an individual harbours *P. vivax* hypnozoites is the single greatest limitation of the MSAT strategy for reducing transmission of vivax malaria. Dormant hypnozoites can seed multiple future relapses and, given their longevity, will be highly prevalent in endemic populations, even if the entomological inoculation rate is relatively low. This claim is indirectly supported by the very high rate of heterologous recurrence following falciparum malaria seen in many regions with *Plasmodium* co-endemicity (Ratcliff et al., 2007; Hasugian et al., 2007; Douglas et al., 2011; Smithuis et al., 2010; Karunajeewa et al., 2008). In cross-sectional treatment campaigns, the only way of ensuring that individuals harbouring hypnozoites who do not have blood-stage infection are treated is to distribute anti-malarial drugs to everybody (MDA).

For the reasons outlined previously, effective hypnozoitocidal therapy is likely to be much more important in mass screening and treatment campaigns than blood schizontocidal treatment (Aguas et al., 2012). Since only a small proportion of people treated in MDA campaigns will have blood-stage infection, the relative importance of hypnozoitocidal therapy in this strategy will be even greater. Between the third and the fifth of November 1981, 70% of the population of Nicaragua (~1.9 million people) received a 3-day course of chloroquine (1500 mg total dose) and primaquine (45 mg total dose) (see Chapter 6 for descriptions of several historical MDA campaigns). This intervention was associated with a subsequent reduction in the incidence of both *P. falciparum* and *P. vivax*. The effect on *P. falciparum* lasted for approximately 9 months, whereas for *P. vivax*, it lasted just 4 months (Garfield and Vermund, 1983). The authors of the analysis suggested that the limited effectiveness against *P. vivax* was likely to relate to the inadequate dose of primaquine administered (Garfield and Vermund, 1983).

In the Ghab region in Syria, a weekly dose of chloroquine and pyrimethamine administered in conjunction with indoor residual spraying of dichlorodiphenyltrichloroethane (DDT) between August and October 1968 resulted in a rapid and marked reduction in the incidence of vivax malaria from 470 cases in 1968 to 114 cases in 1969 (Onori, 1972). Based on the results of concurrent entomological surveillance, the authors concluded that all 114 of the cases in 1969 (92 of which were in individuals who had received presumptive drug treatment) were relapses (Onori, 1972). Without continued aggressive vector control, these relapses are likely to have led to a rebound in vivax malaria incidence, however further spraying of dieldrin in May and August helped to ensure a complete absence of cases in 1970.

On the pacific island of Aneityum in the Vanuatu archipelago (population 718 people) weekly MDA of chloroquine, sulfadoxine + pyrimethamine and primaquine (45 mg) for 9 weeks in conjunction with distribution of insecticide-treated bednets and introduction of larvivorous fish resulted in complete suppression of malaria transmission due to both *P. falciparum* and *P. vivax* (Kaneko et al., 2000; Kaneko, 2010). Periodic malariometric surveillance over the following 9 years showed that, aside from two imported cases of malaria (one due to *P. vivax* and the other due to mixed *P. vivax*/*P. falciparum* infection), malaria was eliminated from the island. The single 45 mg dose of primaquine was given during seven of the nine weekly visits resulting in a total dose of 315 mg (approximately 5.25 mg/kg). Compliance with drug treatment was 88% and the prevalence of bednet ownership was 94%. Although the MDA was only one of several components of the malaria elimination campaign in Aneityum, the treatment schedule will have maximised its benefit. Firstly, a sufficiently long-acting and potent combination of blood schizontocides was used to ensure parasitocidal concentrations of drug persisted for long enough to cover the average lifetime of 5 weeks of *Anopheles farauti* – the major mosquito vector in the region (Belkin et al., 1945). As a result, even mosquitoes that were infected prior to the beginning of the campaign could not have initiated a new human infection. Since sulfadoxine + pyrimethamine has sporontocidal activity, formation of *P. vivax* hypnozoites from inoculations by previously infected mosquitoes should also have been prevented. Secondly, a (presumably) large enough total dose of primaquine was provided to eradicate *P. vivax* hypnozoites. A weekly 45 mg dose of primaquine is thought to be safe in individuals with mild or moderate enzyme deficiency (Leslie et al., 2008). Thirdly, administration of all drug doses was supervised and, therefore, poor adherence was unlikely. Lastly, population coverage was high. Although weekly (or regular) dosing of all individuals in a population may be the ideal means of rapidly reducing *P. vivax* transmission, such a schedule would be very resource intensive and difficult in larger, more dispersed populations. More practical schedules need to be explored.

Considering the logistic and financial difficulties of conducting mass anti-malarial drug administration trials, mathematical models are an attractive, but complicated, means of gaining insight into the likely effectiveness of MDA strategies. Ishikawa et al. used a model of *P. vivax* transmission based on epidemiological data from the Solomon Islands to show that a hypnozoitocidal course of primaquine administered as a weekly dose over 7 weeks along with chloroquine would result in rapid, and almost total, reduction in the *P. vivax* parasite index with maintenance of very low transmission for at least 4 years (Ishikawa et al., 2003). According to the model, a single dose of primaquine and chloroquine would result in a similarly rapid reduction in the parasite index but with a rebound to pre-intervention levels within 4 years (Ishikawa et al., 2003). The longevity of the effect of the latter regimen is likely to have been overestimated since the authors appeared to assume that there is a proportionate reduction in the risk of relapse with each additional dose of primaquine. Data from clinical studies suggest that the effectiveness of primaquine is more likely to be all or nothing depending on whether a certain threshold total dose is reached (Gogtay et al., 1999; Baird and Rieckmann, 2003; Rowland and Durrani, 1999). In another model, Aguas et al. found that an MDA campaign including a blood schizontocidal medication in combination with a wholly effective hypnozoitocidal course of primaquine would reduce the incidence of *P. vivax* as rapidly as *P. falciparum* (Aguas et al., 2012). However, the authors also showed that the re-emergence of *P. vivax* would be much more rapid than *P. falciparum* following reintroduction of foreign strains from nearby regions (Aguas et al., 2012).

Eradicating hypnozoites is clearly critical to the effectiveness of MDA efforts. If only a single round of drug treatment is possible, a full hypnozoitocidal course of primaquine should theoretically be provided. However, given the dangers of daily dosing in G6PD-deficient individuals or pregnant mothers with G6PD deficient foetuses, the G6PD and pregnancy status of each individual should ideally be established before providing primaquine. Given that most individuals in the population will be healthy, adherence to a 2-week course is likely to be very poor, and thus threaten the utility of this drug. If individual G6PD and pregnancy status was known, tafenoquine might be a very useful alternative hypnozoitocidal drug due to its long half-life and resultantly short treatment regimen. Short-course, high-dose primaquine regimens have been effective in clinical studies (Krudsood et al., 2008; Clyde and McCarthy, 1977) and may also be useful for single-visit MDA campaigns.

Young children are at greater risk of high-parasitaemia *P. vivax* infections than adults and are more likely to have patent *P. vivax* relapses (Douglas et al., 2011; Phimpraphi et al., 2008). Ostensibly, this would suggest that targeting young children may maximise the cost-effectiveness of MDA campaigns for reducing *P. vivax* transmission. However, children are also the most likely to experience symptomatic infections and are, therefore, more likely to present passively for treatment than adults (Douglas et al., 2011). The prevalence of hypnozoites in this age group will be lower than in adults. Intermittent preventive treatment in infancy (IPTi) and childhood (IPTc) are two age-limited forms of repeated MDA that have been employed for reducing morbidity associated with malaria, particularly in regions that are highly endemic for *P. falciparum*. Senn et al. recently showed that intermittent preventive treatment of infants with SP + AQ or AS + SP reduced the incidence of clinical vivax malaria episodes in Papua New Guinea by 23% and 4%, respectively (less than the corresponding reductions for *P. falciparum*) (Senn et al., 2012). On the Thai-Myanmar border, monthly or alternate monthly courses of DHA + PIP in healthy adult males at high occupational risk of malaria prevented 93% and 93% of vivax malaria episodes, respectively (Lwin et al., 2012). Both regimens appeared to protect against *P. vivax* gametocytaemia though tests of statistical significance specifically for this species were not given (Lwin et al., 2012). Weekly doses of chloroquine during pregnancy in a Karen population on the Thai-Myanmar border had 100% protective efficacy against *P. vivax* malaria, as detected by light microscopy (Villegas et al., 2007). Protection against clinical vivax malaria episodes or gametocytaemia detected by light microscopy does not necessarily equate to a reduction in transmission of the parasite. Evidence from large scale, cluster-randomised studies with long-term follow-up will be required before the potential of intermittent preventive treatment in infancy, childhood or pregnancy to reduce transmission of *P. vivax* can be established firmly.

Finally, MDA may increase drug pressure on the local parasite population and, therefore, promote the emergence of drug-resistant strains. For this reason, it has been suggested that first-line blood-schizontocidal therapies should be avoided in MDA programmes (Bousema and Drakeley, 2011). Acquired hypnozoite resistance to primaquine has not yet been proved but if it occurred would be particularly disastrous since there is no licensed alternative.

5. CONCLUSIONS

Anti-malarial drugs are an important means of reducing transmission of *P. vivax* infections. In the treatment of symptomatic disease, the primary consideration is prevention of recurrences due to recrudescence and relapses. Short course, potent and relatively slowly eliminated blood schizontocidal regimens administered concurrently with a total dose of 7 mg/kg primaquine (except in areas where there is solid evidence that a total dose of 3.5 mg/kg primaquine is effective) will have the greatest potential to reduce transmissibility of individual *P. vivax* infections. Given the spread of *P. vivax* strains that are resistant to chloroquine, the artemisinin combination therapies dihydroartemisinin + piperaquine and artesunate + mefloquine are the most assured means of preventing recrudescence. Preliminary evidence suggests that, like chloroquine, these combinations potentiate the hypnozoitocidal effect of primaquine, but further supportive evidence is required to solidify this claim. Two factors limit the potential effectiveness of primaquine in endemic regions: poor adherence to the 14-day-treatment regimen and reluctance to prescribe this drug because of the risk of primaquine-induced haemolysis in those with G6PD deficiency. Development of a rapid, accurate and affordable bedside test for G6PD deficiency is one of the highest priorities in the field of vivax malaria research today (Table 5.2). Great priority must also be given to identifying means of improving adherence to primaquine regimens such as shortening the treatment course and direct observation of treatment.

Table 5.2 Future Research Priorities

Development of new, safe, slowly eliminated and effective hypnozoitocidal drugs for preventing *P. vivax* relapses.
Assessment of short-course, high-dose primaquine regimens for improving adherence.
Confirmation of the effectiveness of primaquine when prescribed with blood schizontocides other than chloroquine and quinine.
Development of a rapid, cheap and accurate bedside test for glucose-6-phosphate dehydrogenase deficiency.
Clarification of the threshold level of glucose-6-phosphate dehydrogenase deficiency at which it is no longer safe to give standard courses of primaquine.
Development of other means of improving adherence to primaquine regimens.
Clarification of the relapse pattern of endemic *P. vivax* strains in various geographical locations.
Clarification of the total dose of primaquine required to prevent relapse in various geographical locations.
Assessment of the public health significance of sub-patent *P. vivax* infections.
Development of rapid, cheap and accurate means of diagnosing low-density *P. vivax* infections both in the clinic and in the community.
Development of a single-dose blood-schizontocidal regimen that is effective against both *P. falciparum* and *P. vivax*.
Assessment of the impact of a long period of post-treatment prophylaxis on overall transmission of *P. vivax* in various geographical locations.
Debate and analysis of the risks and benefits of providing hypnozoitocidal courses of primaquine to patients with falciparum malaria in regions with *Plasmodium* co-endemicity.
Mathematical modelling and epidemiological assessment of the most effective dosing regimen for reducing transmission of *P. vivax* in mass drug administration campaigns.
Mathematical modelling and epidemiological assessment of the most appropriate target age groups for mass drug administration campaigns if resources do not allow treatment of all individuals in a population.
Mathematical modelling and epidemiological assessment of the potential impact of mass drug administration campaigns on emergence of drug-resistant *P. vivax* strains.

In the absence of a practicable means of diagnosing G6PD deficiency, many regions opt not to use primaquine in the treatment of symptomatic *P. vivax* infections. In these situations, use of very slowly eliminated blood schizontocidal regimens such as dihydroartemisinin plus piperaquine, artesunate plus mefloquine and chloroquine (where parasites remain fully susceptible) has the potential to suppress the first relapse of tropical strains of *P. vivax*. Where *P. vivax* transmission is intense, this should reduce total life-time parasite carriage. Large, multi-centre randomised trials in which patients are assigned to repeated doses of the same drug are required to prove this speculative assertion.

Given the high rate of *P. vivax* relapse following falciparum malaria in co-endemic areas, there is an argument for using hypnozoitocidal doses of primaquine in those with falciparum malaria in co-endemic regions. This proposition raises important ethical dilemmas related to a dubious risk:benefit ratio. If G6PD testing becomes widely available allowing safe prescription of primaquine, these dilemmas may become redundant.

Asymptomatic *P. vivax* infections are likely to be responsible for a substantial proportion of all *P. vivax* transmission, especially in moderately and highly endemic areas. Two means of treating asymptomatic individuals include mass screening and treatment and mass presumptive drug administration. Since *P. vivax* infections are efficiently transmitted at sub-patent gametocyte densities and the presence or absence of hypnozoites cannot be established in those without blood-stage infection, MDA is likely to have the greater impact of these two approaches. Eradication of hypnozoites will be critical to the effectiveness of these strategies. If G6PD status can be established, a slowly eliminated hypnozoitocidal agent such as tafenoquine may prove to be particularly useful for single-visit campaigns.

P. vivax has side-stepped current malaria control efforts and is becoming a proportionately more common cause of malaria globally. Development of a safe and effective means of eradicating *P. vivax* hypnozoites will be critical to reversing this trend.

REFERENCES

Aguas, R., Ferreira, M.U., Gomes, M.G., 2012. Modeling the effects of relapse in the transmission dynamics of malaria parasites. J. Parasitol. Res. 2012 921715.

Alexandre, M.A., et al., 2010. Severe *Plasmodium vivax* malaria, Brazilian Amazon. Emerg. Infect. Dis. 16 (10), 1611–1614.

Alving, A.S., et al., 1953. Korean vivax malaria. II. Curative treatment with pamaquine and primaquine. Am. J. Trop. Med. Hyg. 2 (6), 970–976.

Alving, A.S., et al., 1955. Potentiation of the curative action of primaquine in vivax malaria by quinine and chloroquine. J. Lab. Clin. Med. 46 (2), 301–306.

Alving, A.S., et al., 1960. Mitigation of the haemolytic effect of primaquine and enhancement of its action against exoerythrocytic forms of the Chesson strain of *Plasmodium vivax* by intermittent regimens of drug administration. Bull. World Health Organ. 22, 621–631.

Anstey, N.M., et al., 2009. The pathophysiology of vivax malaria. Trends Parasitol. 25, 220–227.

Anthony, M., et al., 2012. The global pipeline of new medicines for the control and elimination of malaria. Malar. J. 11 (1), 316.

Arevalo-Herrera, M., Chitnis, C., Herrera, S., 2010. Current status of *Plasmodium vivax* vaccine. Hum. Vaccin. 6 (1), 124–132.

Assefa, A., et al., 2010. Therapeutic efficacy of Artemether/Lumefantrine (Coartem(R)) against *Plasmodium falciparum* in Kersa, South West Ethiopia. Parasit. Vectors 3 (1), 1.

Austin, S.C., Stolley, P.D., Lasky, T., 1992. The history of malariotherapy for neurosyphilis. Modern parallels. JAMA 268 (4), 516–519.

Awab, G.R., et al., 2010. Dihydroartemisinin-piperaquine versus chloroquine to treat vivax malaria in Afghanistan: an open randomized, non-inferiority, trial. Malar. J. 9 (1), 105.

Baird, J.K., Hoffman, S.L., 2004. Primaquine therapy for malaria. Clin. Infect. Dis. 39, 1336–1345.

Baird, J.K., Surjadjaja, C., 2010. Consideration of ethics in primaquine therapy against malaria transmission. Trends Parasitol. 7 (1), 11–16.

Baird, J.K., Rieckmann, K.H., 2003. Can primaquine therapy for vivax malaria be improved?. Trends Parasitol. 19, 115–120.

Baird, J.K., et al., 1995. Treatment of chloroquine-resistant *Plasmodium vivax* with chloroquine and primaquine or halofantrine. J. Infect. Dis. 171 (6), 1678–1682.

Baird, J.K., et al., 1997. In vivo resistance to chloroquine by *Plasmodium vivax* and *Plasmodium falciparum* at Nabire, Irian Jaya, Indonesia. Am. J. Trop. Med. Hyg. 56 (6), 627–631.

Baird, J.K., 2007. Neglect of *Plasmodium vivax* malaria. Trends Parasitol. 23 (11), 533–539.

Baird, J.K., 2009. Resistance to therapies for infection by *Plasmodium vivax*. Clin. Microbiol. Rev. 22 (3), 508–534.

Baird, J.K., 2011. Radical cure: the case for anti-relapse therapy against all malarias. Clin. Infect. Dis. 52 (5), 621–623.

Barcus, M.J., et al., 2007. Demographic risk factors for severe and fatal vivax and falciparum malaria among hospital admissions in Northeastern Indonesian Papua. Am. J. Trop. Med. Hyg. 77 (5), 984–991.

Bassat, Q., Alonso, P.L., 2011. Defying malaria: fathoming severe *Plasmodium vivax* disease. Nat. Med. 17 (1), 48–49.

Beg, M.A., et al., 2008. Comparative features and outcomes of malaria at a tertiary care hospital in Karachi, Pakistan. Int. J. Infect. Dis. 12 (1), 37–42.

Beier, J.C., 1998. Malaria parasite development in mosquitoes. Annu. Rev. Entomol. 43, 519–543.

Belkin, J.N., Knight, K.L., Rozeboom, L.E., 1945. Anopheline mosquitoes of The Solomon Islands and New Hebrides. J. Parasitol. 31 (4), 241–265.

Beutler, E., Vulliamy, T.J., 2002. Hematologically important mutations: glucose-6-phosphate dehydrogenase. Blood Cells Mol. Dis. 28 (2), 93–103.

Bharti, A.R., et al., 2006. Experimental infection of the neotropical malaria vector Anopheles darlingi by human patient-derived *Plasmodium vivax* in the Peruvian Amazon. Am. J. Trop. Med. Hyg. 75 (4), 610–616.

Bousema, T., Drakeley, C., 2011. *Plasmodium falciparum* and *Plasmodium vivax* gametocytes – their epidemiology and infectivity in relation to malaria control and elimination. Clin. Microbiol. Rev. 24 (2), 377–410.

Boyd, M.F., 1938. The threshold of parasite density in relation to clinical activity in primary infections with *Plasmodium vivax*. Am. J. Trop. Med. s1–18 (5), 497–503.

Brueckner, R.P., et al., 1998. First-time-in-humans safety and pharmacokinetics of WR 238605, a new antimalarial. Am. J. Trop. Med. Hyg. 58 (5), 645–649.

Carlton, J.M., Sina, B.J., Adams, J.H., 2011. Why is *Plasmodium vivax* a neglected tropical disease?. PLoS Negl. Trop. Dis. 5 (6), e1160.

Carmona-Fonseca, J., Maestre, A., 2009. Prevention of *Plasmodium vivax* malaria recurrence: efficacy of the standard total dose of primaquine administered over 3 days. Acta Trop. 112 (2), 188–192.

Chen, J.H., et al., 2010. Detection of *Plasmodium vivax* infection in the Republic of Korea by loop-mediated isothermal amplification (LAMP). Acta Trop. 113 (1), 61–65.

Chotivanich, K., et al., 2009. Antimalarial drug susceptibility of *Plasmodium vivax* in the Republic of Korea. Am. J. Trop. Med. Hyg. 80 (6), 902–904.

Clyde, D.F., McCarthy, V.C., 1977. Radical cure of Chesson strain vivax malaria in man by 7, not 14, days of treatment with primaquine. Am. J. Trop. Med. Hyg. 26 (3), 562.

Coatney, G.R., Getz, M.E., 1962. Primaquine and quinocide as curative agents against sporozoite-induced Chesson strain vivax malaria. Bull. World Health Organ. 27, 290–293.

Coatney, G.R., et al., 1950. Studies in human malaria: XVIII. The life pattern of sporozoite-induced St. Elizabeth strain vivax malaria. Am. J. Hyg. 51 (2), 200.

Coatney, G.R., et al., 1953. Korean vivax malaria. Am. J. Trop. Med. Hyg. 2 (6), 985–988.

Collins, W.E., Jeffery, G.M., 1996. Primaquine resistance in *Plasmodium vivax*. Am. J. Trop. Med. Hyg. 55 (3), 243.

Crockett, M., Kain, K.C., 2007. Tafenoquine: a promising new antimalarial agent. Expert Opin. Investig. Drugs 16 (5), 705–715.

Dao, N.V.H., et al., 2007. Vivax malaria: preliminary observations following a shorter course of treatment with artesunate plus primaquine. Trans. R. Soc. Trop. Med. Hyg. 101 (6), 534–539.

Darlow, B., et al., 1982. Sulfadoxine-pyrimethamine for the treatment of acute malaria in children of Papua New Guinea. II. *Plasmodium vivax*. Am. J. Trop. Med. Hyg. 31 (1), 10–13.

Denis, M.B., et al., 2006. Efficacy of artemether-lumefantrine for the treatment of uncomplicated falciparum malaria in northwest Cambodia. Trop. Med. Int. Health 11 (12), 1800–1807.

Douglas, N.M., et al., 2010. Artemisinin combination therapy for vivax malaria. Lancet Infect. Dis. 10 (6), 405–416.

Douglas, N.M., et al., 2011. *Plasmodium vivax* recurrence following falciparum and mixed species malaria: risk factors and effect of antimalarial kinetics. Clin. Infect. Dis. 52 (5), 612–620.

Douglas, N.M., et al., 2012. The anaemia of *Plasmodium vivax* malaria. Malar. J. 11 (1), 135.

Elmes, N.J., et al., 2008. The efficacy and tolerability of three different regimens of tafenoquine versus primaquine for post-exposure prophylaxis of *Plasmodium vivax* malaria in the Southwest Pacific. Trans. R. Soc. Trop. Med. Hyg. 102 (11), 1095–1101.

Ezzet, F., et al., 2000. Pharmacokinetics and pharmacodynamics of lumefantrine (benflumetol) in acute falciparum malaria. Antimicrob. Agents Chemother. 44 (3), 697–704.

Feachem, R.G., et al., 2010. Shrinking the malaria map: progress and prospects. Lancet 376 (9752), 1566–1578.

Flower, B., et al., 2011. Blind, breathless, and paralysed from benign malaria. Lancet 377 (9763), 438.

Gama, B.E., et al., 2011. Chemoresistance of *Plasmodium falciparum* and *Plasmodium vivax* parasites in Brazil: consequences on disease morbidity and control. Mem. Inst. Oswaldo Cruz 106 (Suppl. 1), 159–166.

Garfield, R.M., Vermund, S.H., 1983. Changes in malaria incidence after mass drug administration in Nicaragua. Lancet 2 (8348), 500–503.

Genton, B., et al., 2008. *Plasmodium vivax* and mixed infections are associated with severe malaria in children: a prospective cohort study from Papua New Guinea. PLoS Med. 5 (6), e127.
Gogtay, N.J., et al., 1999. Efficacies of 5- and 14-day primaquine regimens in the prevention of relapses in *Plasmodium vivax* infections. Ann. Trop. Med. Parasitol. 93 (8), 809–812.
Goller, J.L., et al., 2007. Regional differences in the response of *Plasmodium vivax* malaria to primaquine as anti-relapse therapy. Am. J. Trop. Med. Hyg. 76 (2), 203–207.
Grietens, K.P., et al., 2010. Adherence to 7-day primaquine treatment for the radical cure of *P. vivax* in the Peruvian Amazon. Am. J. Trop. Med. Hyg. 82 (6), 1017–1023.
Guerra, C.A., et al., 2010. The international limits and population at risk of *Plasmodium vivax* transmission in 2009. PLoS Negl. Trop. Dis. 4 (8), e774.
Gunasekera, W.M.K.T.d.A.W., et al., 2011. Usefulness of polymerase chain reaction to supplement field microscopy in a pre-selected population with a high probability of malaria infections. Am. J. Trop. Med. Hyg. 85 (1), 6–11.
Hamedi, Y., et al., 2004. Therapeutic efficacy of artesunate in *Plasmodium vivax* malaria in Thailand. Southeast Asian J. Trop. Med. Public Health 35 (3), 570–574.
Hankey, D.D., et al., 1953. Korean vivax malaria. I. Natural history and response to chloroquine. Am. J. Trop. Med. Hyg. 2 (6), 958–969.
Haque, R., et al., 2007. Therapeutic efficacy of artemether-lumefantrine for the treatment of uncomplicated *Plasmodium falciparum* malaria in Bangladesh. Am. J. Trop. Med. Hyg. 76 (1), 39–41.
Harris, I., et al., 2010. A large proportion of asymptomatic *Plasmodium* infections with low and sub-microscopic parasite densities in the low transmission setting of Temotu Province, Solomon Islands: challenges for malaria diagnostics in an elimination setting. Malar. J. 9 (1), 254.
Hasugian, A.R., et al., 2007. Dihydroartemisinin-piperaquine versus artesunate-amodiaquine: superior efficacy and posttreatment prophylaxis against multidrug-resistant *Plasmodium falciparum* and *Plasmodium vivax* malaria. Clin. Infect. Dis. 44, 1067–1074.
Hasugian, A.R., et al., 2009. In vivo and in vitro efficacy of amodiaquine monotherapy for treatment of infection by chloroquine-resistant *Plasmodium vivax*. Antimicrob. Agents Chemother. 53 (3), 1094–1099.
Hay, S.I., et al., 2004. The global distribution and population at risk of malaria: past, present and future. Lancet Infect. Dis. 4, 327–336.
Hill, E., Amatuzio, D.S., 1949. Southwest Pacific vivax malaria: clinical features and observations concerning duration of clinical activity. Am. J. Trop. Med. s1–29 (2), 203–214.
Huh, A.J., et al., 2011. Parasitemia characteristics of *Plasmodium vivax* malaria patients in the Republic of Korea. J. Korean Med. Sci. 26 (1), 42–46.
Hung, T., et al., 2004. Population pharmacokinetics of piperaquine in adults and children with uncomplicated falciparum or vivax malaria. Br. J. Clin. Pharmacol. 57 (3), 253–262.
Hutagalung, R., et al., 2005. A randomized trial of artemether-lumefantrine versus mefloquine-artesunate for the treatment of uncomplicated multi-drug resistant *Plasmodium falciparum* on the western border of Thailand. Malar. J. 4, 46.
Hwang, J., et al., 2011. In vivo efficacy of artemether-lumefantrine against uncomplicated *Plasmodium falciparum* malaria in Central Ethiopia. Malar. J. 10, 209.
Imwong, M., et al., 2003. Novel point mutations in the dihydrofolate reductase gene of *Plasmodium vivax*: evidence for sequential selection by drug pressure. Antimicrob. Agents Chemother. 47 (5), 1514–1521.
Ishikawa, H., et al., 2003. A mathematical model for the transmission of *Plasmodium vivax* malaria. Parasitol. Int. 52 (1), 81–93.
Jeffery, G.M., 1952. The infection of mosquitoes by *Plasmodium vivax* (Chesson strain) during the early primary parasitemias. Am. J. Trop. Med. Hyg. 1 (4), 612–617.

John, G.K., et al., 2012. Primaquine radical cure of *Plasmodium vivax*: a critical review of the literature. Malar. J. 11 (1), 280.

Kaneko, A., et al., 2000. Malaria eradication on islands. Lancet 356 (9241), 1560–1564.

Kaneko, A., 2010. A community-directed strategy for sustainable malaria elimination on islands: short-term MDA integrated with ITNs and robust surveillance. Acta Trop. 114 (3), 177–183.

Karunajeewa, H.A., et al., 2008. A trial of combination antimalarial therapies in children from Papua New Guinea. N. Engl. J. Med. 359, 2545–2557.

Karyana, M., et al., 2008. Epidemiology of multidrug resistant *P. vivax* and *P. falciparum* infection in Southern Papua, Indonesia. Malar. J. 7, 148.

Kinzer, M.H., et al., 2010. Active case detection, treatment of falciparum malaria with combined chloroquine and sulphadoxine/pyrimethamine and vivax malaria with chloroquine and molecular markers of anti-malarial resistance in the Republic of Vanuatu. Malar. J. 9 (1), 89.

Kochar, D.K., et al., 2005. *Plasmodium vivax* malaria. Emerg. Infect. Dis. 11 (1), 132–134.

Kochar, D.K., et al., 2009. Severe *Plasmodium vivax* malaria: a report on serial cases from Bikaner in Northwestern India. Am. J. Trop. Med. Hyg. 80 (2), 194–198.

Krishna, S., White, N.J., 1996. Pharmacokinetics of quinine, chloroquine and amodiaquine: clinical implications. Clin. Pharmacokinet. 30 (4), 263–299.

Krudsood, S., et al., 2006. Safety and tolerability of elubaquine (bulaquine, CDRI 80/53) for treatment of *Plasmodium vivax* malaria in Thailand. Korean J. Parasitol. 44 (3), 221–228.

Krudsood, S., et al., 2007. Clinical efficacy of chloroquine versus artemether-lumefantrine for *Plasmodium vivax* treatment in Thailand. Korean J. Parasitol. 45 (2), 111–114.

Krudsood, S., et al., 2008. High-dose primaquine regimens against relapse of *Plasmodium vivax* malaria. Am. J. Trop. Med. Hyg. 78 (5), 736–740.

LaMontagne, M.P., Blumbergs, P., Strube, R.E., 1982. Antimalarials. 14. 5-(aryloxy)-4-methylprimaquine analogues. A highly effective series of blood and tissue schizonticidal agents. J. Med. Chem. 25 (9), 1094–1097.

Lampah, D.A., et al., 2011. Coma associated with microscopy-diagnosed *Plasmodium vivax*: a prospective study in Papua, Indonesia. PLoS Negl. Trop. Dis. 5 (6), e1032.

Lefevre, G., et al., 2001. A clinical and pharmacokinetic trial of six doses of artemether-lumefantrine for multidrug-resistant *Plasmodium falciparum* malaria in Thailand. Am. J. Trop. Med. Hyg. 64 (5–6), 247–256.

Leslie, T., et al., 2004. Compliance with 14-day primaquine therapy for radical cure of vivax malaria: a randomized placebo-controlled trial comparing unsupervised with supervised treatment. Trans. R. Soc. Trop. Med. Hyg. 98, 168–173.

Leslie, T., et al., 2008. A randomised trial of an eight-week, once weekly primaquine regimen to prevent relapse of *Plasmodium vivax* in Northwest Frontier Province, Pakistan. PLoS One 3 (8), e2861.

Looareesuwan, S., et al., 1987. High rate of *Plasmodium vivax* relapse following treatment of falciparum malaria in Thailand. Lancet 2 (8567), 1052–1055.

Lu, F., et al., 2011. In vitro anti-malarial drug susceptibility of temperate *Plasmodium vivax* from Central China. Am. J. Trop. Med. Hyg. 85 (2), 197–201.

Lux, D., et al., 2003. In-vitro sensitivity testing of *Plasmodium vivax*: response to lumefantrine and chloroquine in Northwestern Thailand. Wien. Klin. Wochenschr. 115 (Suppl. 3), 50–54.

Luxemburger, C., et al., 1996. The epidemiology of malaria in a Karen population on the western border of Thailand. Trans. R. Soc. Trop. Med. Hyg. 90 (2), 105–111.

Lwin, K.M., et al., 2012. Randomized, double-blind, placebo-controlled trial of monthly versus bimonthly dihydroartemisinin-piperaquine chemoprevention in adults at high risk of malaria. Antimicrob. Agents Chemother. 56 (3), 1571–1577.

Macauley, C., 2005. Aggressive active case detection: a malaria control strategy based on the Brazilian model. Soc. Sci. Med. 60 (3), 563–573.
Maneeboonyang, W., et al., 2011. Directly observed therapy with primaquine to reduce the recurrence rate of *Plasmodium vivax* infection along the Thai-Myanmar border. Southeast Asian J. Trop. Med. Public Health 42 (1), 9–18.
Mayxay, M., et al., 2004. Randomized comparison of chloroquine plus sulfadoxine-pyrimethamine versus artesunate plus mefloquine versus artemether-lumefantrine in the treatment of uncomplicated falciparum malaria in the Lao People's Democratic Republic. Clin. Infect. Dis. 39 (8), 1139–1147.
McGready, R., et al., 2008. A randomised controlled trial of artemether-lumefantrine versus artesunate for uncomplicated *Plasmodium falciparum* treatment in pregnancy. PLoS Med. 5 (12), e253.
Medicines for Malaria Venture, 2011. Interactive Science Portfolio. (cited 2012 11th of March 2012); Available from: http://www.mmv.org/research-development/science-portfolio.
Michon, P., et al., 2007. The risk of malarial infections and disease in Papua New Guinean children. Am. J. Trop. Med. Hyg. 76 (6), 997–1008.
Miller, L.H., et al., 1974. Sensitivity of four Central American strains of *Plasmodium vivax* to primaquine. Am. J. Trop. Med. Hyg. 23 (2), 309–310.
Mueller, I., et al., 2009. Key gaps in the knowledge of *Plasmodium vivax*, a neglected human malaria parasite. Lancet Infect. Dis. 9 (9), 555–566.
Nkhoma, E.T., et al., 2009. The global prevalence of glucose-6-phosphate dehydrogenase deficiency: a systematic review and meta-analysis. Blood Cells Mol. Dis. 42 (3), 267–278.
Nosten, F., et al., 2000. Effects of artesunate-mefloquine combination on incidence of *Plasmodium falciparum* malaria and mefloquine resistance in Western Thailand: a prospective study. Lancet 356 (9226), 297–302.
Onori, E., 1972. Experience with mass drug administration as a supplementary attack measure in areas of vivax malaria. Bull. World Health Organ. 47 (5), 543–548.
Phimpraphi, W., et al., 2008. Longitudinal study of *Plasmodium falciparum* and *Plasmodium vivax* in a Karen population in Thailand. Malar. J. 7 (1), 99.
Phyo, A.P., et al., 2011. Dihydroartemisinin-piperaquine versus chloroquine in the treatment of *P. vivax* malaria in Thailand: a randomized controlled trial. Clin. Infect. Dis. 53, 977–984.
Poespoprodjo, J.R., et al., 2009. Vivax malaria: a major cause of morbidity in early infancy. Clin. Infect. Dis. 48 (12), 1704–1712.
Poravuth, Y., et al., 2011. Pyronaridine-artesunate versus chloroquine in patients with acute *Plasmodium vivax* malaria: a randomized, double-blind, non-inferiority trial. PLoS One 6 (1), e14501.
Poschl, B., et al., 2010. Comparative diagnosis of malaria infections by microscopy, nested PCR, and LAMP in Northern Thailand. Am. J. Trop. Med. Hyg. 83 (1), 56–60.
Price, R.N., Douglas, N.M., 2010. Maximising the public health benefit of antimalarials. Lancet Infect. Dis. 10 (10), 654–655.
Price, R.N., et al., 2007. Vivax malaria: neglected and not benign. Am. J. Trop. Med. Hyg. 77, 79–87.
Pukrittayakamee, S., et al., 2004. Therapeutic responses to antimalarial and antibacterial drugs in vivax malaria. Acta Trop. 89, 351–356.
Pukrittayakamee, S., et al., 2008. Effects of different antimalarial drugs on gametocyte carriage in *P. vivax* malaria. Am. J. Trop. Med. Hyg. 79 (3), 378–384.
Pukrittayakamee, S., et al., 2010. A comparison of two short-course primaquine regimens for the treatment and radical cure of *Plasmodium vivax* malaria in Thailand. Am. J. Trop. Med. Hyg. 82 (4), 542–547.

Ratcliff, A., et al., 2007. Two fixed-dose artemisinin combinations for drug-resistant falciparum and vivax malaria in Papua, Indonesia: an open-label randomised comparison. Lancet 369, 757–765.

Reeve, P.A., et al., 1992. Acute intravascular haemolysis in Vanuatu following a single dose of primaquine in individuals with glucose-6-phosphate dehydrogenase deficiency. J. Trop. Med. Hyg. 95 (5), 349–351.

Rijken, M.J., et al., 2011. Chloroquine resistant vivax malaria in a pregnant woman on the western border of Thailand. Malar. J. 10 (1), 113.

Rodriguez, J.C., et al., 2011. Epidemiology and control of malaria in Colombia. Mem. Inst. Oswaldo Cruz 106 (Suppl. 1), 114–122.

Rowland, M., Durrani, N., 1999. Randomized controlled trials of 5- and 14-days primaquine therapy against relapses of vivax malaria in an Afghan refugee settlement in Pakistan. Trans. R. Soc. Trop. Med. Hyg. 93 (6), 641–643.

Russell, B., et al., 2008. Determinants of in vitro drug susceptibility testing of *Plasmodium vivax*. Antimicrob. Agents Chemother. 52 (3), 1040–1045.

Senn, N., et al., 2012. Intermittent preventive treatment for malaria in Papua New Guinean infants exposed to *Plasmodium falciparum* and *P. vivax*: a randomized controlled trial. PLoS Med. 9 (3), e1001195.

Silachamroon, U., et al., 2003. Clinical trial of oral artesunate with or without high-dose primaquine for the treatment of vivax malaria in Thailand. Am. J. Trop. Med. Hyg. 69 (1), 14–18.

Singh, R.K., 2000. Emergence of chloroquine-resistant vivax malaria in South Bihar (India). Trans. R. Soc. Trop. Med. Hyg. 94, 327.

Sirichaisinthop, J., et al., 2011. Evaluation of loop-mediated isothermal amplification (LAMP) for malaria diagnosis in a field setting. Am. J. Trop. Med. Hyg. 85 (4), 594–596.

Smithuis, F., et al., 2010. Effectiveness of five artemisinin combination regimens with or without primaquine in uncomplicated falciparum malaria: an open-label randomised trial. Lancet Infect. Dis. 10 (10), 673–681.

Steenkeste, N., et al., 2010. Sub-microscopic malaria cases and mixed malaria infection in a remote area of high malaria endemicity in Rattanakiri province, Cambodia: implication for malaria elimination. Malar. J. 9 (1), 108.

Stepniewska, K., et al., 2008. *Plasmodium falciparum* gametocyte dynamics in areas of different malaria endemicity. Malar. J. 7, 249.

Stohrer, J.M., et al., 2004. Therapeutic efficacy of artemether-lumefantrine and artesunate-mefloquine for treatment of uncomplicated *Plasmodium falciparum* malaria in Luang Namtha Province, Lao People's Democratic Republic. Trop. Med. Int. Health 9 (11), 1175–1183.

Sumawinata, I.W., et al., 2003. Very high risk of therapeutic failure with chloroquine for uncomplicated *Plasmodium falciparum* and *P. vivax* malaria in Indonesian Papua. Am. J. Trop. Med. Hyg. 68 (4), 416–420.

Sutanto, I., et al., 2009. Resistance to chloroquine by *Plasmodium vivax* at Alor in the Lesser Sundas Archipelago in Eastern Indonesia. Am. J. Trop. Med. Hyg. 81 (2), 338–342.

Takeuchi, R., et al., 2010. Directly-observed therapy (DOT) for the radical 14-day primaquine treatment of *Plasmodium vivax* malaria on the Thai-Myanmar border. Malar. J. 9 (1), 308.

Tan, L.K.K., et al., 2008. Acute lung injury and other serious complications of *Plasmodium vivax* malaria. Lancet Infect. Dis. 8, 449–454.

Tarning, J., et al., 2008. Population pharmacokinetics of piperaquine after two different treatment regimens with dihydroartemisinin-piperaquine in patients with *Plasmodium falciparum* malaria in Thailand. Antimicrob. Agents Chemother. 52, 1052–1061.

Thapa, S., et al., 2007. Comparison of artemether-lumefantrine with sulfadoxine-pyrimethamine for the treatment of uncomplicated falciparum malaria in eastern Nepal. Am. J. Trop. Med. Hyg. 77 (3), 423–430.

Thriemer, K., et al., 2010. Azithromycin combination therapy for the treatment of uncomplicated falciparum malaria in Bangladesh: an open-label randomized, controlled clinical trial. J. Infect. Dis. 202 (3), 392–398.

Tjitra, E., et al., 2002. Therapeutic efficacies of artesunate-sulfadoxine-pyrimethamine and chloroquine-sulfadoxine-pyrimethamine in vivax malaria pilot studies: relationship to *Plasmodium vivax* dhfr mutations. Antimicrob. Agents Chemother. 46, 3947–3953.

Tjitra, E., et al., 2008. Multidrug-resistant *Plasmodium vivax* associated with severe and fatal malaria: a prospective study in Papua, Indonesia. PLoS Med. 5 (6), e128.

Valecha, N., et al., 2001. Comparative antirelapse efficacy of CDRI compound (80/53 (bulaquine) vs primaquine in double blind clinical trial. Curr. Sci. 80 (4), 561–563.

van den Broek, I.V., et al., 2005. Efficacy of chloroquine + sulfadoxine–pyrimethamine, mefloquine + artesunate and artemether + lumefantrine combination therapies to treat *Plasmodium falciparum* malaria in the Chittagong Hill Tracts, Bangladesh. Trans. R. Soc. Trop. Med. Hyg. 99 (10), 727–735.

Van den Eede, P., et al., 2011. *Plasmodium vivax* sub-patent infections after radical treatment are common in Peruvian patients: results of a 1-year prospective cohort study. PLoS One 6 (1), e16257.

van Vugt, M., et al., 1998. Randomized comparison of artemether-benflumetol and artesunate-mefloquine in treatment of multidrug-resistant falciparum malaria. Antimicrob. Agents Chemother. 42 (1), 135–139.

van Vugt, M., et al., 1999. Efficacy of six doses of artemether-lumefantrine (benflumetol) in multidrug-resistant *Plasmodium falciparum* malaria. Am. J. Trop. Med. Hyg. 60 (6), 936–942.

Villegas, L., et al., 2007. Chloroquine prophylaxis against vivax malaria in pregnancy: a randomized, double-blind, placebo-controlled trial. Trop. Med. Int. Health 12 (2), 209–218.

White, N., 1999. Antimalarial drug resistance and combination chemotherapy. Philos. Trans. R. Soc. Lond. B 354, 739–749.

White, N.J., 2008. The role of anti-malarial drugs in eliminating malaria. Malar. J. 7 (Suppl. 1), S8.

White, N.J., 2011. Determinants of relapse periodicity in *Plasmodium vivax* malaria. Malar. J. 10, 297.

Wilairatana, P., et al., 1999. Efficacy of primaquine regimens for primaquine-resistant *Plasmodium vivax* malaria in Thailand. Am. J. Trop. Med. Hyg. 61 (6), 973–977.

World Health Organization, et al., 2009. Malaria Rapid Diagnostic Test Performance: Results of WHO Product Testing of Malaria RDTs: Round 1 (2008). World Health Organization, Geneva.

World Health Organization, 2010. Guidelines for the Treatment of Malaria. World Health Organization.

CHAPTER SIX

Control and Elimination of *Plasmodium vivax*

G. Dennis Shanks[*,†]
*Army Malaria Institute, Enoggera, Queensland, Australia
†Department of Zoology, University of Oxford, Oxford, UK

Contents

Advances in Parasitology, Volume 80
ISSN 0065-308X, http://dx.doi.org/10.1016/B978-0-12-397900-1.00006-2

Abstract

Plasmodium vivax represents a special challenge to malaria control because of the ability of a single infection to relapse over months to years. *P. vivax* is more tolerant of low temperatures than *P. falciparum*, which spreads its potential range far beyond the tropics into sub-Arctic areas. Ordinary malaria control measures such as residual insecticide spraying and impregnated bed nets are effective for *P. vivax*, but long-lasting (up to 3 years) residual hepatic parasites (hypnozoites) mean that even well-executed malaria control programs must maintain maximal efforts for an extended period in order to eliminate indigenous infections. Hypnozoites are only eliminated by using an 8-aminoquinoline (currently only primaquine), which requires compliance with a long regimen as well as care to avoid those at risk of haemolysis due to the common genetic polymorphism, glucose-6-phosphate dehydrogenase deficiency. Risk of reintroduction of *P. vivax* into areas without malaria but still containing competent *Anopheles* vectors is enhanced as persons carrying hypnozoites are undetectable until they become symptomatic from activation of the quiescent liver parasite. Mass drug administration using drug combinations including primaquine have successfully eliminated malaria from small islands demonstrating proof of principal as a potential elimination method. It will be very difficult to maintain adequate malaria surveillance measures for years after malaria has ceased to be a public health problem, which will clearly be required to eliminate relapsing malaria such as *P. vivax*. New interventions will likely be required to eliminate vivax malaria; highly desirable new products include transmission-blocking vaccines, new drug combinations to treat chloroquine resistant strains and a safe, long-lasting 8-aminoquinoline.

1. WHY CONTROLLING *P. VIVAX* IS DIFFICULT

Malaria control means to reduce the public health burden of malaria below an acceptable threshold; malaria elimination is the geographically defined extinction of malaria parasites in the human population (Moonen et al., 2010). Recently, malaria control and especially elimination has been reinvigorated by new resources and attention (Hall and Fauci, 2009; Feachem and Sabot, 2008). Malaria is one of the few infectious diseases whose potential for elimination has been demonstrated by its successful removal from large areas such as Europe and North America. Research goals for malaria elimination have been described, many of which apply to vivax malaria (A research agenda for malaria eradication: drugs, 2011;

Greenwood, 2008b). The first task of a malaria control program is to stop deaths due to malaria, which necessarily means an emphasis on falciparum malaria at the onset (Tatem et al., 2010). However, as has been repeatedly shown in earlier chapters in this issue, vivax causes mortality and substantial morbidity across the world (Genton et al., 2008; Oh et al., 2001). Importantly, vivax malaria has several attributes that make it difficult to control and very difficult to eliminate. Once modest success has been achieved in malaria control, especially in countries outside of Africa, vivax and other forms of relapsing malaria (e.g. *Plasmodium ovale*) will increasingly become important until malaria control is primarily if not exclusively focused on vivax malaria (Rajakaruna et al., 2010; Aguas et al., 2008). In some parts of Asia and Latin America, this has already occurred. In sub-Saharan Africa, malaria control will, for the foreseeable future, be concerned primarily with falciparum malaria, however, substantial populations of relapsing malaria exist in Africa despite their clinical invisibility to most physicians and will become increasingly important as established forms of malaria control using rapid diagnosis and treatment as well as insecticide-treated bed nets (ITN) continue to lower the disease impact of all forms of malaria.

1.1. Geographic Range and Temperature Tolerance

Vivax malaria has a broader geographic range and temperature tolerance than falciparum malaria. The classic success stories from the 1950s–1960s Global Malaria Eradication Program (GMEP) were largely in areas of seasonal malaria transmission away from the equator, which are the regions inhabited by vivax malaria (Bruce-Chwatt, 1977; Colbourne, 1962). The historical distribution of malaria transmission, originally mapped by Lysenko [EDITOR = SEE REFS in CHPT 1; VOL 80], has been digitised by the Malaria Atlas Project (MAP). (Chapter 1, Figure 1.3). In nearly half the countries of the world shown on the map, malaria has been eliminated over the twentieth century, which has considerably shrunk today's distribution of vivax malaria (see chapter 1, volume 80, and Gething et al., 2012). However, about 100 countries, both tropical and sub-tropical, currently have some degree of malaria transmission (Tatem et al., 2010). Although often regarded as due to exceptional circumstances, vivax malaria has been transmitted during the summer months in sub-Arctic regions of the Northern hemisphere. The most important single determinant of malaria transmission in such cold climates is the amount of time when the temperatures remain above a mean of 18 °C (Gething et al., 2010), as described in chapter 1 of this volume. Besides seasonality, temperatures also vary by altitude which can produce highland malaria during brief warm periods above 2000 m. If such warmth is maintained for a few weeks, then those mosquitoes carrying malaria parasites can support development of infective sporozoites. This is true of areas in high latitudes as well as tropical countries with mountains. Transmission can also occur in warmer than ambient conditions such as inside houses or animal shelters which can result in highly focal transmission as occurred in England before elimination. Despite global warming, reintroduction of malaria into previously eliminated areas is not primarily dependant on temperature changes (Gething et al., 2010). Movement of gametocyte-carrying individuals is much more important than climate in terms of reintroduction of vivax malaria into eliminated areas such as the United States and Europe, where the Anopheline mosquito vectors have never been absent. Rare cases of autochthonous malaria transmission continue to be reported in the USA and Europe but represent little threat to malaria control if they remain isolated instances without secondary cases (Danis et al., 2011).

1.2. Different Types of Vivax Malaria

Whether *P. vivax* consists of separate species or not will be left to those with greater taxonomic knowledge and interest. From a public health viewpoint, however, different vivax types certainly exist and the variability of relapses and drug sensitivity have important control and elimination implications (White, 2011; Baird, 2009). As explained in chapter 2, volume 80, tropical vivax is much more likely to repeatedly relapse as well as demonstrate greater resistance to chloroquine treatment than those types found away from the equator, such as in China and the Korean peninsula. This is probably due respectively to lack of evolutionary drive to adapt to limited transmission seasons and wide-spread use of chloroquine in tropical areas. The circumstances that encourage chloroquine resistance are continuously circulating blood parasites under consistent drug pressure, which occurs mostly in tropical areas. Institution of polymerase chain reactions (PCR) for malaria diagnosis has determined that the number of mixed falciparum/vivax infections as well as asymptomatic parasitaemic persons is much larger than was appreciated from traditional microscopic diagnosis (Mayxay et al., 2004). A malaria control program that depends largely on symptomatic persons coming to a clinic for microscopic diagnosis may be adequate for malaria control, but it certainly will not be for malaria elimination as there will remain a reservoir of parasites to continue transmission in the absence of ill persons (Harris et al., 2010). It is unclear if molecular diagnostic methods can be practically transferred to the field situation, but the requirement of a much more complete degree of case finding for malaria elimination as opposed to control necessitates diagnostic methods that are more sensitive as well as easier to perform than microscopy if we are to successfully eliminate relapsing malaria in the tropics (Tanner and de Savigny, 2008). Field deployable PCR systems as well as loop mediated isothermal PCR (LAMP), which does not require a thermocycler, gives hope that technology will eventually evolve to enable the sensitive detection of vivax malaria parasites in the blood of asymptomatic persons.

1.3. Hypnozoites and Implications for Elimination

The residual hepatic forms of the vivax parasite known as hypnozoites have been explained in chapter 2, volume 80. This important evolutionary adaptation allows vivax malaria to remain in the liver waiting for favourable transmission conditions that usually coincide with rains producing mosquito habitat (Shortt and Garnham, 1948). Hypnozoites have profound implications for malaria control and elimination. The clinically silent reservoir of parasites which can, at unpredictable and lengthy intervals, reinfect the bloodstream gives vivax malaria a great advantage against any malaria elimination program. In endemic areas, it is not possible to distinguish new infections from old relapses as they are clinically equivalent, however, the public health implications are quite different. Ongoing transmission can be defeated using current methods such as ITN, but the chain of transmission will not be ended until the residual liver parasites are eliminated along with the risk of internal re-establishment of malaria transmission. The only method of eliminating hypnozoites is the use of 8-aminoquinoline drugs such as primaquine. Hypnozoite elimination has been possible at the community-level in large military populations where mass drug administration can be enforced but has been accomplished only very rarely in endemic civilian populations (Hill and Amatuzio, 1949; Kaneko, 2010). Vivax malaria elimination in large populations will require some form of mass drug administration. This will involve either determining how to safely use the potentially toxic drug primaquine or discovery of another drug capable of killing hypnozoites without the risk of haemolysis in those with a genetic polymorphism leading to glucose-6-phosphate dehydrogenase (G6PD) deficiency (see chapter 4, volume 81).

The public health challenge presented by P. vivax hypnozoites to malaria elimination can be summarized as one of program length and completeness. The tools exist today to control malaria even in areas of high endemicity if sufficient resources and political will can be applied to rapid diagnosis and treatment with effective drugs coupled with distribution of ITN to break the malaria transmission cycle (Greenwood, 2008a). A short intense program (attack phase) of residual insecticide spray during the GMEP was often sufficient to nearly eliminate falciparum in areas without year-round transmission. Vivax malaria's ability to reappear from the liver following intense control efforts, often meant control but not elimination was achieved. This window of vulnerability for malaria elimination failure during which programs must maintain intensive anti-transmission measures is uncertain but is at least 3 years and probably as long as 5 years. Over time, the failure to achieve the promised eradication of malaria despite impressive decreases in malaria mortality and morbidity doomed the GMEP. This historical lesson is important to current elimination efforts such that initial success does not create unrealistic expectations that malaria elimination will be easy or cheap.

Furthermore, relapses of vivax malaria from hypnozoites make it much more likely that it will be this species which is reintroduced into eliminated areas. This was seen in the USA when autochthonous cases occurred probably from persons travelling from Mexico and Central America while carrying parasites either in their blood or liver (Rodriguez et al., 2009; Zucker, 1996). This was also demonstrated by vivax malaria epidemics in the southern (now independent) republics of the former Soviet Union likely from soldiers returning from Afghanistan (Sergiev et al., 1993).

1.4. *P. vivax* in Africa

One does not associate *P. vivax* with Africa because the population is largely Duffy antigen negative (Howes et al. 2011), which deprives the parasite of one of its most important cellular receptors on the erythrocyte. The related parasite *P. ovale* is often seen as a vivax equivalent in Africa due to its ability to cause mixed infections with falciparum and relapse, presumably from hypnozoites. These two factors mean that in much of sub-Saharan Africa no vivax malaria is reported (further discussed in chapter 1, volume 80), but the parasite is not absent as travellers do on occasion return from Africa with documented vivax infections (Blossom et al., 2005; Guerra et al. 2010; PLoS NTDs). When groups of people with Duffy antigen have been brought to Africa, they have brought *P. vivax* with them. An example would be the Indian labourers used to construct the Mombasa to Lake Victoria railway in the early part of the twentieth century. Peoples in the Horn of Africa (Somalia, Ethiopia) have Duffy antigen on their erythrocytes so vivax malaria is quite common in that corner of Africa (Ashton et al., 2011). This includes highland areas as well as tourist destinations such as the Omo River. Duffy negativity is not an absolute bar to vivax infection as demonstrated in both Papua New Guinea and Africa (Mendes et al., 2011; Ryan et al., 2006). It is likely to be some years before anyone will consider relapsing malaria a major problem in Africa, but it will be important in terms of elimination programs when and if the burden of falciparum malaria is reduced to levels where non-falciparum species make up most of the malaria infections. In the meantime, better reporting and surveillance of non-falciparum species in sub-Saharan Africa is important in order to develop data for future programs (Culleton et al., 2009). These issues are further discussed in chapter 2 of volume 81.

2. ESTABLISHED ANTI-TRANSMISSION MEASURES

Doing simple things well is very important in malaria control. Although there is an appropriate research emphasis on finding new tools with which to control malaria, the lack of better drugs or vaccines is not the most common reason for persistent malaria endemicity. Many malarious countries have marginal health systems, which are under-resourced in terms of both money and personnel. Organization and logistics failures including administrative corruption, which degrades all disease control efforts, are the most prevalent reasons that bed nets are not distributed or houses not sprayed. Therefore, the consistent application of what is known to work is the best approach with the hope that in the future, we will be able to eventually achieve elimination with better interventions developed during the journey from control to elimination.

2.1. Residual Insecticide Spraying of Houses

When the mosquito transmission of malaria was first discovered by Ronald Ross at the end of the nineteenth century, malaria control was felt to be a tractable problem. Stop the mosquitoes and the malaria will go away. This is, of course, true, but far from simple given the *Anopheles* mosquito's long co-evolution with both *P. vivax* and *Homo sapiens*. Mosquitoes have immense reproductive capacity, so only in a few areas has elimination of the vector been possible, usually in geographically isolated areas such as Corsica, but also achieved in parts of Brazil (Toty et al., 2010; Killeen, 2003). The more common approach is to kill enough mosquitoes to break the transmission cycle, typically aiming at only those female mosquitoes feeding on humans at night. *Anophelines* differ greatly in their suitability as malaria vectors (see chapter 1, volume 80), which is not evident unless careful field entomologic studies are carried out. So the basic question in classical malaria control has always been how to kill night feeding mosquitoes most effectively.

Insecticide residual spraying (IRS) of house walls was the basis of the GMEP and noted many successes in areas of seasonal transmission. Once adopted, IRS became a grand logistics problem, but not a conceptual one. If one could just deliver adequate amounts of insecticide (DDT in the case of GMEP) to all the houses in a defined area, malaria transmission would go down to levels that the parasite could no longer reproduce itself ($R_0 < 1$). Malaria elimination then became another organizational problem of how to maintain an IRS program long enough for the malaria to go away. Given the usually stated limit on malaria relapses of 3 years post infection, 3–5-year intensive programs offered certainly malaria control and possibly elimination. It is very difficult to outwait the parasite especially vivax as it can reappear years later from hypnozoites to restart transmission. Practical problems dogged all malaria control programs, but in both China and India, countries with immense human and parasite (both falciparum and vivax) populations, a very high degree of control was achieved with IRS, resulting in the prevention of many deaths and associated disease (Cui et al., 2011; Kidson and Indaratna, 1998). The perception of failure only appeared later after malaria resurged once control measures were abandoned prior to achieving elimination.

Turning control into elimination depends on both achieving a high degree of control and then maintaining it for years with a good surveillance system to detect the few remaining cases. Falciparum was removed before vivax, thus one of the first positive signs of malaria control was elimination of malaria mortality. Removal of malaria from a population had multiple positive health effects indicating the parasite's role in combination with many other non-malarial infections. Once mortality and much malaria morbidity were removed, however, the motivation to continue the program markedly decreased. This lack of interest may have been aided by discounting vivax malaria as being of a less serious nature than falciparum malaria, a misunderstanding of the morbidity caused by vivax as described in chapter 3 of volume 80. As resources migrated to other more pressing health problems, malaria control programs gradually lost community support and then disintegrated. Health concerns about the long-acting insecticide DDT, which was the basis of the GMEP, removed critical community acceptance and left the program without any fall-back options. Over a period of years, sometimes with intermittent epidemics, malaria returned to previous endemic levels.

2.2. Insecticide Impregnated Bed Nets

The evolution of an alternative to IRS took time, but eventually giving the householders control over the insecticide by placing it into a portable bed net was developed. The desired effect was the same as IRS, killing female *Anophelines* as they attempted to bite humans inside their houses. Vivax malaria was, to some extent, ignored in the testing of bed nets as the key indicator was reduction of malaria mortality (which has historically been mistakenly not associated with P. vivax). That control of vivax malaria would follow from methods proven to work for falciparum malaria was assumed but not extensively documented by field studies (Luxemburger et al., 1994; Rowland et al., 1996; Sutanto et al., 1999). Such field studies are very difficult for the same reason that eliminating vivax is difficult, the ability of vivax malaria to reappear for years after transmission ends. Following the eventual realization that malaria control could not simply be another program put upon poorly functioning primary care systems, malaria control became a means to deliver ITN to endemic populations often in conjunction with other health interventions such as prenatal care or childhood immunization (Over et al., 2004). Problems previously experienced by the GMEP again appeared, especially the unsolved issue of how to move from reasonably good malaria control to elimination of the parasite.

2.3. Example of Solomon Islands

From the beginning of colonial administration in the Solomon Islands, medical officers noted that 'Malaria is one of the greatest factors in the heavy death-rate among infants and children' (Crichlow, 1929). The arrival of the US and Japanese Armies in the early 1940s provided huge scope for the parasite with large populations of previously unexposed persons and breakdown of the civilian health care system. A malaria survey done by Black in 1952 showed that malaria was found throughout the Solomon Islands ranging from holoendemic on Guadalcanal and Savo Islands to much lower levels on some of the outlying coral islands with largely Polynesian as opposed to Melanesian populations (Black, 1952). The GMEP came late to the Solomon Islands with trials of DDT spraying in the early 1960s, a pre-eradication program starting in the mid-1960s and expansion to the entire country by 1970 (Avery, 1973). Considerable success was experienced with a high degree of control resulting in markedly decreased malariometric parameters. The notable problem areas were the north coast of Guadalcanal because of the large transient populations that often cooperated poorly with the spray teams, Nggela islands where transmission was never interrupted probably because of large brackish water mosquito breeding sites, and a transplanted Polynesian population on New Georgia that were left living in undesirable swamp land. Progress was sufficient for the director of the country's malaria eradication program to entitle his MD thesis submitted to the University of Sheffield in 1977 'The Epidemiology of Disappearing Malaria in the Solomon Islands'. (Avery, 1973).

Such progress was not maintained, however, as funding for malaria control fell with the major external funding agencies cutting resources in the approach to independence largely because of the perceived failure that the GMEP which could not deliver eradication in a time-limited program. The vector mosquitoes did not cooperate by evolving behaviours that avoided DDT spray in houses by feeding earlier while people were still outside their homes. Permethrin-impregnated bed nets were tested in the highly malarious area of the north coast of Guadalcanal and shown to be effective (Kere et al., 1993). Malaria control, however, disintegrated in the face of ethnic conflict between the islands of Guadalcanal and Malaita in the 1990s resulting in the Solomon Islands having some of the highest reported malaria rates in Asia.

Malaria control was re-established particularly in the wake of restoration of civil order by the Regional Assistance Mission to the Solomon Islands (RAMSI) consisting largely of Australian soldiers and policemen. The Global Fund for AIDS, TB and Malaria (GFATM) enabled funding for malaria control which largely consisted of the country-wide distribution of ITN (Over et al., 2004). Over time, with additional resources and assistance from the Australian Government's AusAID sponsored Pacific Malaria Initiative, effective malaria control again reached all of the Solomon Islands. Such control again begged the question of whether elimination was possible at least on some islands. The isolated islands of the Santa Cruz group were shown to have low endemicity and remarkably little morbidity that could be assigned to malaria infection (Harris et al., 2010). The island of Isabel was a remarkably successful indigenous program based on excellent staff from the National Vector Borne Disease Control Program and highly effective participation through chiefs, church and community. Of particular note is that most of the persons found to have malaria are asymptomatic with vivax parasitemia only detected by PCR (Harris et al., 2010). Problems experienced were those that would have been familiar during the GMEP, including, for instance, those brought about by transient oil palm plantation workers on the north coast of Guadalcanal and Nggella Island (Avery, 1973). If the Solomon Islands is not to repeat this cycle of control with subsequent loss of program focus following reinstitution of control, then both government and funding agencies will have to recognize that malaria, particularly vivax malaria, is a stubborn aspect of tropical ecology and will require programs measured in decades and not years.

3. DRUGS FOR CONTROL OF RELAPSING MALARIA

Elimination of malaria through the administration of medication to an entire population has the dual attractions of being conceptually simple and time-limited. Unfortunately, for practical reasons such as the inability to achieve complete compliance and inherent limitations of the drugs currently available, examples of vivax malaria elimination using drugs as a primary intervention are extremely rare. Given the renewed interest and resources committed to malaria elimination through the GFATM and others recently, mass drug administration (MDA) is being reconsidered as a potential means to achieve elimination in some areas with particularly favourable epidemiology (A research agenda for malaria eradication: drugs, 2011). Which drugs to use and under what circumstances they should be employed is, however, very uncertain.

3.1. Mass Drug Administration as Public Health Tool

Malaria elimination cannot occur in a vacuum. Unless and until the malaria burden is decreased to a very small percentage of the population, discussion of malaria elimination is futile especially in areas which currently experience year-round transmission (Hay et al., 2008). The basic components of Roll Back Malaria (RBM) are rapid treatment of malaria cases with effective drugs and distribution of ITN to lower both malaria mortality and morbidity. Adequate treatment in this context means not just decreasing the parasitaemia in order to abate symptoms but actual cure meaning killing the last parasite. All antimalarial programs have to include the task of locating and adequately treating those persons infected with malaria. Elimination of malaria is a further development of this concept when areas under extremely good malaria control are advanced into an elimination phase with the goal of killing the last parasite not in the individual, but in the entire population. A large pulse of antimalarial medication through an entire area's population in order to eliminate residual parasites and block transmission until all infective mosquitoes have died could be the final phase in a long campaign prior to long-term surveillance primarily aimed at excluding new imported infections.

Other quite justified uses of antimalarial drugs must be differentiated from medication used to treat clinical cases during malaria control or elimination activities. Full treatment regimens of antimalarial medication have been successfully used to decrease malaria episodes when given to children with their standard immunizations during the first year of life (intermittent preventive treatment of infants or IPTi) (Greenwood, 2007) or to protect pregnant women during their last two trimesters (intermittent preventive treatment for pregnancy or IPTp) (ter Kuile et al., 2007). Rapid drug treatment of febrile school children by their teachers has also been shown to reduce mortality in some African settings. Such presumptive treatments to prevent mortality or severe morbidity are not a form of malaria elimination, but a means to decrease the malaria burden in an area still experiencing high levels of transmission. Concerted and resource demanding efforts to stop malaria transmission can really only be done from a base of excellent malaria control with a functional health system already able to deliver curative treatment to those infected or highly likely to be infected with malaria.

3.2. Historical Use of Drugs for Malaria Control/Elimination

There are many instances of antimalarial drugs being used within specific populations such as rubber tappers (Kingsbury and Amies, 1931), banana growers (Macphail, 1927), tea estate workers (Ray, 1948; Strangeways-Dixon, 1950), villagers (Clark et al., 1942; Kondrashin and Sanyal, 1985; Dapeng et al., 1996; Escudie et al., 1962), school children (Miller, 1955, 1956; Clyde, 1962), soldiers (Shanks et al., 1992, 1995a, 1995b; Malaria in the army in India, 1934) and miners (Gunther, 1951; Gunther et al., 1952; Shanks et al., 1993) as a means of controlling malaria, where the aim was protection of an occupational group rather than elimination in a geographic area. Antimalarial drug use for MDA focused on falciparum malaria has been reviewed (Greenwood, 2004; von Seidlein et al., 2003). The Italian elimination program was ultimately successful after 60 years despite many setbacks largely due the social disruption of two World Wars (Snowden, 2006). It arose from a social concern for rural farm labourers whose poverty exposed them to lethal malaria as an occupational hazard in certain high-transmission areas. The technologic advance that made the quinine distribution program of the Italian malaria control program possible was the greatly expanded cultivation of cinchona trees in the then Dutch East Indies rather than its native South America. The price of quinine, then the only specific drug known for malaria, dropped dramatically making its wide-spread distribution affordable and feasible. An Italian state quinine monopoly was set up to distribute medication to the rural poor as a means of social justice and an attempt to raise agricultural output. This well-intentioned effort largely failed because of limitations inherent to quinine and lack of other forms of socioeconomic progress. Quinine was bad-tasting and in sufficient doses to be effective induced unpleasant adverse events such as tinnitus. Heroic amounts of quinine were distributed and when it was actually ingested, did prevent a great deal of mortality but did nothing to stop vivax relapses thus discrediting the drug for a purpose it could not achieve (Snowden, 2006). Quinine cured some persons of falciparum but not vivax and failed to stop transmission. A single drug program in the absence of broader forms of medical care and other means to reduce poverty such as improved housing did not eliminate malaria and warns against a single intervention based only on medication in an impoverished population with minimal health services against a parasite capable of an evolutionary response.

3.3. Primaquine and Other 8-Aminoquinolines

As 8-aminoquinolines kill all stages of the malaria parasite, they are particularly well-suited for malaria elimination efforts. Unfortunately, the entire class of compounds also has characteristics that have made MDA difficult to achieve such as haemolysis in persons with the genetic defect G6PD deficiency and limited gastrointestinal tolerance especially when taken on an empty stomach (Cappellini and Fiorelli, 2008). Some trials with precursor drugs to primaquine such as plasmoquine demonstrated the potential of 8-aminoquinolines but were too toxic for wide-spread use (Kingsbury and Amies, 1931; Baeza, 1939; Barber et al., 1929, 1932; Walker, 1955; Field et al., 1940). Primaquine was first synthesized in the USA during the Second World War and subsequently further developed during the Korean War when many soldiers were infected with *P. vivax* with long relapse periods (Alving et al., 1952). The standard dosing regimen of 15 mg daily for 14 days that was used reflected a compromise between efficacy which was improved at 30 mg daily and toxicity particularly from haemolysis in black soldiers with G6PD deficiency (Clayman et al., 1952; Hockwald et al., 1952; Garrison et al., 1952; Coatney et al., 1953). Unlike most medications, the efficacy of primaquine is dependent on the total dose of drug given regardless of how frequently the medication is given, such that an alternative regimen of 45 mg weekly for 8 weeks was also instituted (Alving et al., 1960). Primaquine and G6PD deficiency are further discussed in chapter 4, volume 81.

Primaquine has been given to a vast number of people starting with large groups of US soldiers returning from the Korean War (Archambeault, 1954); in G6PD normal individuals, it is well tolerated. Persons with modest enzyme deficiency such as blacks in the USA (usually 5–10% of normal G6PD activity) only rarely have major haemolytic episodes (Hockwald et al., 1952). Unfortunately, that is not true of severely deficient individuals who can experience major haemolytic episodes usually 2–3 days after starting primaquine that can, on occasion, lead to severe anaemia and acute renal failure (Reeve et al., 1992; Ganczakowski et al., 1995; Kaneko et al., 1998; Hill et al., 2006). Primaquine is also contraindicated in persons with methaemoglobin reductase deficiency (Hill et al., 2006). In areas with essentially no G6PD deficiency as in parts of Latin America and China, primaquine can be given confidently without blood testing. It was discovered prior to the use of primaquine for MDA that G6PD deficiency did not exist in the island of Aneityum, but that was untrue of other nearby islands in Vanuatu (Kaneko et al., 1998). There are other G6PD variants that are severely enzyme deficient, particularly in persons who are descended from groups that originally lived around the Mediterranean Sea (Italy, Greece and North Africa) (Cappellini and Fiorelli, 2008). Blood testing for G6PD deficiency must be done to avoid drug-induced haemolytic events in most populations prior to the use of primaquine. Pregnant women should not receive primaquine as even if they test normally for the enzyme, it is not possible to test the foetus. Women with normal G6PD status who require primaquine can safely receive it after delivery of their child. Having a drug that requires a specific blood test prior to use does not invalidate a public health indication such as malaria elimination, but it does make it extremely difficult especially under the limited circumstances common in developing countries (A research agenda for malaria eradication: drugs, 2011; Kuwahata et al., 2010).

The record of using primaquine MDA for malaria control/elimination has been mixed in terms of it public health effect (Table 6.1). In 1968, in the island of Zanzibar, >120,000 persons (estimated 83% of total population) received amodiaquine and primaquine to supplement the DDT spray program. Although no severe toxic events were documented, the malaria parasitemia rate did not change much and the huge effort involved in drug distribution was felt to have been wasted (Dola, 1974). One of the largest uses of MDA including primaquine was in Nicaragua in 1981 (Garfield and Vermund, 1983). Close to 2 million persons or 70% of the total population received a 3-day course of chloroquine/primaquine in a nation-wide campaign. After a short period of decreased parasitaemia, malaria rates returned to baseline. Occasional haemolytic events were reported in Nicaragua, but it is unclear how much surveillance might have been done for such adverse events. Malaria transmission was not interrupted although an estimated 9200 malaria cases were prevented (Garfield and Vermund, 1983).

3.4. Role of MDA in Future Malaria Elimination Efforts

The objective of using MDA for malaria elimination is to kill the human reservoir of parasites responsible for malaria transmission (hypnozoites which subsequently produce mature gametocytes) not decreasing disease burden

Table 6.1 Selected Historical Examples of 8-Aminoquinoline Drugs Used for Malaria Control/Elimination

Location and Date	Drugs and Results	References
Villagers in Taiwan, early 1960s	Chloroquine/primaquine Elimination of residual areas of transmission	Department of Health, Executive Yuan, 1991; Chen, 1991
Nissan Island, Papua New Guinea, 1962	Amodiaquine/primaquine 12 weeks of drug ended transmission for at least 2 years in population of 2000	Rieckmann et al., 1968; Rieckmann, 1963
Zanzibar, 1968	80% of population received amodiaquine and primaquine to supplement DDT spray program	Dola, 1974
Nicaragua, 1981	Chloroquine/primaquine 70% national coverage gave temporary decrease in parasitemia, did not end transmission	Garfield et al., 1989
Villagers in Malaysia, 1984–1985	SP/primaquine Reduction in parasitemia for two months	Hii et al., 1987
Villagers in Indonesia, 1987–1989	SP/primaquine 8 month reduction in parasitemia	Doi et al., 1989
Villagers in China 1980s–1990s	Chloroquine/primaquine Malaria control as part of integrated system involving bed nets, insecticide spray	Dapeng et al., 1996
Island of Aneityum, 1992–2000	SP/primaquine Transmission ended after 9 weeks of medication and maintained by surveillance	Kaneko et al., 2000
Vivax epidemic in Kyrgyzstan, 2007	Actual regimen uncertain but appears to have been treatment doses of chloroquine/primaquine	Usenbaev et al., 2006, 2008

(Kouznetsov, 1979; Bruce-Chwatt, 1972). Unless and until the malaria burden is minimized, any pre-mature attempts at elimination are likely to expend resources including the critical factor of community participation without achieving lasting results. MDA is currently a difficult and very labour-intensive process that holds several risks beyond the possible failure to stop transmission.

Geographically isolated areas and/or those with seasonal transmission may be more suited to MDA. Regardless, MDA should not be considered as an addition to a failing bed net/insecticide spray campaign (Avery, 1973). It is unlikely to redeem such a failing program just as adding another new drug to a failing medication is unlikely to give an adequate combination. There is a critical need not to engender additional drug resistance that could limit the already very limited chemotherapeutic choices of malaria treatment. Although artemisinin combination therapy (ACT) does block malaria transmission under some circumstances, their use in MDA programs must be carefully considered especially given what appears to be the genesis of artemisinin resistance by falciparum malaria in some areas of the world (White, 2010). Vivax malaria is much less likely to develop drug resistance due to its hypnozoite reservoir which maintains a group of parasites that are sequestered away from antimalarial drug effect due to their metabolic inactivity. This was demonstrated by the acquisition of chloroquine resistance in vivax malaria some 30 years following falciparum malaria's acquisition of such drug resistance. Although primaquine resistance in vivax malaria has not been directly demonstrated in field studies, tropical strains of vivax clearly tolerate much more primaquine prior to curing any individual patient by killing the last hypnozoite. Malaria control programs always must be alert to the possibility of the evolution of drug resistance especially since there is no apparent replacement for primaquine if such resistance occurs.

Ideally, a drug for MDA would not be a major treatment option in the same region, which is one reason why 8-aminoquinolines are at the top of the drug list for malaria elimination purposes (A research agenda for malaria eradication: drugs, 2011). Drugs must be well tolerated to be used in entire communities or even countries. Any serious adverse event especially one that can clearly be linked to the medication would end a community's participation no matter what the long-term risk benefit equation indicated. A new drug or drug combination for MDA malaria control will have to have at least as good a safety profile as those previously used such as pyrimethamine or chloroquine. Community participation and motivation is critical to MDA. Without very high levels of compliance, residual parasites will be left in non-participants and reinfect the community later. In order to get to the point where an MDA program might stop transmission, one must minimize the malaria burden. Practically, this means that from the community's viewpoint, malaria will appear as a minor problem in a long list of unmet health needs. MDA might be considered when a malaria epidemic occurs (Kouznetsov, 1979; Bruce-Chwatt, 1972), but that is a different public health indication that seeks to minimize morbidity and mortality, not eliminate malaria. Relapsing malaria is a particular challenge. Because of *P. vivax*'s adaptation to a wide-ranging human host, one can expect malaria relapses for 3–5 years and perhaps longer even after a near-perfectly conducted elimination program (Kaneko et al., 2000). See Table 6.2 for characteristics distinguishing vivax malaria control from elimination.

Better ways to deliver the drug to entire communities must be found. Simple, easy-to-use solutions make a huge difference in being able to

Table 6.2 Differences Distinguishing Vivax Malaria Control from Elimination Programs are Described; This Applies to Public Health Applications rather than the Individual Patient

Intervention	Vivax Malaria Control	Vivax Malaria Elimination
Purpose of program	Stopping vivax malaria transmission	Destroying the last remaining parasite within a large population
Use of ITN	Main intervention to break current malaria transmission by killing night biting *Anopheles*	Maintained in order to prevent reintroduction of vivax by immigrants with hypnozoites
Use of mass drug administration of primaquine	Not appropriate until transmission largely ended	Difficult but critical to final elimination of residual parasites from population
Disease surveillance	Relatively easy as laboratory staff often find parasites in blood smears	Difficult as vast majority of blood smears will be negative
Length of control program	Usually 1–2 years of intense operations	Probably more than 5 years of maintaining program with few if any symptomatic malaria cases
Importance of asymptomatic persons	Unimportant compared to those sick with vivax	Extremely important to identify and treat this residual parasite reservoir
Community participation	Relatively easy as progress visible by fewer sick persons	Difficult as elimination occurs when there is little if any active disease
Political will and resources	Relatively easy to maintain as one can demonstrate visible progress	Very difficult due to competition from other higher burden diseases

conduct large-scale field operations such as MDA (A research agenda for malaria eradication: drugs, 2011). Skin patch or other simple drug-delivery technology may offer another solution especially if one can develop forms that can be used by non-medical personal and monitored to document compliance. A suggested product profile is shown in Table 6.3 about what an idealized MDA might look like for malaria elimination.

Table 6.3 Proposed Product Profile for Mass Administration of a Drug to Stop Malaria Transmission in a Defined Area

Parameter	Minimum Essential	Ideal
Efficacy	Sustained (2–4 weeks) transmission blocking within a defined area after MDA	3 months transmission blocking within a defined area after MDA
Duration of action*	2 weeks	3 months
Bioavailability / effect of food	Little	None
Dosing regimen	Daily doses over < 2 weeks	Single dose
Safety†	SAEs and AEs no worse than chloroquine	No SAEs; minimal AEs
Safety in pregnancy‡	Requires screening out of pregnant women	Does not require any screening for pregnancy in community
Compatibility with EPI vaccines	Serological responses not significantly affected	Serological responses not affected
Formulations	Scored tablets resistant to humidity	Paediatric version available; Sustained release single tablet resistant to humidity
Cost of treatment	≤ \$4.00	*circa* \$1.00
Shelf life	≥2 years	6 years
Susceptibility to loss of efficacy due to acquired resistance§	Very limited if any	Severely limited by sporontocidal activity of active compound

*Although shorter duration of activity may be possible in highly seasonal transmission areas by giving MDA at nadir of transmission, pulse of drug into a community by MDA is meant to stop transmission long enough to push area under excellent control into actual elimination of parasite.
†Tolerance and safety are key issues in any MDA strategy. Safety in G6PD-deficient individuals must be sufficient to not require G6PD screening (Kuwahata et al., 2010). Methylene blue appears to have less haemolytic potential than primaquine (Meissner et al., 2005; Coulibaly et al., 2009).
‡Uncertainty as to how 8-aminoquinoline would affect G6PD-deficient foetus within a G6PD-normal mother may require urine pregnancy testing of women of child bearing age.
§8-aminoquinoline likely to block transmission for substantial periods of time both in drug treated individuals and in mosquitoes who feed on drug treated persons.

3.5. China's Use of MDA

The People's Republic of China has progressively controlled malaria during the twentieth century, such that a country that was once very malarious is now within sight of elimination (Cui et al., 2011). Except for the border with Burma, all endemic malaria in China is vivax following the recent disappearance of falciparum malaria from Hainan Island. However, the geographic extent and population numbers still within areas of malaria transmission in China are enormous even if individual cases are very infrequent outside of focal epidemics (Dapeng et al., 1996; Zhou et al., 2011; Hui et al., 2009; Xu et al., 2006). In the 1970s, China experienced a massive upsurge of vivax malaria with an estimated 30 million malaria cases occurring of which 95% were vivax (Zhang and Deng, 1989). In response, China employed the mass administration of primaquine in treatment dosages (22.5 mg per day in adults for 8 days) (Sleigh et al., 1998). Literally, tens of millions of individuals in central China (e.g. Anhui Province) were treated with primaquine to stop transmission and kill hypnozoites in the early part of the year prior to when seasonal relapses would be expected. Public health interventions of such size are difficult to evaluate as one can only use historical controls, but clearly the transmission rates of vivax malaria in China have continued to drop. Except for some ethnic minority groups, Chinese people have very low frequencies of G6PD and thus a low risk of haemolysis when receiving primaquine. Despite the low risk, the large numbers of people treated with primaquine resulted in more than a handful of persons with acute haemolysis triggered by primaquine. Order of magnitude estimates would indicate one severe adverse event per 10,000 persons receiving MDA. Currently MDA with primaquine is used in China focally for entire villages with vivax malaria prevalence >3% and only in neighbouring households to known cases when prevalence is 1–3%. Combined with re-locating the previous year's vivax cases and re-treating them with primaquine during the spring, this extensive use of primaquine has lowered the amount of vivax malaria in China (Prof Gao Qi, Director Jiangsu Institute of Parasitology, Wuxi, China personal communication, 2012). It is unclear if this type of strategy could be applied in other areas of the world with different populations having a greater prevalence of G6PD deficiency (Shekalaghe et al., 2010; Howes et al. 2012).

3.6. Kyrgyzstan's Use of MDA

Central Asia, specifically, the republics of the former Soviet Union, was considered one of the GMEP success stories with malaria eliminated from the area during the 1950s–1960s. There was no malaria in Kyrgyzstan until the early 1980s when regional population movements associated with the war in Afghanistan reintroduced vivax malaria to the Fergana Valley (Razakov Sh, 2000). By 2002, 2744 cases of malaria (98% vivax) were registered in Kyrgyzstan and malaria control programs were being re-instituted. During 2002, about 5000 persons were given preventive therapy with primaquine in a region bordering Tajikistan (Usenbaev et al., 2006). In 2007, a localized peri-urban epidemic led to a decision to give MDA with primaquine in addition to ordinary case finding and treatment. From a target population of nearly 9000 persons, it was decided to give primaquine treatment to 6897 persons and 90% actually received the intended medication (Usenbaev et al., 2008). About 2% were unable to tolerate primaquine and stopped the medication; no haemolytic reactions were reported. In conjunction with other environmental antimalarial measures such as larvicides, the vivax cases fell to three by 2009.

4. DIFFICULTY GOING FROM LOW TO NO MALARIA TRANSMISSION

Disease elimination programs are different in that the unit of treatment is the population and not the individual. Established anti-transmission methods have repeatedly been shown to reduce malaria transmission to low levels when used properly and with sufficient vigour. The difficulty is going from low malaria, when it is of little public health consequence, to no malaria, where the parasite is indeed absent. Diminishing returns are obtained as fewer parasites are killed only by applying increasingly greater efforts. In both Taiwan and the USA, vivax was the primary parasite and certainly the last one to be eliminated by successful programs in 1965 and 1952, respectively (Yip, 2000; 1991; Williams, 1963; Lisansky, 1958).

4.1. Malaria Elimination in Taiwan

The outstanding success of island elimination was Taiwan and it is instructive to note the scale of the operations required to eliminate both falciparum and vivax malaria from Taiwan (Yip, 1998, 2000; Department of Health, Executive Yuan, 1991; Chen, 1991). Taiwan conducted a classic GMEP campaign largely dependent on DDT IRS where MDA was ancillary to the use of insecticides. Advantages that the Taiwan program had in order to achieve elimination included being an island, absence of drug resistance and strong governmental commitment to elimination. Problems encountered during the program included the need to extend basic health services into rural areas where malaria was being transmitted, residual foci of transmission in ethnic minority and other difficult to reach groups and difficulty in maintaining accurate diagnoses by microscopy when very few positives were seen. Following a 6-year DDT spray program starting in 1952, >20 residual foci of transmission were dealt with using intensive IRS and MDA of entire populations usually numbered in the hundreds with treatment courses of chloroquine/primaquine. Finding and eliminating these residual foci was a massive effort of malaria surveillance involving >5 million blood slides taken from July 1958 to December 1964 in order to find and treat the last 1023 malaria infections. During this consolidation phase, the program had to examine >5000 slides in order to find a single true positive, which is a considerably lower rate than the false-positive rate from even expert microscopists. When Taiwan was certified as eradicated of malaria on 25 November 1965, it was the conclusion of a 20-year program that at one point involved over 7000 staff, a full research institute, as well as a large logistical establishment (Department of Health, Executive Yuan, 1991).

4.2. Malaria Elimination in Mexico

Mexico has not eliminated malaria but has made great progress towards that goal with most transmission being restricted to vivax malaria in southern states such as Oaxaca (Rodriguez et al., 2009). Still largely based on DDT IRS, in Mexico case finding and investigation plays a large role in trying to achieve elimination goals. An integrated epidemiological tracking system capable of mapping all cases and coupled with knowledge of specific malaria risk factors has allowed focussing antimalarial efforts on remaining transmission foci. Although the primaquine regimen to kill hypnozoites is well established, the multiple doses required do not promote compliance. In Mexico, it was found that changing the primaquine regimen to intermittent monthly medication that was directly observed greatly improved compliance and decreased relapses (Rodriguez et al., 2009). Since effectiveness of primaquine is dependent only on the total dose, ingested alternatives to daily dosing seem to work well. As a country approaches elimination, intensive work with the few remaining cases as an indicator of where transmission has been occurring as well as a means of stopping future transmission seems to be a successful formula (Danis-Lozano et al., 2007).

5. SURVEILLANCE AS KEY INTERVENTION FOR MALARIA ELIMINATION

One of the greatest differences between a malaria control program and elimination is the degree of epidemiological surveillance required (Moonen et al., 2010). If one can identify persons with parasites, then the parasites can be eliminated, but identification of those persons can be difficult especially with vivax hypnozoites (Kaiser, 1966). As malaria cases become fewer, more effort is required to find them. Surveillance and its degree of completeness must increase as one progresses from a control program to elimination. Surveillance must be seen as a key intervention that distinguishes elimination from malaria control.

5.1. Korea

South Korea was certified malaria free by the WHO in 1985. In the early 1990s, indigenous vivax malaria cases began to reappear in both Korean and US soldiers stationed in the Demilitarized Zone on the border with North Korea (Park et al., 2009; Klein et al., 2009; Jun et al., 2009). By 2000, 4142 cases (all vivax) were reported and transmission had expanded from a border problem to involve many other areas of South Korea (Kim et al., 2009). Chemoprophylaxis with chloroquine was done initially in 16,000 South Korean soldiers during 1997 and then increasing to 200,000 soldiers in 2005 with decline in case numbers; unfortunately, it was very difficult to obtain compliance with primaquine eradication courses following the soldiers, release from military service (Yeom et al., 2005). Vivax relapses often occurred in ex-soldiers following their disbursement throughout the civilian population of South Korea, especially given the tendency for Korean vivax to relapse the year following initial infection. Climate restricts vivax malaria transmission to the months of June to September in Korea. Following reduction of annual cases to <1000 in 2004, cases have again increased above 2000 in 2007 (Park et al., 2009). Surveillance and detailed geographic tracking of vivax malaria cases in Korea has been done with hopes of returning to the pre-1990 situation of no malaria (Choi et al., 2010). Malaria elimination target date is now 2015 but 1317 cases were reported in 2009 indicating a great deal of work remains in order to interrupt transmission. Given the unclear situation in North Korea and the inability to do any antimalarial activities in the Demilitarized Zone, careful case finding coupled with complete primaquine treatment seems to be the main course of action for South Korea.

5.2. Aneityum

The most unambiguous success of MDA using chloroquine/sulfadoxine pyrimethamine (SP)/primaquine for malaria control remains the island of Aneityum, a small island at the southern end of Vanuatu and the extreme southern boundary of global malaria transmission (Fig. 6.1). Nearly 90% compliance with the medication over 9 weeks added to insecticide-treated bed nets, a favourable epidemiology and good post-MDA surveillance for new cases managed to eliminate malaria from this island of about 800 persons (Kaneko et al., 2000). For more than 20 years, malaria surveillance was continued on Aneityum and visitors were not allowed onto the main island without having a blood smear taken. During 2002 a large religious gathering resulting in an unusual number of visitors to the island, vivax malaria was reintroduced and rapidly became established in all villages on the island (Kaneko, 2010). Additional rounds of MDA including primaquine were required in order to remove the parasite again. Such a large concentration of malaria control resources cannot be generally applied as on the small island of Aneityum, so one has to find more practical approaches that can be applied to larger and less isolated areas.

Malaria elimination is favoured in isolated geographic areas such as islands when the population can be easily accessed and introduction of new persons potentially with new parasites can be strictly monitored at a limited number of entry points. In Sardinia, malaria eradication was aimed at mosquito elimination which failed despite the eventual elimination of malaria (Brown, 1998; Marchi and Munstermann, 1987). Besides the southernmost island of Aneityum (Kaneko et al., 2000), malaria elimination using 9–12 weeks of chloroquine/SP/primaquine MDA was attempted on other isolated islands in Melanesia, specifically the Santa Cruz Islands in the Solomon Islands as well as Nissan Island in Papua New Guinea (Avery, 1977; Rieckmann et al., 1968; Rieckmann, 1963). These programs were in addition to DDT house spraying and usually were attempts to eliminate residual malaria transmission that was not disappearing in the face of concerted residual insecticide use. MDA would reduce parasite rates but only in a few cases such as Aneityum and Nissan was transmission actually suspended. Failure of these elimination programs could usually be traced to either continuing foci of transmission or reintroduction of parasites from visitors from neighbouring islands.

6. PROBLEMS OF MALARIA REINTRODUCTION INTO ELIMINATED AREAS

The world is not static, especially given modern transportation systems that can move a person to nearly anywhere on the globe within a day.

Figure 6.1 The island of Aneityum in Vanuatu is where malaria was eliminated by a combined program of mass drug administration and vector control during the 1990s. For colour version of this figure, the reader is referred to the online version of this book.

Some of those moving persons have vivax parasites either in their blood or liver, thus they are able to infect suitable vector mosquitoes. Even isolated island nations have to cooperate with their neighbours in order to minimize reintroduction of malaria into eliminated areas. One such example is the island of Mauritius where vivax malaria was re-established following a typhoon that devastated the island in 1975 (Bruce-Chwatt and Bruce-Chwatt, 1974). It is likely that some of the reconstruction staff arriving after the typhoon brought malaria parasites from the Indian subcontinent triggering an epidemic in 1980 and necessitating a long, expensive program of elimination (Tatarsky et al., 2011). Given our dynamic world situation, it is best to consider malaria reintroduction issues early and plan for them even before malaria elimination is achieved within any particularly geographic area.

6.1. Trans-national Malaria Control

International borders challenge all malaria control programs as countries rarely operate identical public health programs, thus allowing malaria to defuse across borders into areas that have largely eliminated malaria. Vivax malaria is especially adept at movement across borders because of asymptomatic carriage of hypnozoites. The GMEP allowed nearby countries to substantially decrease their malaria transmission rates simultaneously, as in Southern Europe in the 1950s, thus border control for malaria purposes was not a major problem. This is unlikely to be true today. Population movements have recently caused malaria reintroductions in France, Greece and Spain although these autochthonous cases are unlikely to endanger their malaria elimination status (Danis et al., 2011; Toty et al., 2010; Armengaud et al., 2008; Santa-Olalla Peralta et al., 2010). Several trans-national cooperative projects to deal with cross border malaria issues have been set up including the Asia Pacific Malaria Elimination Network (APMEN) and the Elimination Eight (E8) Regional Initiative in eight countries of Southern Africa (Hsiang et al., 2010; Moss et al., 2011).

6.2. USA Post-Second World War and Korean Conflict

By the beginning of the twentieth century, the USA's malaria problem was predominantly due to vivax malaria and largely limited to rural southern areas, particularly river valleys and coastal areas (Goodwin, 1950; Andrews et al., 1950). Given the very limited state of public health infrastructure at the time, malaria endemicity was progressively measured by what was thought to be the insensitive but specific means of counting death certificates listing cause of death as malaria. Unfortunately, this method was completely undermined by insurance companies' definition of insurable disease excluding tuberculosis and venereal diseases, which gave family members and physicians a powerful financial incentive to certify deaths due to other infections as being caused by malaria (Faust et al., 1947). Malaria was actually being largely extinguished by economic development, which moved families out of marginal housing in river valleys to more urban areas into houses with window screening. This epidemiological shift was not appreciated until the Malaria Control in War Areas (direct forerunner of the US Centers for Disease Control) was established in Atlanta to deal with the expected malaria epidemics that would be caused by putting large military training camps in the southern USA during the Second World War (Williams, 1963; Lisansky, 1958; Andrews et al., 1950). By eliminating the use of death certificates noting malaria and clinical diagnoses of malaria through the use of blood smear microscopy, actual malaria endemicity was found to be much less than predicted. An IRS program over 5 years following the end of the Second World War largely completed malaria eradication in the USA (Andrews et al., 1950).

Of particular concern in the USA was the possibility that soldiers returning either from Italy or the Southwest Pacific would reinfect mosquitoes from vivax malaria relapses (Garrison et al., 1952; Archambeault, 1954). This did not happen in spite of primaquine being still under development and, thus, unavailable in 1945, resulting in the large numbers of ex-soldiers with relapsing vivax malaria in the USA. This potential problem was repeated during the Korean War when >10,000 vivax malaria infections were documented in USA soldiers returning from Korea to the USA. Still relatively few autochthonous infections were seen. The most likely explanation for this lack of transmission from returning soldiers is that improved housing left relatively few ex-soldiers exposed to night mosquitoes in formerly endemic areas. A compulsory program of chloroquine/primaquine given during the return sea voyage from Korea to the West Coast of the USA killed most hypnozoites, stopped the epidemic of Korean malaria relapsing in the USA and eliminated the risk of reintroduction of vivax malaria (Coatney et al., 1953; Archambeault, 1954).

6.3. Former Soviet Union Post Afghan War

Failure of established malaria control programs often occurs during the population movements and social disruptions associated with war. Vivax malaria was almost entirely eliminated from the southern republics of the Soviet

Union by the 1960s. War and political instability in Afghanistan and the resulting population movements reintroduced malaria back into the areas of Uzbekistan, Tajikistan, Kyrgyzstan, Turkmenistan and Kazakhstan (Razakov Sh, 2000; Usenbaev et al., 2006; Bismil'din et al., 2001; Amangel'diev et al., 2000; Matthys et al., 2008; Baranova and Sergiev, 2000). Malaria transmission was promoted by soldiers returning from the Afghan war with a total of 7683 known malaria cases imported to the Soviet Union, the vast majority being vivax malaria with only scattered cases of falciparum (Sergiev et al., 1993). Of the malaria relapses documented, 98% occurred within the first year after leaving Afghanistan. Many cases appeared to be due to failure to take appropriate post-deployment primaquine meant to kill hypnozoites. This reintroduction of vivax malaria has necessitated a large control/elimination effort by reconstituting discontinued malaria programs. The case of Kyrgyzstan has already been discussed above. Turkmenistan has been certified free of malaria in 2010 following sporadic local transmission in the 1990s including a 108-case epidemic of vivax in 1998 (Turkmenistan certified malaria-free, 2010). Uzbekistan appears to import most of its malaria cases from Tajikistan, of which 95% is due to vivax malaria (Razakov Sh, 2000). Following a post-independence civil war, endemic malaria was reestablished in southern Tajikistan with 30,000 cases per year during the peak year of 1997. This has since been controlled with the Tajik government aiming to eliminate malaria by 2015 (Matthys et al., 2008). Nearly half of the malaria cases documented in Kazakhstan are in military personnel who have been serving on the Tajik–Afghan border (Bismil'din et al., 2001). Although the political landscape of the former Soviet Union is very different, the malaria problem they are facing is analogous to that of South Korea.

7. PROSPECTS FOR IMPROVED DRUGS FOR CONTROL/ELIMINATION

Although the antimalarial drugs available today do work well when used correctly, they are far from ideal, particularly in terms of use in public health programs for entire populations. The ideal circumstance envisaged by malERA was a single encounter radical cure and prophylaxis (SERCaP), which had few adverse events and was inexpensive (A research agenda for malaria eradication: drugs, 2011). Although SERCaP is certainly a good goal to work towards, the number of new drugs already in clinical testing or ways to adapt old drugs is quite limited.

7.1. Tafenoquine

One future possibility is to develop a better 8-aminoquinoline than primaquine. This may already exist in tafenoquine, which is an investigational, long-acting primaquine analogue that has been already been given to >2000 persons, mostly Australian soldiers (Nasveld et al., 2002; Shanks et al., 2001; Lell et al., 2000). Major benefit is that this high dose may be possible to administer in a single dose, which would exclude problems of compliance currently seen with long primaquine dosing schedules. (Shanks et al., 2001). Tafenoquine has been used in combination with chloroquine to treat acute vivax malaria, to eliminate residual liver parasites to prevent relapses and as chemoprophylaxis for ongoing exposure in endemic areas (Shanks et al., 2001). There is good laboratory evidence that tafenoquine will kill parasites in the mosquito (Ponsa et al., 2003). Tafenoquine is not markedly more efficacious than primaquine, just much easier to administer which in itself might qualify it for a public health use if there were no questions about severe adverse events. Tafenoquine, however, is known to cause haemolysis in G6PD-deficient persons although even severe haemolytic events have not continued past a single episode despite tafenoquine's long half-life (Shanks et al., 2001). Currently, studies are underway to discover if there is an effective dosage or delivery system for tafenoquine that can be given safely to large populations without screening each individual for G6PD deficiency.

7.2. Methylene Blue

Methylene blue is a dye that was the earliest synthesized antimalarial drug having first been tested at the end of the nineteenth century. Used in combination with other medications such as amodiaquine, it has been shown to be well tolerated and effective for falciparum malaria in African adults and children (Bountogo et al., 2010; Zoungrana et al., 2008; Meissner et al., 2005, 2006). While conducting these African field trials, methylene blue was noted to be an excellent gametocide, which has been confirmed *in vitro* with falciparum (Coulibaly et al., 2009; Adjalley et al., 2011). The ability of vivax malaria to produce gametocytes from the beginning of blood infection unlike falciparum, which does not produce sexual stages capable of mosquito infection for some days to weeks after blood infection is achieved, indicates that methylene blue might have an anti-transmission indication for both species. Methylene blue has haemolytic potential in persons with G6PD but it appears to be less haemolytic than primaquine. As a dye, methylene blue also turns urine and other body fluids blue temporarily, so those receiving the drug need to be warned to avoid unnecessary alarm.

7.3. Other Medications Still in Pre-clinical Stages

Better drugs capable of killing all phases of plasmodium's life-cycle are needed which includes parasites in the blood, liver and mosquito. Traditionally, this has meant 8-aminoquinolines and perhaps tafenoquine can be adapted to fit a public health indication such as MDA for elimination. Certainly tafenoquine is the only new drug that is already in advanced clinical testing, which might possibly be adapted for MDA. New molecular classes of antimalarial drugs to block transmission and kill hypnozoites are needed urgently. A least one of the discovery projects sponsored by the Medicines for Malaria Venture (www.mmv.org) is a new class of compounds described as imidazolidinediones that kill liver parasites in a non-human primate model (Guan et al., 2007; Meister et al., 2011). Early investigations of such compounds are promising, but one must always be aware of the number of drugs which fall by the wayside because of a variety of pharmacological problems (Meister et al., 2011; Derbyshire et al., 2011; Borrmann and Matuschewski, 2011). One can certainly hope that some of the promising new compounds will eventually make it out of the laboratory and into clinical testing, but even with a decades-long view of malaria elimination, it is unlikely that completely new drugs will play a role in malaria elimination for many years to come.

8. PROSPECTS FOR IMPROVED ANTI-TRANSMISSION MEASURES

Despite the great progress that has been achieved in many countries using established control methods, malaria elimination will almost certainly require new interventions, especially in areas with year-round transmission (Chanda et al., 2011). The danger in this realization is that waiting for a perfect solution negates the progress that can be achieved now with ITN, IRS and case finding with effective drug therapy. Research on new antimalarial solutions is required in order to respond to a plastic parasite known to be capable of evolutionary change to thwart control/elimination programs (mal ERACGoVC, 2011). When GMEP ran into problems, there were no alternatives that had already been developed such that the program collapsed as donors realized that quick success could not be achieved (Brown, 1998). It is important that current demands are balanced against future needs in malaria control and that malaria research remain an ongoing part of control/elimination programs.

8.1. Anti-transmission Vaccine

The smallest number of malaria parasites occurs when gametocytes from the blood develop into oocysts in the mosquito gut. Killing the relatively few parasites within the mosquito would block transmission at what is thought to be the parasite's most evolutionarily vulnerable point (Polley et al., 2004). Although this can be done with primaquine, an anti-transmission vaccine has been sought based on natural immunity to vivax showing blocking of mosquito infection in man in a variety of Asian and American settings (Kappe et al., 2010; Plowe et al., 2009). Because anti-transmission measures are aimed at a population and not an individual, there has been little interest until recently in what has been called an *altruistic vaccine*. Antigens have been identified, which have been successfully used *in vitro* and in animal models. Initial clinical testing of a combined falciparum and vivax transmission-blocking vaccine has occurred indicating that such vaccines are safe and immunogenic. One can be hopeful that transmission-blocking vaccines will progress further towards field trials, however, it should be remembered that the supposedly simple blood stage vaccine RTS,S has been in development for more than 20 years as an indication of how difficult it is to field test vaccines against tropical diseases (Vogel and Roberts, 2011).

8.2. Improved Anti-mosquito Measures

Anopheline mosquitoes are fully capable of evolving insecticide resistance and may also change their behaviour in ways to make them less exposed to ITN. The future prospects for new insecticide compounds, especially any completely new chemical entities, are not good as their development largely depends on agricultural needs and not human health. The evolution of mosquito enzymes capable of imparting insecticide resistance has been documented in many field situations, particularly in Africa where the main concern is falciparum malaria, but resistance threatens all malaria control programs regardless of species (Abilio et al., 2011). Particularly worrying are malaria programs that doubly depend on pyrethroid compounds both to impregnate bed nets and also spray house walls as this provides multiple opportunities for mosquito evolution of insecticide resistance (Kaiser et al., 2010). IRS with DDT, particularly in Southern Africa, has been used as an alternative to pyrethroids; despite environmental concerns, the limited use of DDT for human health as opposed to agricultural production is safe

(Brooke et al., 2002). It is difficult to imagine any modern malaria control/elimination program without ITN and/or IRS, so the very limited number of insecticides needs to be managed in such a way (e.g. alternation) to minimize the evolution of resistance in vector mosquitoes.

8.3. Better Diagnostic Devices

Even though the biology of the man–mosquito–parasite triad cannot be changed, it is possible that new tools might shift the balance in favour of malaria elimination. High on the list of desirable items for malaria elimination are improved diagnostic tests especially for low parasitaemias and non-falciparum species (Harris et al., 2010). Vivax malaria does not have the falciparum protein HRP-2, which is the basis of most malaria rapid detection tests (Ly et al., 2010). As a consequence, vivax detection using RDT depends on other proteins such as lactic acid dehydrogenase and vivax detection is less sensitive than HRP-2-based tests for falciparum malaria (Senn et al., 2011; Ashton et al., 2010; Ashley et al., 2009). As malaria programs evolve from controlling malaria disease burden towards the elimination of the parasite, finding the increasingly few infected persons and adequately treating those often asymptomatic persons will become increasingly important. The microscopic examination of stained blood smears still remains the best way to quickly find low-level parasitaemias. New technology will hopefully change this century-old paradigm, but to date, that has not happened despite the invention and on-going improvement of immunochromatographic detection strips.

9. CONCLUSIONS

For the first time in at least a generation, malaria elimination is firmly on the world development pathway as not only a desirable goal but also a program for global implementation (Moonen et al., 2010; Feachem and Sabot, 2008). Although the context of such discussions usually implies that the main concern is falciparum malaria in Africa, vivax remains the most prevalent parasite with the widest geographic range. Vivax will also be the most difficult parasite to eliminate because of hypnozoites causing relapses long after initial infection. The main strategy for malaria elimination is control in areas of high endemicity and progressive elimination from the current borders of malaria transmission. Those countries that will be eliminating malaria in the foreseeable future, with the possible exception of Haiti, are all dealing with vivax malaria (Tatem et al., 2010). Presence of residual falciparum malaria cases is a reasonable marker that a country is still in the pre-elimination phase of malaria control. Attitudes towards vivax malaria as a minor problem are only slowly changing as the true disease burden of this relapsing parasite is better understood. The basic tools of malaria control/elimination already exist – what is in doubt is whether public health resources and political will can be remained focused on malaria for the decades long effort required to eliminate malaria, particularly vivax (Feachem et al., 2010; Sabot et al., 2010).

Better organization and cooperation will also be needed if malaria elimination is to be of more than theoretical importance. Programs such as the Expanded Program of Immunization (EPI) and lymphatic filariasis eradication have learned many valuable lessons about conducting disease elimination programs under the resource-constrained circumstances of developing countries (Breman et al., 2011). More examples of synergistic cooperation with such experienced organizations will be required. The GMEP did not fail to eliminate malaria for lack of effort or good intentions. A new drug (chloroquine) and a new technology (DDT spraying of house walls) had made malaria elimination seem to be an achievable goal for the twentieth century, which then faded into a mirage as biology and politics interacted to stop well-functioning malaria programs short of eradication. It is well to remember now in the twenty-first century that as the world once again puts malaria elimination back on the global agenda that malaria elimination will be a very long effort that hopefully will be made possible with improved drugs, insecticides and diagnostics to find and kill the plastic *P. vivax* parasite.

ACKNOWLEDGEMENTS

The author thanks Drs David Saunders and Alan Magill for critically reading and commenting on the manuscript and Professor Gao Qi, Director Jiangsu Institute of Parasitology, Wuxi, China for permission to refer to unpublished information on mass drug administration in China.

REFERENCES

Abilio, A.P., Kleinschmidt, I., Rehman, A.M., Cuamba, N., Ramdeen, V., Mthembu, D.S., et al., 2011. The emergence of insecticide resistance in central Mozambique and potential threat to the successful indoor residual spraying malaria control programme. Malar. J. 10, 110.

Adjalley, S.H., Johnston, G.L., Li, T., Eastman, R.T., Ekland, E.H., Eappen, A.G., et al., 2011. Quantitative assessment of *Plasmodium falciparum* sexual development reveals potent transmission-blocking activity by methylene blue. Proc. Natl. Acad. Sci. U S A 108 (47), E1214–E1223.

Aguas, R., White, L.J., Snow, R.W., Gomes, M.G., 2008. Prospects for malaria eradication in sub-Saharan Africa. PloS One 3 (3), e1767.

Alving, A.S., Arnold, J., Robinson, D.H., 1952. Mass therapy of subclinical vivax malaria with primaquine. J. Am. Med. Assoc. 149 (17), 1558–1562.

Alving, A.S., Johnson, C.F., Tarlov, A.R., Brewer, G.J., Kellermeyer, R.W., Carson, P.E., 1960. Mitigation of the haemolytic effect of primaquine and enhancement of its action against exoerythrocytic forms of the Chesson strain of *Piasmodium vivax* by intermittent regimens of drug administration: a preliminary report. Bull. World Health Organ. 22, 621–631.

Amangel'diev, K.A., Morozova, K.V., Medalieva, D.O., 2000. The epidemic situation with malaria in Turkmenistan. Meditsinskaia Parazitologiia I Parazitarnye Bolezni 2, 29–32 Epidemicheskaia situatsiia po maliarii v Turkmenistane.

Andrews, J.M., Quinby, G.E., Langmuir, A.D., 1950. Malaria eradication in the United States. Am. J. Public Health Nations Health 40 (11), 1405–1411.

Archambeault, C., 1954. Mass antimalarial therapy in veterans returning from Korea. J. Am. Med. Assoc. 154 (17), 1411–1415.

Armengaud, A., Legros, F., D'Ortenzio, E., Quatresous, I., Barre, H., Houze, S., et al., 2008. A case of autochthonous *Plasmodium vivax* malaria, Corsica, August 2006. Travel Med. Infect. Dis. 6 (1–2), 36–40.

Ashley, E.A., Touabi, M., Ahrer, M., Hutagalung, R., Htun, K., Luchavez, J., et al., 2009. Evaluation of three parasite lactate dehydrogenase-based rapid diagnostic tests for the diagnosis of falciparum and vivax malaria. Malar. J. 8, 241.

Ashton, R.A., Kefyalew, T., Tesfaye, G., Counihan, H., Yadeta, D., Cundill, B., et al., 2010. Performance of three multi-species rapid diagnostic tests for diagnosis of *Plasmodium falciparum* and *Plasmodium vivax* malaria in Oromia Regional State, Ethiopia. Malar. J. 9, 297.

Ashton, R.A., Kefyalew, T., Tesfaye, G., Pullan, R.L., Yadeta, D., Reithinger, R., et al., 2011. School-based surveys of malaria in Oromia Regional State, Ethiopia: a rapid survey method for malaria in low transmission settings. Malar. J. 10 (1), 25.

Avery, J.G., 1973. A review of the malaria eradication programme in the British Solomon Islands 1970-72. P. N. G. Med. J. 17 (1), 50–67.

Avery, J., 1977. The Epidemiology of Disappearing Malaria in the Solomon Islands. University of Sheffield, Sheffield UK.

Baeza, J., 1939. Partial control of malaria in a localized area by plasmoquine. J. Malaya Branch Br. Med. Assoc. 2 (4), 239–243.

Baird, J.K., 2009. Resistance to therapies for infection by *Plasmodium vivax*. Clin. Microbiol. Rev. 22 (3), 508–534.

Baranova, A.M., Sergiev, V.P., 2000. The malaria situation in the Russian Federation (1997–1999). Meditsinskaia Parazitologiia I Parazitarnye Bolezni (2), 22–25. Situatsiia po maliarii v Rossiiskoi Federatsii (1997–1999 gody).

Barber, M., Komp, W., Newman, B., 1929. The effect of small doses of Plasmochin on the viability of gametocytes of malaria as measured by mosquito infection experiments. Public Health Rep. 44, 1409–1420.

Barber, M., Rice, J., Brown, O., 1932. Malaria studies on the Fireston rubber plantation in Liberia, West Africa. Am. J. Hyg. 15, 601–633.

Bismil'din, F.B., Shapieva, Z., Anpilova, E.N., 2001. Current malaria situation in the Republic of Kazakhstan. Meditsinskaia parazitologiia i parazitarnye bolezni (1), 24–33.

Black, R., 1952. Malaria in the British Solomon Islands Protectorate. Noumea: South Pacific Comm, Tech paper 33.

Blossom, D., King, C.H., Armitage, K., 2005. Occult *Plasmodium vivax* infection diagnosed by a polymerase chain reaction-based detection system: a case report. Am. J. Trop. Med. Hyg. 73 (1), 188–190.

Borrmann, S., Matuschewski, K., 2011. Targeting Plasmodium liver stages: better late than never. Trends Mol. Med. 17 (9), 527–536.

Bountogo, M., Zoungrana, A., Coulibaly, B., Klose, C., Mansmann, U., Mockenhaupt, F.P., et al., 2010. Efficacy of methylene blue monotherapy in semi-immune adults with uncomplicated falciparum malaria: a controlled trial in Burkina Faso. Trop. Med. Int. Health:TM & IH 15 (6), 713–717.

Breman, J.G., de Quadros, C.A., Dowdle, W.R., Foege, W.H., Henderson, D.A., John, T.J., et al., 2011. The role of research in viral disease eradication and elimination programs: lessons for malaria eradication. PLoS Med. 8 (1), e1000405.

Brooke, B.D., Hunt, R.H., Chandre, F., Carnevale, P., Coetzee, M., 2002. Stable chromosomal inversion polymorphisms and insecticide resistance in the malaria vector mosquito *Anopheles gambiae* (Diptera: Culicidae). J. Med. Entomol. 39 (4), 568–573.

Brown, P.J., 1998. Failure-as-success: multiple meanings of eradication in the Rockerfeller Foundation Sardinia project, 1946–1951. Parassitologia 40, 117–130.

Bruce-Chwatt, L.J., Bruce-Chwatt, J.M., 1974. Malaria in Mauritius–as dead as the dodo. Bull. N.Y. Acad. Med. 50 (10), 1069–1080.

Bruce-Chwatt, L.J., 1972. The role of drugs in a malaria program. Am. J. Trop. Med. Hyg. 21 (5), 731–735.

Bruce-Chwatt, L.J., 1977. Malaria eradication in Portugal. Trans. R. Soc. Trop. Med. Hyg. 71 (3), 232–240.

Cappellini, M.D., Fiorelli, G., 2008. Glucose-6-phosphate dehydrogenase deficiency. Lancet 371, 64–74.

Chanda, E., Hemingway, J., Kleinschmidt, I., Rehman, A.M., Ramdeen, V., Phiri, F.N., et al., 2011. Insecticide resistance and the future of malaria control in Zambia. PloS One 6 (9), e24336.

Chen, W.I., 1991. Malaria eradication in Taiwan, 1952–1964–some memorable facts. Gaoxiong Yi Xue Ke Xue Za Zhi (The Kaohsiung Journal of Medical Sciences) 7 (5), 263–270.

Choi, Y.K., Choi, K.M., Park, M.H., Lee, E.G., Kim, Y.J., Lee, B.C., et al., 2010. Rapid dissemination of newly introduced *Plasmodium vivax* genotypes in South Korea. Am. J. Trop. Med. Hyg. 82 (3), 426–432.

Clark, H., Komp, W., Jobbins, D., 1942. A tenth year's observations on malaria in Panama, with reference to the occurrence of variations in the parasite index, during continued treatment with atabrine and plasmochine. Am. J. Trop. Med. 22, 191–216.

Clayman, C.B., Arnold, J., Hockwald, R.S., Yount Jr., E.H., Edgcomb, J.H., Alving, A.S., 1952. Toxicity of primaquine in Caucasians. J. Am. Med. Assoc. 149 (17), 1563–1568.

Clyde, D.F., 1962. Mass administration of an antimalarial drug combining 4-aminoquinoline and 8-aminoquinoline in Tanganyika. Bull. World Health Organ. 27, 203–212.

Coatney, G.R., Alving, A.S., Jones Jr., R., Hankey, D.D., Robinson, D.H., Garrison, P.L., et al., 1953. Korean vivax malaria. V. Cure of the infection by primaquine administered during long-term latency. Am. J. Trop. Med. Hyg. 2 (6), 985–988.

Colbourne, M.J., 1962. Prospects for malaria eradication, with special reference to the Western Pacific. Trans. R. Soc. Trop. Med. Hyg. 56, 179–201.

Coulibaly, B., Zoungrana, A., Mockenhaupt, F.P., Schirmer, R.H., Klose, C., Mansmann, U., et al., 2009. Strong gametocytocidal effect of methylene blue-based combination therapy against falciparum malaria: a randomised controlled trial. PloS One 4 (5), e5318.

Crichlow, N., 1929. The prevalent diseases of the British Solomon Islands. Trans. R. Soc. Trop. Med. Hyg. 23 (2), 179–184.

Cui, L., Yan, G., Sattabongkot, J., Chen, B., Cao, Y., Fan, Q., et al., 2011. Challenges and prospects for malaria elimination in the greater Mekong Subregion. Acta. Trop.

Culleton, R., Ndounga, M., Zeyrek, F.Y., Coban, C., Casimiro, P.N., Takeo, S., et al., 2009. Evidence for the transmission of *Plasmodium vivax* in the Republic of the Congo, West Central Africa. J. Infect. Dis. 200 (9), 1465–1469.

Danis, K., Baka, A., Lenglet, A., Van Bortel, W., Terzaki, I., Tseroni, M., et al., 2011. Autochthonous *Plasmodium vivax* malaria in Greece. Euro Surveillance: Bulletin Europeen Sur Les Maladies Transmissibles (European Communicable Disease Bulletin) 16 (42).

Danis-Lozano, R., Rodriguez, M.H., Betanzos-Reyes, A.F., Hernandez-Avila, J.E., Gonzalez-Ceron, L., Mendez-Galvan, J.F., et al., 2007. Individual risk factors for *Plasmodium vivax* infection in the residual malaria transmission focus of Oaxaca, Mexico. Salud Publica De Mexico 49 (3), 199–209.

Dapeng, L., Leyuan, S., Xili, L., Xiance, Y., 1996. A successful control programme for falciparum malaria in Xinyang, China. Trans. R. Soc. Trop. Med. Hyg. 90 (2), 100–102.

Department of Health, Executive Yuan, 1991. Malaria Eradication in Taiwan. Taipei, Taiwan.

Derbyshire, E.R., Mota, M.M., Clardy, J., 2011. The next opportunity in anti-malaria drug discovery: the liver stage. PLoS Pathog. 7 (9), e1002178.

Doi, H., Kaneko, A., Panjaitan, W., Ishii, A., 1989. Chemotherapeutic malaria control operation by single dose of Fansidar plus primaquine in North Sumatra, Indonesia. Southeast Asian J. Trop. Med. Public Health 20 (3), 341–349.

Dola, S., 1974. Mass drug administration as a supplementary attack measure in malaria eradication programme. East Afr. Med. J. 51 (7), 529–531.

Escudie, A., Hamon, J., Schneider, J., Mar–Apr 1962. Resultats d'une chimioprohylaxie antipaludique de masse par l'association amino-4-quinoleine/amino-8-quinoleine en milieu rural Africain de la region de Bobo-Dioulasso. Med. Trop. 22, 268–305.

Faust, E.C., Scott, J.A., Mc, D.G., 1947. Malaria mortality and morbidity in the United States for the year 1945. J. Natl. Malar. Soc. 6 (3), 184–191.

Feachem, R., Sabot, O., 2008. A new global malaria eradication strategy. Lancet.

Feachem, R.G., Phillips, A.A., Targett, G.A., Snow, R.W., 2010. Call to action: priorities for malaria elimination. Lancet 376 (9752), 1517–1521.

Field, J., Niven, J., Hodgkin, E., Mitchell, J., 1940. Field observations on the effects of prophylactic plasmoquine on the incidence, course, severity and transmission of *P vivax* malaria. Kuala Lumpur: Contract No.: 3.

Ganczakowski, M., Town, M., Bowden, D.K., Vulliamy, T.J., Kaneko, A., Clegg, J.B., et al., 1995. Multiple glucose 6-phosphate dehydrogenase-deficient variants correlate with malaria endemicity in the Vanuatu archipelago (southwestern Pacific). Am. J. Hum. Genet. 56 (1), 294–301.

Garfield, R.M., Vermund, S.H., 1983. Changes in malaria incidence after mass drug administration in Nicaragua. Lancet 2 (8348), 500–503.

Garfield, R.M., Prado, E., Gates, J.R., Vermund, S.H., 1989. Malaria in Nicaragua: community-based control efforts and the impact of war. Int. J. Epidemiol. 18 (2), 434–439.

Garrison, P.L., Hankey, D.D., Coker, W.G., Donovan, W.N., Jastremski, B., Coatney, G.R., et al., 1952. Cure of Korean vivax malaria with pamaquine and primaquine. J. Am. Med. Assoc. 149 (17), 1562–1563.

Genton, B., D'Acremont, V., Rare, L., Baea, K., Reeder, J.C., Alpers, M.P., et al., 2008. *Plasmodium vivax* and mixed infections are associated with severe malaria in children: a prospective cohort study from Papua New Guinea. PLoS Med. 5 (6), e127.

Gething, P.W., Smith, D.L., Patil, A.P., Tatem, A.J., Snow, R.W., Hay, S.I., 2010. Climate change and the global malaria recession. Nature 465 (7296), 342–345.

Goodwin Jr., M.H., 1950. Observations on the recession of malaria in an area of southwestern Georgia. Am. J. Trop. Med. Hyg. 30 (5), 651–661.

Greenwood, B., 2004. The use of anti-malarial drugs to prevent malaria in the population of malaria-endemic areas. Am. J. Trop. Med. Hyg. 70 (1), 1–7.

Greenwood, B., 2007. Intermittent preventive antimalarial treatment in infants. Clin. Infect. Dis. 45 (1), 26–28.

Greenwood, B., 2008a. Progress in malaria control in endemic areas. Travel Med. Infect. Dis. 6 (4), 173–176.

Greenwood, B.M., 2008b. Control to elimination: implications for malaria research. Trends Parasitol. 24 (10), 449–454.
Guan, J., Wang, X., Smith, K., Ager, A., Gettayacamin, M., Kyle, D.E., et al., 2007. Malaria causal prophylactic activity of imidazolidinedione derivatives. J. Med. Chem. 50 (24), 6226–6231.
Guerra, C.A., Howes, R.E., Patil, A.P., Gething, P.W., Van Boeckel, T.P., Temperley, W.H., et al., 2010. The international limits and population at risk of *Plasmodium vivax* transmission in 2009. PLoS Negl. Trop. Dis. 4 (8), e774.
Gunther, C., Fraser, N., Wright, W., 1952. Proguanil and malaria among non-tolerant New Guinea natives. Trans. R. Soc. Trop. Med. Hyg. 46, 185–200.
Gunther, C.E., 1951. Proguanil hydrochloride (paludrine) in the prevention and treatment of malaria in New Guinea. Trans. R. Soc. Trop. Med. Hyg. 44 (4), 473–478.
Hall, B.F., Fauci, A.S., 2009. Malaria control, elimination, and eradication: the role of the evolving biomedical research agenda. J. Infect. Dis. 200 (11), 1639–1643.
Harris, I., Sharrock, W.W., Bain, L.M., Gray, K.A., Bobogare, A., Boaz, L., et al., 2010. A large proportion of asymptomatic *Plasmodium* infections with low and sub-microscopic parasite densities in the low transmission setting of Temotu Province, Solomon Islands: challenges for malaria diagnostics in an elimination setting. Malar. J. 9, 254.
Hay, S.I., Smith, D.L., Snow, R.W., 2008. Measuring malaria endemicity from intense to interrupted transmission. Lancet Infect. Dis 8(6), 369–378.
Hii, J.L., Vun, Y.S., Chin, K.F., Chua, R., Tambakau, S., Binisol, E.S., et al., 1987. The influence of permethrin-impregnated bednets and mass drug administration on the incidence of *Plasmodium falciparum* malaria in children in Sabah, Malaysia. Med. Vet. Entomol. 1, 397–407.
Hill, E., Amatuzio, D., 1949. Southwest Pacific vivax malaria; clinical features and observations concerning duration of clinical activity. Am. J. Trop. Med. Hyg. 29 (2), 203–214.
Hill, D.R., Baird, J.K., Parise, M.E., Lewis, L.S., Ryan, E.T., Magill, A.J., 2006. Primaquine: report from CDC expert meeting on malaria chemoprophylaxis I. Am. J. Trop. Med. Hyg. 75 (3), 402–415.
Hockwald, R.S., Arnold, J., Clayman, C.B., Alving, A.S., 1952. Toxicity of primaquine in Negroes. J. Am. Med. Assoc. 149 (17), 1568–1570.
Howes, R.E., Piel, F.B., Patil, A.P., Nyangiri, O.A., Gething, P.W., Dewi, M., et al., 2010. G6PD deficiency prevalence and estimates of affected populations in malaria endemic countries: a geostatistical model-based map. PLoS Med. (in press).
Howes, R.E., Patil, A.P., Piel, F.B., Nyangiri, O.A., Kabaria, C.W., Gething, P.W., et al., 2011. The global distribution of the Duffy blood group. Nat. Comms. 2, 266.
Hsiang, M.S., Abeyasinghe, R., Whittaker, M., Feachem, R.G., 2010. Malaria elimination in Asia-Pacific: an under-told story. Lancet 375 (9726), 1586–1587.
Hui, F.M., Xu, B., Chen, Z.W., Cheng, X., Liang, L., Huang, H.B., et al., 2009. Spatio-temporal distribution of malaria in Yunnan Province, China. Am. J. Trop. Med. Hyg. 81 (3), 503–509.
Jun, G., Yeom, J.S., Hong, J.Y., Shin, E.H., Chang, K.S., Yu, J.R., et al., 2009. Resurgence of *Plasmodium vivax* malaria in the Republic of Korea during 2006–2007. Am. J. Trop. Med. Hyg. 81 (4), 605–610.
Kaiser, M.L., Koekemoer, L.L., Coetzee, M., Hunt, R.H., Brooke, B.D., 2010. Staggered larval time-to-hatch and insecticide resistance in the major malaria vector *Anopheles gambiae* S form. Malar. J. 9, 360.
Kaiser, R.L., 1966. Epidemiology of malaria eradication. 3. The role of surveillance in a malaria eradication program. Am. J. Public Health Nations Health 56 (1), 90–93.
Kaneko, A., Taleo, G., Kalkoa, M., Yaviong, J., Reeve, P.A., Ganczakowski, M., et al., 1998. Malaria epidemiology, glucose 6-phosphate dehydrogenase deficiency and human settlement in the Vanuatu Archipelago. Acta. Trop. 70 (3), 285–302.

Kaneko, A., Taleo, G., Kalkoa, M., Yamar, S., Kobayakawa, T., Bjorkman, A., 2000. Malaria eradication on islands. Lancet 356 (9241), 1560–1564.

Kaneko, A.A., 2010. Community-directed strategy for sustainable malaria elimination on islands: short-term MDA integrated with ITNs and robust surveillance. Acta. Trop. 114 (3), 177–183.

Kappe, S.H., Vaughan, A.M., Boddey, J.A., Cowman, A.F., 2010. That was then but this is now: malaria research in the time of an eradication agenda. Science 328 (5980), 862–866.

Kasehagen, L.J., Mueller, I., Kiniboro, B., Bockarie, M.J., Reeder, J.C., Kazura, J.W., et al., 2007. Reduced *Plasmodium vivax* erythrocyte infection in PNG Duffy-negative heterozygotes. PloS One 2 (3), e336.

Kere, N.K., Parkinson, A.D., Samrawickerema, W.A., 1993. The effect of permethrin impregnated bednets on the incidence of *Plasmodium falciparum*, in children of north Guadalcanal, Solomon Islands. Southeast Asian J. Trop. Med. Public Health 24 (1), 130–137.

Kidson, C., Indaratna, K., 1998. Ecology, economics and political will: the vicissitudes of malaria strategies in Asia. Parassitologia 40 (1–2), 39–46.

Killeen, G.F., 2003. Following in Soper's footsteps: northeast Brazil 63 years after eradication of *Anopheles gambiae*. Lancet Infect. Dis. 3 (10), 663–666.

Kim, H.C., Pacha, L.A., Lee, W.J., Lee, J.K., Gaydos, J.C., Sames, W.J., et al., 2009. Malaria in the Republic of Korea, 1993–2007. Variables related to re-emergence and persistence of *Plasmodium vivax* among Korean populations and U.S. forces in Korea. Mil. Med. 174 (7), 762–769.

Kingsbury, A., Amies, C., 1931. A field experiment on the value of plasmoquine in the prophylaxis of malaria. Trans. R. Soc. Trop. Med. Hyg. 25 (3), 159–172.

Klein, T.A., Pacha, L.A., Lee, H.C., Kim, H.C., Lee, W.J., Lee, J.K., et al., 2009. *Plasmodium vivax* malaria among U.S. forces Korea in the Republic of Korea, 1993–2007. Mil. Med. 174 (4), 412–418.

Kondrashin, A.V., Sanyal, M.C., 1985. Mass drug administration in Andhra Pradesh in areas under *Plasmodium falciparum* containment programme. J. Commun. Dis. 17 (4), 293–299.

Kouznetsov, R.L., 1979. The place of chemotherapy in antimalaria programmes. WHO Chron. 33 (1), 9–12.

Kuwahata, M., Wijesinghe, R., Ho, M.F., Pelecanos, A., Bobogare, A., Landry, L., et al., 2010. Population screening for glucose-6-phosphate dehydrogenase deficiencies in Isabel Province, Solomon Islands, using a modified enzyme assay on filter paper dried bloodspots. Malar. J. 9, 223.

Lell, B., Faucher, J.F., Missinou, M.A., Borrmann, S., Dangelmaier, O., Horton, J., et al., 2000. Malaria chemoprophylaxis with tafenoquine: a randomised study. Lancet 355 (9220), 2041–2045.

Lisansky, E., 1958. The eradication of malaria as an endemic disease in the United States. Ann. Intern. Med. 48 (2), 428–438.

Luxemburger, C., Perea, W., Delmas, G., Pruja, C., Pecoul, B., Moren, A., 1994. Permethrin-impregnanted bed nets for the prevention of malaria in schoolchildren on the Thai-Burmese border. Trans. R. Soc. Trop. Med. Hyg. 88 (2), 155–159.

Ly, A.B., Tall, A., Perry, R., Baril, L., Badiane, A., Faye, J., et al., 2010. Use of HRP-2-based rapid diagnostic test for *Plasmodium falciparum* malaria: assessing accuracy and cost-effectiveness in the villages of Dielmo and Ndiop, Senegal. Malar. J. 9, 153.

Macphail, N., 1927. Plasmochin in malaria, 16th Annual Report of Medical Dept. United Fruit Company, New York.

malERA, 2011. A research agenda for malaria eradication: drugs. PLoS Med. 8 (1), e1000402.

malERA, 2011. A research agenda for malaria eradication: vector control. PLoS Med. 8 (1), e1000401.

Malaria in the army in India. 1934. Lancet 223, 802.

Marchi, A., Munstermann, L.E., 1987. The mosquitoes of Sardinia: species records 35 years after the malaria eradication campaign. Med. Vet. Entomol. 1 (1), 89–96.

Matthys, B., Sherkanov, T., Karimov, S.S., Khabirov, Z., Mostowlansky, T., Utzinger, J., et al., 2008. History of malaria control in Tajikistan and rapid malaria appraisal in an agro-ecological setting. Malar. J. 7, 217.

Mayxay, M., Pukrittayakamee, S., Newton, P.N., White, N.J., 2004. Mixed-species malaria infections in humans. Trends Parasitol. 20 (5), 233–240.

Meissner, P.E., Mandi, G., Witte, S., Coulibaly, B., Mansmann, U., Rengelshausen, J., et al., 2005. Safety of the methylene blue plus chloroquine combination in the treatment of uncomplicated falciparum malaria in young children of Burkina Faso [ISRCTN27290841]. Malar. J. 4, 45.

Meissner, P.E., Mandi, G., Coulibaly, B., Witte, S., Tapsoba, T., Mansmann, U., et al., 2006. Methylene blue for malaria in Africa: results from a dose-finding study in combination with chloroquine. Malar. J. 5, 84.

Meister, S., Plouffe, D.M., Kuhen, K.L., Bonamy, G.M., Wu, T., Barnes, S.W., et al., 2011. Imaging of *Plasmodium* liver stages to drive next-generation antimalarial drug discovery. Science 334 (6061), 1372–1377.

Mendes, C., Dias, F., Figueiredo, J., Mora, V.G., Cano, J., de Sousa, B., et al., 2011. Duffy negative antigen is no longer a barrier to *Plasmodium vivax* – molecular evidences from the African West Coast (Angola and Equatorial Guinea). PLoS Negl. Trop. Dis. 5 (6), e1192.

Miller, M., 1955. Suppression of malaria by monthly drug administration. Am. J. Trop. Med. Hyg. 4, 790–799.

Miller, M., 1956. Further studies on malaria suppression by monthly drug administration. Am. J. Trop. Med. Hyg. 5, 624–637.

Moonen, B., Cohen, J.M., Snow, R.W., Slutsker, L., Drakeley, C., Smith, D.L., et al., 2010. Operational strategies to achieve and maintain malaria elimination. Lancet 376 (9752), 1592–1603.

Moss, W.J., Norris, D.E., Mharakurwa, S., Scott, A., Mulenga, M., Mason, P.R., et al., 2011. Challenges and prospects for malaria elimination in the Southern Africa region. Acta. Trop.

Nasveld, P., Kitchener, S., Edstein, M., Rieckmann, K., 2002. Comparison of tafenoquine (WR238605) and primaquine in the post-exposure (terminal) prophylaxis of vivax malaria in Australian Defence Force personnel. Trans. R. Soc. Trop. Med. Hyg. 96 (6), 683–684.

Oh, M.D., Shin, H., Shin, D., Kim, U., Lee, S., Kim, N., et al., 2001. Clinical features of vivax malaria. Am. J. Trop. Med. Hyg. 65 (2), 143–146.

Over, M., Bakote'e, B., Velayudhan, R., Wilikai, P., Graves, P.M., 2004. Impregnated nets or DDT residual spraying? Field effectiveness of malaria prevention techniques in Solomon Islands, 1993–1999. Am. J. Trop. Med. Hyg. 71 (Suppl. 2), 214–223.

Park, J.W., Jun, G., Yeom, J.S., 2009. *Plasmodium vivax* malaria: status in the Republic of Korea following reemergence. Korean J. Parasitol. (Suppl. 47), S39–S50.

Plowe, C.V., Alonso, P., Hoffman, S.L., 2009. The potential role of vaccines in the elimination of falciparum malaria and the eventual eradication of malaria. J. Infect. Dis. 200 (11), 1646–1649.

Polley, S.D., McRobert, L., Sutherland, C.J., 2004. Vaccination for vivax malaria: targeting the invaders. Trends Parasitol. 20 (3), 99–102.

Ponsa, N., Sattabongkot, J., Kittayapong, P., Eikarat, N., Coleman, R.E., 2003. Transmission-blocking activity of tafenoquine (WR-238605) and artelinic acid against naturally circulating strains of *Plasmodium vivax* in Thailand. Am. J. Trop. Med. Hyg. 69 (5), 542–547.

Rajakaruna, R.S., Alifrangis, M., Amerasinghe, P.H., Konradsen, F., 2010. Pre-elimination stage of malaria in Sri Lanka: assessing the level of hidden parasites in the population. Malar. J. 9, 25.

Ray, A., 1948. Prophylactic use of paludrine in a tea estate. Indian J. Malariol. 2 (1/2), 35–66.
Razakov Sh, A., 2000. The epidemic situation with malaria in Uzbekistan. Meditsinskaia Parazitologiia I Parazitarnye Bolezni (2), 32–34. (Epidemicheskaia situatsiia po maliarii v Uzbekistane).
Reeve, P.A., Toaliu, H., Kaneko, A., Hall, J.J., Ganczakowski, M., 1992. Acute intravascular haemolysis in Vanuatu following a single dose of primaquine in individuals with glucose-6-phosphate dehydrogenase deficiency. J. Trop. Med. Hyg. 95 (5), 349–351.
Rieckmann, K.H., McNamara, J.V., Frischer, H., Stockert, T.A., Carson, P.E., Powell, R.D., 1968. Gametocytocidal and sporontocidal effects of primaquine and of sulfadiazine with pyrimethamine in a chloroquine-resistant strain of *Plasmodium falciparum*. Bull. World Health Organ. 38 (4), 625–632.
Rieckmann, K., 1963. Malaria Eradication Using Mass Drug Administration on Nissan Island in Papua New Guinea. University of Sydney, Sydney.
Rodriguez, M.H., Betanzos-Reyes, A.F., Hernandez-Avila, J.E., Mendez-Galvan, J.F., Danis-Lozano, R., Altamirano-Jimenez, A., 2009. The participation of secondary clinical episodes in the epidemiology of vivax malaria during pre- and post-implementation of focal control in the state of Oaxaca, Mexico. Am. J. Trop. Med. Hyg. 80 (6), 889–895.
Rowland, M., Bouma, M., Ducornez, D., Durrani, N., Rozendal, J., Schapira, A., et al., 1996. Pyrethroid-impregnated bed nets for personal protection against malaria for Afghan refugees. Trans. R. Soc. Trop. Med. Hyg. 90 (4), 357–361.
Ryan, J.R., Stoute, J.A., Amon, J., Dunton, R.F., Mtalib, R., Koros, J., et al., 2006. Evidence for transmission of *Plasmodium vivax* among a duffy antigen negative population in Western Kenya. Am. J. Trop. Med. Hyg. 75 (4), 575–581.
Sabot, O., Cohen, J.M., Hsiang, M.S., Kahn, J.G., Basu, S., Tang, L., et al., 2010. Costs and financial feasibility of malaria elimination. Lancet 376 (9752), 1604–1615.
Santa-Olalla Peralta, P., Vazquez-Torres, M.C., Latorre-Fandos, E., Mairal-Claver, P., Cortina-Solano, P., Puy-Azon, A., et al., 2010. First autochthonous malaria case due to *Plasmodium vivax* since eradication, Spain, October 2010. Euro Surveillance: Bulletin Europeen Sur Les Maladies Transmissibles (European Communicable Disease Bulletin) 15 (41), 19684.
Senn, N., Rarau, P., Manong, D., Salib, M., Siba, P., Robinson, L.J., et al., 2011. Rapid diagnostic test-based Management of malaria: an effectiveness study in Papua New Guinean infants with *Plasmodium falciparum* and *Plasmodium vivax* malaria. Clin. Infect. Dis.
Sergiev, V.P., Baranova, A.M., Orlov, V.S., Mihajlov, L.G., Kouznetsov, R.L., Neujmin, N.I., et al., 1993. Importation of malaria into the USSR from Afghanistan, 1981–89. Bull. World Health Organ. 71 (3–4), 385–388.
Shanks, G.D., Edstein, M.D., Suriyamongkol, V., Timsaad, S., Webster, H.K., 1992. Malaria chemoprophylaxis using proguanil/dapsone combinations on the Thai-Cambodian border. Am. J. Trop. Med. Hyg. 46 (6), 643–648.
Shanks, G.D., Edstein, M.D., Kereu, R.K., Spicer, P.E., Rieckmann, K.H., 1993. Postexposure administration of halofantrine for the prevention of malaria. Clin. Infect. Dis. 17 (4), 628–631.
Shanks, G.D., Barnett, A., Edstein, M.D., Rieckmann, K.H., 1995a. Effectiveness of doxycycline combined with primaquine for malaria prophylaxis. Med. J. Aust. 162 (6), 306–307 9-10.
Shanks, G.D., Roessler, P., Edstein, M.D., Rieckmann, K.H., 1995b. Doxycycline for malaria prophylaxis in Australian soldiers deployed to United Nations missions in Somalia and Cambodia. Mil. Med. 160 (9), 443–445.
Shanks, G.D., Oloo, A.J., Aleman, G.M., Ohrt, C., Klotz, F.W., Braitman, D., et al., 2001. A new primaquine analogue, tafenoquine (WR 238605), for prophylaxis against *Plasmodium falciparum* malaria. Clin. Infect. Dis. 33 (12), 1968–1974.
Shekalaghe, S., ter Braak, R., Daou, M., Kavishe, R., van den Bijilaardt, W., van der Bosch, S., et al., 2010. Tanzania, hemolysis after a single dose of primaquine coadministered with an artemisinin is not restricted to glucose-6-phosphate deyhrogenase-deficient individuals. Antimicrob. Agents Chemother. 54 (5), 1762–1768.

Shortt, H.E., Garnham, P.C., 1948. Demonstration of a persisting exo-erythrocytic cycle in *Plasmodium cynomolgi* and its bearing on the production of relapses. Br. Med. J. 1 (4564), 1225–1228.
Sleigh, A.C., Liu, X.L., Jackson, S., Li, P., Shang, L.Y., 1998. Resurgence of vivax malaria in Henan Province, China. Bull. World Health Organ. 76 (3), 265–270.
Snowden, F.M., 2006. The Conquest of Malaria: Italy, 1900–1962. Yale University Press, New Haven.
Strangeways-Dixon, D., 1950. Paludrine (proguanil) as a malarial prophylactic amongst African labour in Kenya. East Afr. Med. J. 28 (3), 1–4.
Sutanto, I., Pribadi, W., Purnomo, Bandi R., Rusmiarto, S., Atmosoedjonon, S., et al., 1999. Efficacy of permethrin-impregnated bed nets on malaria control in a hyperendemic area in Irian Jaya, Indonesia: influenza of seasonal rainfall fluctuations. Southeast Asian J. Trop. Med. Public Health 30 (3), 440–446.
Tanner, M., de Savigny, D., 2008. Malaria eradication back on the table. Bull. World Health Organ. 86 (2), 82.
Tatarsky, A., Aboobakar, S., Cohen, J.M., Gopee, N., Bheecarry, A., Moonasar, D., et al., 2011. Preventing the reintroduction of malaria in Mauritius: a programmatic and financial assessment. PloS One 6 (9), e23832.
Tatem, A.J., Smith, D.L., Gething, P.W., Kabaria, C.W., Snow, R.W., Hay, S.I., 2010. Ranking of elimination feasibility between malaria-endemic countries. Lancet 376 (9752), 1579–1591.
ter Kuile, F.O., van Eijk, A.M., Filler, S.J., 2007. Effect of sulfadoxine-pyrimethamine resistance on the efficacy of intermittent preventive therapy for malaria control during pregnancy: a systematic review. J. Am. Med. Assoc. 297 (23), 2603–2616.
Toty, C., Barre, H., Le Goff, G., Larget-Thiery, I., Rahola, N., Couret, D., et al., 2010. Malaria risk in Corsica, former hot spot of malaria in France. Malar. J. 9, 231.
Turkmenistan certified malaria-free. 2010. Releve epidemiologique hebdomadaire/Section d'hygiene du Secretariat de la Societe des Nations (Weekly epidemiological record/Health Section of the Secretariat of the League of Nations) 85 (47), 461–463.
Usenbaev, N.T., Ezhov, M.N., Zvantsov, A.B., Annarbaev, A., Zhoroev, A.A., Almerekov, K., 2006. An outbreak of *Plasmodium vivax* malaria in Kyrghyzstan. Meditsinskaia Parazitologiia I Parazitarnye Bolezni (1), 17–20.
Usenbaev, N.T., Baranova, A.M., Anarbaev, A.A., Almerekov, K., 2008. Experience in sanitizing an urban focus of vivax malaria (Tashkumyr, Kyrghyzstan). Meditsinskaia Parazitologiia I Parazitarnye Bolezni (3), 45–46.
Vogel, G., Roberts, L., 2011. Malaria. Vaccine trial meets modest expectations, buoys hopes. Science 334 (6054), 298–299.
von Seidlein, L., Walraven, G., Milligan, P.J., Alexander, N., Manneh, F., Deen, J.L., et al., 2003. The effect of mass administration of sulfadoxine-pyrimethamine combined with artesunate on malaria incidence: a double-blind, community-randomized, placebo-controlled trial in The Gambia. Trans. R. Soc. Trop. Med. Hyg. 97 (2), 217–225.
Walker, A.J., 1955. Potentialities of monthly doses of camoquin and a gametocidal drug in malaria control. Trans. R. Soc. Trop. Med. Hyg. 49 (4), 351–355.
White, N.J., 2010. Artemisinin resistance–the clock is ticking. Lancet 376 (9758), 2051–2052.
White, N.J., 2011. Determinants of relapse periodicity in *Plasmodium vivax* malaria. Malar. J. 10, 297.
Williams Jr., L.L., 1963. Malaria eradication in the United States. Am. J. Public Health Nations Health 53, 17–21.
Xu, B.L., Su, Y.P., Shang, L.Y., Zhang, H.W., 2006. Malaria control in Henan Province, People's Republic of China. Am. J. Trop. Med. Hyg. 74 (4), 564–567.
Yeom, J.S., Ryu, S.H., Oh, S., Choi, D.H., Song, K.J., Oh, Y.H., et al., 2005. Evaluation of anti-malarial effects of mass chemoprophylaxis in the Republic of Korea army. J. Korean Med. Sci. 20 (5), 707–712.

Yip, K., 1998. Antimalarial work in China: a historical perspective. Parassitologia 40 (1–2), 29–38.
Yip, K., 2000. Malaria eradication: the Taiwan experience. Parassitologia 42, 117–126.
Zhang, X.B., Deng, D., 1989. An epidemiological study of the residual vivax malaria cases in Huanghuai plain. Zhongguo Ji Sheng Chong Xue Yu Ji Sheng Chong Bing Za Zhi (Chinese Journal of Parasitology & Parasitic Diseases) 7 (1), 22–27.
Zhou, H.Y., Wang, W.M., Cao, J., Zhu, G.D., Gao, Q., 2011. Epidemiological analysis of malaria situation in Jiangsu Province in 2009. Zhongguo Xue Xi Chong Bing Fang Zhi Za Zhi (Chinese Journal of Schistosomiasis Control) 23 (4), 402–405.
Zoungrana, A., Coulibaly, B., Sie, A., Walter-Sack, I., Mockenhaupt, F.P., Kouyate, B., et al., 2008. Safety and efficacy of methylene blue combined with artesunate or amodiaquine for uncomplicated falciparum malaria: a randomized controlled trial from Burkina Faso. PloS One 3 (2) e1630.
Zucker, J., 1996. Changing patterns of autochthonous malaria transmission in the United States: a review of recent outbreaks. Emerg. Infect. Dis. 2 (1), 37–43.

INDEX

Note: Page numbers with "f" denote figures; "t" tables.

D

K

L

M

Q

R

S

W

Y

Kolibri 360 GmbH
Ferdinand-Jühlke-Straße 7
99095 Erfurt, Deutschland
produktsicherheit@kolibri360.de